Michael L. Morrison
William M. Block
M. Dale Strickland
Bret A. Collier
Markus J. Peterson

Wildlife Study Design

Second Edition

Dr. Michael L. Morrison
Texas A&M University
College Station,TX
USA
mlmorrison@ag.tamu.edu

Dr. William M. Block
Rocky Mountain Research Station
USDA Forest Service
Flagstaff, AZ
USA

Dr. M. Dale Strickland
Western EcoSystems Technology, Inc.
Cheyenne, WY
USA

Dr. Bret A. Collier
Texas A&M University
College Station, TX
USA

Dr. Markus J. Peterson
Texas A&M University
College Station, TX
USA

ISBN: 978-1-4419-2594-7 e-ISBN: 978-0-387-75528-1
DOI: 10.1007/978-0-387-75528-1

Printed on acid-free paper

9 8 7 6 5 4 3 2 1

springer.com

SPRINGER SERIES ON

ENVIRONMENTAL MANAGEMENT

BRUCE N. ANDERSON
ROBERT W. HOWARTH
LAWRENCE R. WALKER

Series Editors

Springer Series on Environmental Management
Volumes published since 1989

The Professional Practice of Environmental Management (1989)
R.S. Dorney and L. Dorney (eds.)

Chemical in the Aquatic Environment: Advanced Hazard Assessment (1989)
L. Landner (ed.)

Inorganic Contaminants of Surface Water: Research and Monitoring Priorities (1991)
J.W. Moore

Chernobyl: A Policy Response Study (1991)
B. Segerstahl (ed.)

Long-Term Consequences of Disasters: The Reconstruction of Friuli, Italy, in Its International Context, 1976–1988 (1991)
R. Geipel

Food Web Management: A Case Study of Lake Mendota (1992)
J.F. Kitchell (ed.)

Restoration and Recovery of an Industrial Region: Progress in Restoring the Smelter-Damaged Landscape near Sudbury, Canada (1995)
J.M. Gunn (ed.)

Limnological and Engineering Analysis of a Polluted Urban Lake: Prelude to Environmental Management of Onondaga Lake, New York (1996)
S.W. Effler (ed.)

Assessment and Management of Plant Invasions (1997)
J.O. Luken and J.W. Thieret (eds.)

Marine Debris: Sources, Impacts, and Solutions (1997)
J.M. Coe and D.B. Rogers (eds.)

Environmental Problem Solving: Psychosocial Barriers to Adaptive Change (1999)
A. Miller

Rural Planning from an Environmental Systems Perspective (1999)
F.B. Golley and J. Bellot (eds.)

Wildlife Study Design (2001)
M.L. Morrison, W.M. Block, M.D. Strickland, and W.L. Kendall

Selenium Assessment in Aquatic Ecosystems: A Guide for Hazard Evaluation and Water Quality Criteria (2002)
A.D. Lemly

Quantifying Environmental Impact Assessments Using Fuzzy Logic (2005)
R.B. Shephard

Changing Land Use Patterns in the Coastal Zone: Managing Environmental Quality in Rapidly Developing Regions (2006)
G.S. Kleppel, M.R. DeVoe, and M.V. Rawson (eds.)

The Longleaf Pine Ecosystem: Ecology, Silviculture, and Restoration (2006)
S. Jose, E.J. Tokela, and D.L. Miller (eds.)

Linking Restoration and Ecological Succession (2007)
L.R. Walker, J. Walker, and R.J. Hobbs (eds.)

Preface

We developed the first edition of this book because we perceived a need for a compilation on study design with application to studies of the ecology, conservation, and management of wildlife. We felt that the need for coverage of study design in one source was strong, and although a few books and monographs existed on some of the topics that we covered, no single work attempted to synthesize the many facets of wildlife study design.

We decided to develop this second edition because our original goal – synthesis of study design – remains strong, and because we each gathered a substantial body of new material with which we could update and expand each chapter. Several of us also used the first edition as the basis for workshops and graduate teaching, which provided us with many valuable suggestions from readers on how to improve the text. In particular, Morrison received a detailed review from the graduate students in his "Wildlife Study Design" course at Texas A&M University. We also paid heed to the reviews of the first edition that appeared in the literature.

As for the first edition, we think this new edition is a useful textbook for advanced undergraduate and graduate students and a valuable guide and reference for scientists and resource managers. Thus, we see this book being used by students in the classroom, by practicing professionals taking workshops on study design, and as a reference by anyone interested in this topic. Although we focus our examples on terrestrial vertebrates, the concepts provided herein have applicability to most ecological studies of flora and fauna.

We approached this book from both a basic and applied perspective. The topics we cover include most of the important areas in statistics, but we were unable to go into great detail regarding statistical methodology. However, we included sufficient details for the reader to understand the concepts. Actual application might require additional reading. To facilitate additional research on the topics, we included extensive literature reviews on most of the areas covered.

A primary change in the second edition was division of the original Chap. 1 into two new chapters. Chapter 1 now focuses on philosophical issues as they relate to science. The philosophy of science provides a logical framework for generating meaningful and well-defined questions based on existing theory and the results of previous studies. It also provides a framework for combining the results of one's study into the larger body of knowledge about wildlife and for generating new questions, thus completing the feedback loop that characterizes science. The new Chapter 2 retains many of the elements present in the first chapter of the original edition, but has been fully revised. In this new Chapter 2, we focus on the concept of basic study design, including variable classification, the necessity of randomization and replication in wildlife study design, and the three major types of designs in decreasing order of rigor (i.e., manipulative experiments, quasi-experiments, and observational studies).

Throughout the remaining chapters we expanded our use of examples and the accompanying literature. In particular, we added considerable new material on detection probabilities, adaptive cluster methods, double sampling, sampling of rare species, and effect size and power. We expanded our coverage of impact assessment with recent literature on disturbance and recovery. One of the changes highlighted by student reviewers of the first edition was the need for more material on what to do "when things go wrong." That is, what can one do to recover a study when the wonderful design put down on paper cannot be fully implemented in the field, or when some event (e.g., natural catastrophe or just plain bad luck) reduces your sample size? We also added a glossary to assist in reviewing key terminology used in study design, as requested by student reviewers.

We thank Janet Slobodien, Editor, Ecology and Environmental Science, Springer Science + Business Media, for guiding both editions through to publication; and also Tom Brazda of Springer for assisting with the final compilation and editing of the book. Joyce Vandewater is thanked for patiently working with us to create and standardize the graphics. We thank the reviewers selected by Springer for providing valuable comments that strengthened this edition. Angela Hallock, Texas A&M University, completed the task of securing copyright permissions for material used in the text. Nils Peterson, Damon Hall, and Tarla Rai Peterson provided incisive reviews of Chapter 1 that greatly improved the final version.

We also thank those who assisted with the first edition, because the valuable comments they made were retained through to this new edition: Rudy King, Rocky Mountain Research Station, US Forest Service; Lyman McDonald, Western EcoSystems Technology, Inc. In particular we thank first edition co-author William L. Kendall for his valuable contributions.

Contents

Glossary

Format of glossary: Key terms used throughout the text are listed below with a brief definition; cross-reference to associated terms is provided where appropriate. The number(s) following each term refers to the chapter(s) in which the term is defined or otherwise discussed.

Abductive Reasoning (1) For our purposes, see *retroductive reasoning*.

Accuracy (2, 5) Combination of bias and precision that measures the conformity of a quantity to its true value. See also *bias*, *precision*.

Adaptive management (7) Planned series of events including monitoring the effects of implementing land management activities on key resources, and then using monitoring results as a basis for modifying those activities when warranted.

Adaptive sampling (4, 5) Sampling procedure in which the proba bility for selecting units to be included in the sample depends upon the values and locations of the variables of interest during the survey period.

Aesthetics (1) The branch of Western philosophy dealing with the nature of beauty, art, and taste, and also the creation and appreciation of beauty. See *axiology*.

Analysis of covariance (3) An analysis using the concepts of analysis of variance and regression that considers the added influence of variables having a measurable influence on the dependent variable when analyzing the dependent variables response to a treatment.

Analysis of variance (2, 3) An analysis of the variation in the outcomes of an experiment to assess the contribution of each variable to the variation.

Anthropocentric (1) Human centeredness; interpreting the world in terms of human values and experiences.

Anthropogenic (1) Of, relating to, or resulting from the influence of human beings on nature.

Area of interest (3) Area to which statistical and deductive inferences will be made.

Attributable risk (6) Defined as the proportional increase in the risk of injury or death attributable to the external factor.

Axiology (1) The branch of Western philosophy that studies the nature, types, and criteria of value and quality (includes value judgments, aesthetics, and ethics).

BACI (2, 3, 6) The before–after/control–impact, or BACI, design is the standard upon which many current designs are based. In the BACI design, a sample is taken before and another sample is taken after a disturbance, in each of the putatively disturbed (impacted) sites and an undisturbed (control) site.

BACIP (3, 6) BACI design with paired sampling, or BACIP. The BACIP design requires paired (simultaneous or nearly so) sampling several times before and after the impact at both the control and impacted site.

Before–after design (3) A relatively weak design appropriate when measurements on the study area before the treatment area are compared with measurements on the same area following the treatment and independent control or reference data are lacking.

Bias (2, 3, 5) Difference between estimator's expectation and the true value of a parameter being estimated. Tendency of replicated parameters to differ systematically from the true parameter value. See also *precision*, *accuracy*.

Blocking (3) Partitioning of variance.

Biodiversity (1) In its most general sense, biodiversity refers to all aspects of variety in the living world. More specifically, the term may be used to describe the number of species

(species richness), the amount of genetic variation, or the number of biotic community types present in an area of interest.

Bioequivalence testing (3) An alternative paradigm for data analysis that reverses the burden of proof so that a treatment is considered biologically significant until evidence suggests otherwise.

Biological population (1, 3) A group of individuals of one species in an area at a given time that potentially could interbreed. The size and nature of the area is defined, often arbitrarily, for the purposes of the study being undertaken.

Biological resources (6) Quantifiable components of the systems such as organisms, populations, species, and communities.

Biometry (9) See *biostatistics*.

Biostatistics (9) Biostatistics in general focuses on statistical applications to biological and ecological data ("ecostsatistics" is not generally used). Many biostatics textbooks are written so as to not require mathematical education beyond elementary algebra, or are written in a "nonmathematical" manner.

Biotic community (1) The assemblage of species populations that occur together in space and time.

Capture–recapture (2, 7) Method used to estimate size and vital rates of ecological populations using the rates of capture and recaptures for unique individuals.

Case study (6, 3, 8) Case study is work that focuses on a specific location or species and is often short term in duration. There are situations in which a biological study was too localized or too brief in duration to warrant a full research article and can be focused as a case study.

Central limit theorem (2) Statistical theory which states that n random variables will assume a normal distribution as the number of variables collected goes to infinity.

Census (1) The process of obtaining information about every individual at a specific time and place.

Chronosequence (7) A way to examine time (short to long) effects of a disturbance or activity over a sequence of time without having to track a set of plots through time. Done by locating various plots representative of conditions at different times post-disturbance. By sampling enough areas you could draw inferences as to possible short- and long-term effects on wildlife.

Cluster sampling (4) Probabilistic sample in which sampling units are selected based on the distribution of organisms within the sampling frame.

Completely randomized design (3) The random application of two treatments to a group of experimental units.

Compliance monitoring (7) Done when mandated by law or statute to ensure that actions are in compliance with existing legal direction. An example of compliance is monitoring established within a biological opinion provided by the US Fish and Wildlife Service during interagency consultation under the Endangered Species Act.

Community metrics (7) Indices of quantitative values that are related to numbers, degree of association, diversity, and evenness of species.

Concomitant variable (3) See *covariate*.

Conceptual model (1, 3, 7) A theoretical construct that represents the system of interest; it includes a set of variables and logical qualitative and sometimes quantitative relationships among them.

Confidence intervals (3) When estimated with the data for an observed effect size, a CI represents the likely range of numbers that cannot be excluded as possible values of the true effect size if the study were repeated infinitely into the future with probability $1 - \alpha$.

Confounding variables (3) Random variables that are likely to influence the response variable in a manor typically out of the control of the biologist including landscape issues (e.g., large-scale habitat variables), biological issues (e.g., variable prey species abundance), land use issues (e.g., rapidly changing crops and pest control), weather, study area access, etc.

Control (controlling variable) (2, 3) A standard for comparison in ecological studies. Controls are typically experimental units on which no treatments have been assigned so that treatment effects can be evaluated on other experimental units. Control can also be achieved by standardization of related variables.

Covariate (3, 4) Random variable collected during the course of a research study which the researcher hypothesizes influences the response variable.

Cross-over design (3) The random assignment of two or more treatments to a study population during the first study period and then the treatments are switched during subsequent study periods so that all study units receive all treatments in sequence.

Deductive reasoning (1) The form of inference where the conclusion about particulars follows necessarily from general or universal premises.

Design/data-based studies (3) Studies where basic statistical inferences concerning the study areas or study populations are justified by the design of the study and data collected.

Detectability (2) A parameter that describes the probability that an individual organism will be detected (seen or captured) during some specific time or place.

Disturbing variables (2, 3) Extraneous variable that can bias the results of a study.

Dose-response regression (3, 6) Analysis in which dose is a measure of exposure to the impact and response is a measure of the biological system. See also *gradient design*.

Dynamic equilibrium (6) Incorporates both temporal and spatial variation, where natural factors and levels of resources usually differ between two or more areas being compared, but the differences between mean levels of the resource remain similar over time. Contrast with *steady-state system* and *spatial equilibrium*.

Effect size (3, 7) A measure of the difference among groups. From a statistical point of view, it is the difference between the null and alternative hypotheses.

Effectiveness monitoring (7) Used to evaluate whether or not a management action or decision met its stated objective.

Element (2) Basic unit on which some measurement is taken in ecological studies.

Empiricism (1) A theory that all knowledge originates in experience; discounts the notion of innate ideas.

Environmental impact studies (3) Field studies that look at the environmental response to unplanned perturbations, as opposed to manipulative experiments, although manipulative experiments and smaller observational studies aid understanding of the mechanism of impact.

Epidemiology (3, 6) The study of the occurrence of disease, injury, or death, usually in reference to human populations.

Epistemology (1) The branch of Western philosophy that studies the nature and scope of knowledge.

Estimator (2, 3, 4) A function of observed sample data that is used to estimate some unknown population parameter.

Ethics (1) The branch of Western philosophy dealing with values and norms of a person or group; addresses concepts such as right and wrong, good and evil, and moral responsibility. See *axiology*.

Experimental units (1, 3) The basic units, such as individual plots and organisms, upon which experimental data could be collected and which determine sample size.

Expert opinion (9) Expert opinion can be formalized into a process that seeks the council of many individuals with expertise in the area of interest. Contrast with *personal opinion*.

Expected value (2, 3) A mathematical expectation of a random variable that equals the sum (or integral) of the values that are possible for it, each multiplied by its probability.

Experimental design (3) The combination of a design structure, treatment structure, and the method of randomization included in an experiment.

Explanatory variable (2, 3) A variable that is used in a statistical relationship to explain or predict changes in the value of another variable.

Factorial experiments (3) Multiple-factor experiments where all possible combinations of factors of interest are tested and these tests are possibly replicated a number of times.

Finite (4) Bounded or limited in magnitude, spatial, or temporal extent.

Gradient design (3, 6) This class of designs analyze an impact along a continuous scale and use regression techniques to test for an association between level of impact and response by the animal. See also *dose-response regression and response gradient design.*

Gray literature (8) The graduate thesis or dissertation, a final report to an agency, in-house agency papers, and the like are not publications per se and are termed gray literature because they are usually not readily available, and they do not usually receive independent peer review.

Habitat (1, 3, 7) The physical space within which an organism lives, and the abiotic and biotic entities (e.g., resources) it uses and selects in that space. Because habitat is organism-specific, it relates the presence of a species, population, or individual (animal or plant) to an area's physical and biological characteristics.

Hypothesis (1, 3) See *research hypothesis*, *scientific hypothesis*, and *statistical hypothesis*.

Hypothetico-deductive model of science (1) The model of science popularized by Karl R. Popper that argues scientists should formulate one or more hypotheses to explain an observed phenomenon, then deductively derive a number of explicit predictions that should be observed as a consequence of each hypothesis. Observations contrary to those predictions lead the researcher to conclusively falsify the hypothesis which then should immediately be rejected. Observations in agreement with the deductive predictions only imply that the hypothesis is still viable.

Impact (3, 6) Impact is a general term used to describe any change that perturbs the current system, whether it is planned or unplanned, human-induced, or an act of nature.

Impact assessment (1, 3) Studies designed to determine the influence of a change that perturbs the current state of a system of interest; these impacts can be planned or unplanned, human-induced, or acts of nature.

Impact-reference design (3, 6) The basic design mimics a classical experimental treatment and control design, where random samples are taken from sites within the disturbed area and from other nondisturbed reference sites.

Implementation monitoring (7) Used to assess whether a directed management activity has been carried out as designed.

Incomplete block design (3) An experiment that uses blocking of variance, although each block has less than a full complement of treatments.

Independent data (3) Data that are neither contiguous in time or space.

Index (2, 7) A numerical value used to compare values collected over time or between areas.

Indicator species (7) Index or represent specific environmental conditions or the population status of other ecologically similar species.

Inductive reasoning (1) The form of inference where a generalized conclusion is reached based on a collection of particular facts or instances.

Inventory (1, 3, 7) Studies designed to determine the distribution and composition of wildlife and/or wildlife habitats.

Knowledge (1) In philosophy, knowledge generally was defined as "justified true belief" from classical times (Plato) until the 1960s. Since then, philosophers have been unable to agree on a single definition for various reasons. In general, knowledge now can be considered society's accepted portrayal of a proposition under consideration. Thus, for the society of scientists, knowledge still remains justified true belief.

Latin square design (3) An extension of the randomized block design to control for multiple sources of variation.

Level-by-time interaction (6) The term "level" refers to the fact that specific categories (levels) of the impact are designated; used in a level-by-time design. Contrast with *trend-by-time interaction.*

Local extinction probability (1) The probability that a species currently present in a biotic community will not be present by the next time period.

Logical empiricism (1) See *logical positivism*; reflects the affinity of later members of this movement for the writings of Locke, Berkeley, and Hume.

Logical positivism (1) An early twentieth century philosophical movement that holds that all meaningful statements are either (1) analytic (e.g., mathematical equations) or (2) conclusively verifiable or at least confirmable by observation and experiment, and that all other statements are therefore cognitively meaningless.

Longitudinal studies (3) Repeated measures experiment common in wildlife telemetry studies, environmental impact studies, habitat use and selection studies, studies of blood chemistry, and many other forms of wildlife research, where logistics typically leads to repeated measures of data from study plots or study organisms.

Long-term study (5, 3, 7) A study that continues "…for as long as the generation time of the dominant organism or long enough to include examples of the important processes that structure the ecosystem under study... the length of study is measured against the dynamic speed of the system being studied" (Strayer et al. 1986).

Levels (6) Levels are measures of a resource such as abundance, diversity, community structure, and reproductive rates. Hence, levels are quantifiable on an objective scale and can be used to estimate means and variance and to test hypotheses.

Magnitude of anticipated effect (3) The magnitude of the perturbation or the importance of the effect to the biology of the species, which often determines the level of concern and the required level of precision.

Manipulative studies (3) Studies that include control of the experimental conditions; there are always two or more treatments with different experimental units receiving different treatments and random application of treatments.

Matched pair design (3, 6) This design reduces the confounding of factors across sites. Under this design, sites within the impacted area are randomly selected and nonrandomly matched with similar reference sites.

Mechanism (2) A physical or chemical process involving how a natural phenomenon works.

Mensurative studies (3) Studies involving making measurements of uncontrolled events at one or more points in space or time with space and time being the only experimental variable or treatment.

Meta-analysis (3) Analysis of a series of independent studies addressing a specific research question.

Metaphysics (1) The branch of Western philosophy concerned with explaining the system of principles underlying a particular subject or discipline. Recently, in common parlance, the term often refers to topics beyond the physical world.

Metapopulation (1) A population subdivided into segments occupying patches of habitat in a fragmented landscape. Individual patches are separated by an environment hostile to the species of interest, and movement and presumably gene flow between patches is inhibited, but still exists.

Model-based studies (3) Studies that predict the outcome of experiments using models. In the extreme case of model-based analysis where no new data are available, all inferences are justified by assumption, are deductive, and are subject to counterarguments.

Monitoring (1, 3, 7) Studies designed to determine rates of change or the influence of management practices on wildlife population dynamics and/or habitats.

Multiple-factor designs (3) Experiments when one or more classes of treatments are combined with one or more classifications of experimental units.

Multivariate analysis (3) Analysis that considers several related random variables simultaneously, each one being considered equally important at the start of the analysis.

Natural factors (6) Physical and chemical features of the environment that affect the level of a resource at a given time and location, such as temperature, substrate, dissolved oxygen, and total organic carbon.

Nested experimental design (3) A design that uses replication of experimental units in at least two levels of a hierarchy.

Nonresponse error (5) Occurs when one fails to record or observe an individual or unit that is part of the selected sample.

Normal science (1) A term employed by Thomas S. Kuhn that characterizes periods where there is general consensus within a scientific community regarding theory, methods, terminology, and types of experiments likely to contribute useful insights – the articulation of a paradigm.

Nuisance parameters (3, 4) A parameter estimated by a statistic, which is not needed except for the calculation of the parameter of interest.

Number of colonizing species (1) The number of species currently in the community that were absent during the last time period.

Observer drift (8) Part of *quality assurance*, observer drift refers to the gradual change in the way observers collect data through time and can affect all studies regardless of the precautions taken during observer selection and training.

Observational studies (3, 4) See *mensurative studies*. Studies that have no specific sampling design, and the researcher has little or no control over how observations on the population were obtained.

One-factor experiment (3) An experiment that uses one type of treatment or one classification factor in the experimental units in the study, such as all the animals in a specific area or all trees of the same species in a management unit.

Ontology (1) The branch of metaphysics (see above) that studies the nature of reality, being, or existence.

Optimal study design (6) If you know what type of impact will occur, when and where it will occur, and have the ability to gather pretreatment data, you are in an optimal situation to design the study. Contrast with *suboptimal study design*.

Overdispersion (3, 4) A statistical occurrence when the observed variance of the data is larger than the predicted variance. Fairly common in analysis using Poisson and Binomial regression techniques.

Paired study design (3) A study that typically evaluates changes in study units paired for similarity.

***P*-value (2, 3)** Probability of obtaining a test statistic at least as extreme at the observed conditional on the null hypothesis being true.

Personal opinion (9) Personal opinion implies a decision based on personal biases and experiences. Contrast with *expert opinion*.

Panmictic populations (1) Populations where interactions between individuals, including potential mating opportunities, are relatively continuous throughout the space occupied by the population.

Paradigm (1, 3) A term employed by Thomas S. Kuhn that characterizes a scientific tradition, including its philosophy, theory, experiments, methods, publications, and applications. Paradigms govern what he called normal science (see above). The term also has come to describe a given world-view in common parlance.

Parameter (2, 3) Quantities that define certain characteristics of an ecological system or population.

Pilot study (1, 2, 5, 8) A pilot study is a full-scale dress rehearsal of the study plan and includes data collection, data processing, and data analyses, thus allowing thorough evaluation of all aspects of the study including initial sample size and power analyses. A pilot study is often done with a much larger sample than a *pretest period*. Such studies are especially useful when initiating longer-term studies.

Population (1, 3) See *biological population*, *sampled population*, and *target population*.

Postmodernism (1) It is a truism that postmodernism is indefinable. It can be described as a cultural zeitgeist of crisis, desperation, anxiety, schizophrenia, nostalgia, pastiche, apocalyptic millennialism, and lassitude. The quintessential postmodern utterance is "Whatever?." The more radical social constructionists (see below) often are called postmodernists.

Postpositivism (1) The stance, based on the writings of Karl R. Popper and others, that human knowledge is not based on unchallengeable empirical foundations as argued by the logical positivists (see above), but is to some degree conjectural. Further, while we do have warrants for asserting beliefs and conjectures, based on the hypothetico-deductive model of science, they can be modified or withdrawn based on further investigation.

Pragmatism (1) An American movement in philosophy founded by Charles Saunders Peirce and popularized by William James and others that is marked by the tenets that (1) the meaning of concepts should be sought in their practical bearings, (2) the function of thought is to guide action, and (3) truth is preeminently to be tested by the practical consequences of belief.

Precision (2, 3) Degree of mutual agreement between individual measurement or the amount of variation between sample estimates arising from the sample sampling process. See also *bias*, *accuracy*.

Press disturbance (6) Press disturbances are those that are sustained beyond the initial disturbance. Contrast with *pulse disturbance*; see also *temporal variance, disturbances affecting*.

Pretesting period (8) Initial field sampling should include tests of data collection procedures; this is often called the pretesting period. Pretesting allows for redesign of data forms and sampling protocols. Pretesting sampling should cover as much of the range of conditions that will be encountered during the study. Some, but seldom all, of the data collected during pretesting might be suitable for inclusion with the final data set. Contrast with *pilot study*.

Preventable fraction (6) The proportion of deaths removed by a preventive step is termed the preventable fraction and is defined as the proportion of injuries or deaths that would be removed if all birds were able to take advantage of the preventive intervention. See also *prevented fraction*.

Prevented fraction (6) Is the actual reduction in mortality that occurred because of the preventive intervention. See also *preventable fraction*.

Preventive intervention (6) Steps taken to prevent an impact (injury or death); used in the context of epidemiological studies. See also *preventable fraction* and *prevented fraction*.

Process variation (2) Variation in population growth irrespective of the methods used to determine population parameters. See also *sampling variation*.

Proportional mortality (6) The proportion of the animals killed.

Pulse disturbance (6) Pulse disturbances are those that are not sustained after the initial disturbance; the effects of the disturbance may be long lasting. Contrast with *press disturbance*; see also *temporal variance, disturbances affecting*.

Quality assurance (5, 8) The purpose of quality assurance (also called quality assurance/quality control, or QA/QC) is to ensure that the execution of the plan is in accordance with the study design. As such it is a process to produce reliable research data with respect to its precision, completeness, comparability, and accuracy. It is important to the successful completion of the study that a formal program of QA/QC is instituted on both the data collection and data processing components.

Quality control (5) The routine application of procedures (such as calibration or maintenance of instruments) to reduce random and systematic errors, and to ensure that data are generated, analyzed, interpreted, synthesized, communicated, and used within acceptable limits.

Quasi-experiments (3) Observational studies where strict adherence to Fisher's requirements for the design of true experiments is impossible or impractical, although adherence to fundamental statistical principles as much as possible is essential and conclusions concerning cause-and-effect relationships are limited.

Randomization (2, 3) The process of selecting a random sample of an ecological population on which to perform a treatment or to take observations.

Randomization tests (3) Computer intensive tests that, for example, involve the repeated sampling of a randomization distribution (say 5,000 times) to determine if a sample statistic is significant at a certain level.

Randomized complete block design (3) An experiment where blocking of variance is used and each treatment is randomly assigned within each block.

Rationalism (1) A theory that reason is in itself a source of knowledge superior to and independent of sense perceptions.

Recovered (6) When natural factors have regained their influence over the biological resource(s) being assessed. See also *recovery*.

Recovery (6) A temporal process in which impacts progressively lessen through natural processes and/or active restoration efforts. See also *recovered*.

Repeated measure designs (3) Experiments where several comparable measurements are taken on each experimental unit.

Replication (2, 3) The process of repeating a study multiple times under similar conditions to confirm findings.

Research hypothesis (1) A tentative explanation for how some process in a system of interest works; a proposed explanation for an observed phenomenon in a given system. Also see *scientific hypothesis* and *statistical hypothesis*.

Response-gradient design (3) Study design useful for quantifying treatment effects when a response is expected to vary relative to the distance or time from the application of the treatment (gradient of response).

Retroductive reasoning (1) The form of inference proposed by Charles Saunders Peirce where a hypothesis is developed, which would, if true, best explain a particular set of observations.

Retrospective power analysis (3) A power analysis that is conducted after the study is completed, the data have been collected and analyzed, and the outcome is known.

Retrospective study (3) An observational study that looks backwards in time.

Revolutionary science (1) See *scientific revolution*.

Sample (1, 3, 4) A subset of a population randomly selected based on some probabilistic scheme on which measurements regarding the population of interest will be made.

Sampling distribution (2, 3) The frequency distribution of a statistic obtained from a large number of random samples drawn from a specific ecological population.

Sampling bias (3, 5) A systematic bias where a parameter is consistently under- or overestimated.

Sampling intensity (3, 5) Refers to how many, how long, and how often units should be sampled.

Sampled population (1, 3) The subset of the target population that is accessible to sampling.

Sample size (3, 4) Number of samples that must be taken to meet some a priori specified level of precision in the resulting parameter estimates.

Sampling variation (2) Variation that is contributed to the methods used to determine population parameters. See also *process variation*.

Sampling units (1, 3, 4) A unique collection of elements (e.g., plots or organism) on which sample data are collected. See also *element*.

Scientific hypothesis (1, 3) A universal proposition explaining an observed phenomenon. For example, the hypothesis of density-dependent population regulation in ecology. Also see *research hypothesis* and *statistical hypothesis*.

Scientific revolution (1) A term employed by Thomas S. Kuhn that refers to an interruption in normal science (see above), where a shift in paradigm (see above) occurs. Darwin's work, for example, led to a scientific revolution in zoology–ecology (e.g., Linnaean vs. Darwinian paradigms).

Sequential study designs (3) Unique study designs in which the sample size is not fixed before the study begins and there are now three potential statistical inferences, namely accept, reject, or uncertainty (more data are needed).

Similar (6) In the context of replicated study sites, concerns matching the basic environmental conditions of sites.

Simple random sampling (2, 3) A basic sampling technique where we select a group of subjects (a sample) for study from a larger group (a population). Each individual is chosen entirely by chance and each member of the population has an equal chance of being included in the sample.

Size bias (4) The propensity for organisms of a larger size or grouped together to be detected and sampled at a higher rate than organisms of a smaller size or group.

Social constructionism (1) A theory based on the work of Hegel and others that holds knowledge ultimately is at least in part created through the combined perceptions of society.

Spatial equilibrium (6) Occurs when 2 or more sampling areas, such as impact and reference, have similar natural factors and, thus, similar levels of a resource. Contrast with *steady-state system* and *dynamic equilibrium*.

Species diversity (1, 7) Indices of community diversity that take into account both species richness and the relative abundance of species.

Species richness (1, 7) The number of species in the biotic community at a given time.

Split-plot designs (3) A form of nested factorial design where the study area is divided into blocks that are then divided into relatively large plots called main plots, which are then subdivided into smaller plots called split plots, resulting in an incomplete block treatment structure.

Stage-based matrices (6) Used to analyze population growth for species in which it is difficult to age individuals, or where it is more appropriate to classify them into life stages or size classes rather than by age.

Statistics (2, 3) Mathematical procedure used to measure attributes of a population based on data collected from a sample of that population.

Statistical hypothesis (1, 3) A deductively derived prediction, based on the research hypothesis, of a specific result that can be tested against data using a statistical algorithm. Also see *scientific hypothesis* and *research hypothesis*.

Statistical power (3, 7) The probability that you will reject a null hypothesis when it is false, i.e., the experiment has a small probability of making a Type II error.

Steady-state system (6) Typified by levels of resources, and the natural factors controlling them, show a constant mean through time. Contrast with *dynamic system* and *spatial equilibrium*.

Strata (3, 4) Division in an organized system based on the characteristics of that system.

Stratified sampling (4, 3) A sampling method used to divide a population into homogenous subgroups or blocks of experimental units that are then sampled individually. See also *strata*.

Stressors (7) Natural and anthropogenic events that affect resource distribution or abundance.

Suboptimal study design (6) When no or little pretreatment data are available and the treatment (impact) has not or can not be replicated. Contrast with *optimal study design*.

Syllogism (1) A deductive formal argument where a major and minor premise necessitates a conclusion (e.g., all animals are mortal, northern bobwhites are animals, therefore bobwhites are mortal). See *deductive reasoning*.

Systematic sampling (3, 4) A sampling method in which samples are collected from a population systematically, or by selecting 1 unit of every 10 in order from a random starting point.

Target population (1, 2, 3) A clear and precise definition of the spatial and temporal aspects of the study area as well as a detailed description of the resource on which information is wanted.

Time of interest (3) The period of interest for statistical and deductive inferences will be made, e.g., diurnal, nocturnal, seasonal, or annual.

Take monitoring (7) Assesses whether an activity adversely affects the occupancy or habitat of a threatened or endangered species.

Temporal variance, disturbances affecting (6) Disturbances affecting temporal variance are those that do not alter the mean abundance, but change the magnitude of the oscillations between sampling periods. See also *press disturbance* and *pulse disturbance*.

Theory (1) There are two distinct uses of "theory" in natural science:

a. A proposed description, explanation, or model capable of predicting future occurrences of the same type that can potentially be evaluated empirically.

b. An integrated and hierarchical set of empirical hypotheses that together explain a significant portion of scientific observations. Most ecologists would argue that (1) the theory of evolution through natural selection and (2) *perhaps* the theory of island biogeography are the only theories of ecology under this definition.

Thresholds/trigger points (7) Pre-determined levels of a response variable that when exceeded will lead to an action or correction.

Time-series design (3, 6) In this design it is expected that the response of the animals to the disturbance will decrease over time; the animals are sampled at the same sites over time.

Treatment (2, 3) Any method, technique, or process that is designed to change the way a physical process works.

Trend (2, 7) A change in the trajectory of an ecological population over time.

Trend-by-time interaction (6) Here, continuous variables are used (rather than distinct levels) to compare trends between measures of the resource and levels of change (or impact) over time; used in trend-by-time interaction design. Contrast with *level-by-time interaction*.

Type I error (2, 3)	Error that occurs by rejecting the null hypothesis when it is true. See also *Type II error.*
Type II error (2, 3)	Error that occurs by accepting a null hypothesis when the alternative hypothesis is true. See also *Type I error.*
Unequal probability (3, 4)	A sampling procedure wherein samples are selected based on probabilities that are tied to the characteristics of the organisms' size or location. See also *simple random sampling.*
Unpaired study design (3)	A study design that estimates the effect of a treatment by examining the difference in the population mean for a selected parameter in a treated and control population.
Validation monitoring (7)	Used to evaluate whether established management direction (e.g., National Forest Plans) provides guidance to meet its stated objectives.

Chapter 1
Concepts for Wildlife Science: Theory

1.1 Introduction

We conduct wildlife studies in the pursuit of knowledge. Therefore, an understanding of what knowledge is and how it is acquired is foundational to wildlife science. Adequately addressing this topic is a daunting challenge for a single text because wildlife science is a synthetic discipline that encompasses aspects of a vast array of other academic disciplines. For example, many vibrant wildlife science programs include faculty who study molecular biology, animal physiology, biometrics, systems analysis, plant ecology, animal ecology, conservation biology, and environmental sociology, humanities, education, economics, policy, and law. The primary emphasis of this text is the design of wildlife-related field studies. Those addressing other aspects of wildlife science should find the text useful, but will undoubtedly require additional sources on design. For example, those interested in learning how to design quantitative or qualitative studies of how humans perceive wildlife-related issues will find the excellent texts by Dillman (2007) and Denzin and Lincoln (2005) useful.

The process of designing, conducting, and drawing conclusions from wildlife field studies draws from several disciplines. This process begins and ends with expert biological knowledge that comes from familiarity with the natural history of the system being studied. This familiarity should inspire meaningful questions about the system that are worth pursuing for management purposes or for the sake of knowledge alone. During the design and implementation of studies, this familiarity helps the researcher identify what is feasible with respect to practicality and budget. When the study is completed and the results are analyzed, this familiarity provides the researcher with perspective in drawing conclusions. Familiarity can, however, lead to tunnel vision when viewing the system and thus misses alternative explanations for observed phenomena. Therefore, to conduct wildlife science as objectively as possible, it is usually necessary to temper expert knowledge that comes from familiarity with principles drawn from other academic disciplines. We incorporate these concepts in later chapters that discuss sampling and specific study designs.

In this chapter, we begin by discussing philosophical issues as they relate to science. After all, it makes little sense to begin collecting data before clearly

M.L. Morrison et al., *Wildlife Study Design*.

understanding the nature of the entity being studied (*ontology*), what constitutes knowledge and how it is acquired (*epistemology*), and why one thinks the research question is valuable, the approach ethical, and the results important (*axiology*). Moreover, the philosophy of science provides a logical framework for generating meaningful and well-defined questions based on existing theory and the results of previous studies. It provides also a framework for combining the results of one's study into the larger body of knowledge about wildlife and for generating new questions, thus completing the feedback loop that characterizes science. For these reasons, we outline how scientific methodology helps us acquire valuable knowledge both in general and in specific regarding wildlife. We end the chapter with a brief discussion of terminology relevant to the remaining chapters.

1.2 Philosophy and Science

1.2.1 The Science Wars

In 1987, physicists Theo Theocharis and Michael Psimopoulos (1987) published an essay in *Nature,* where they referred to the preeminent philosophers of science Karl R. Popper, Thomas Kuhn, Imre Lakatos, and Paul Feyerabend as "betrayers of the truth" and Feyerabend as "currently the worst enemy of science" (pp. 596–597). According to Theocharis and Psimopoulos, by admitting an unavoidable social dimension to science, and that human perceptions of reality are to some degree social constructions, these and other philosophers of science working in the 1960s had enabled an avalanche of "erroneous and harmful … epistemological antitheses" of science (p. 595). They argued

> The problem is that although the epistemological antitheses are demonstrably untenable, inherently obscurantist and possibly dangerous, they have become alarmingly popular with the public, and even worse, with the communities of professional philosophers and scientists. (p. 598)

The result, Theocharis and Psimopoulos feared, was that "having lost their monopoly in the production of knowledge, scientists have also lost their privileged status in society" and the governmental largess to which they had become accustomed (p. 597).

Since the 1960s, entire academic subdisciplines devoted to critiquing science, and refereed journals associated with these endeavors, have become increasingly common and influential. The more radical members of this group often are called *postmodernists*. It is probably fair to say, however, that most scientists either were blissfully unaware of these critiques, or dismissed them as so much leftwing academic nonsense.

By the 1990s, however, other scientists began to join Theocharis and Psimopoulos with concerns about what they perceived to be attacks on the validity and value of science. Paul R. Gross and Norman Levitt (1994), with *Higher Superstition: The Academic Left and Its Quarrels with Science*, opened a frontal attack on critical studies of science. They argued that scholars in critical science studies knew little about science and used sloppy scholarship to grind political axes. Both the academic and mainstream press gave *Higher Superstition* substantial coverage, and "the science wars" were on.

In 1995, the New York Academy of Sciences hosted a conference entitled "The Flight from Science and Reason" (see Gross et al. (1997) for proceedings). These authors, in general, were also highly critical of what they perceived to be outrageous, politically motivated postmodern attacks on science. *Social Text*, a critical theory journal, prepared a special 1996 issue titled "Science Wars" in response to these criticisms. Although several articles made interesting points, if the essay by physicist Alan D. Sokal had not been included, most scientists and the mainstream media probably would have paid little attention. Sokal (1996b) purportedly argued that quantum physics supported trendy postmodern critiques of scientific objectivity. He simultaneously revealed elsewhere that his article was a parody perpetrated to see whether the journal editors would "publish an article liberally salted with nonsense if (a) it sounded good and (b) it flattered the editors' ideological preconceptions" (Sokal 1996a, p. 62). The "Sokal affair," as the hoax and its aftermath came to be known, brought the science wars to the attention of most scientists and humanists in academia through flurries of essays and letters to editors of academic publications. A number of books soon followed that addressed the Sokal affair and the science wars from various perspectives and with various degrees of acrimony (e.g., Sokal and Bricmont 1998; Koertge 1998; Hacking 1999; Ashman and Barringer 2001). At the same time, the public, no doubt already somewhat cynical about academic humanists and scientists alike, read their fill about the science wars in the mainstream media. Like all wars, there were probably no winners. A more relevant question is the degree to which all combatants lost.

A student of wildlife science might well ask, "How can Karl Popper be one of the more notorious enemies of science" and "If science is an objective, rational enterprise addressing material realities, how can there be any argument about the nature of scientific knowledge, let alone the sometimes vicious attacks seen in the science wars?" These are fair questions. Our discussion of ontology, epistemology, and axiology in science, making up the remainder of Sect. 1.2, should help answer these and related questions and simultaneously serve as a brief philosophical foundation for the rest of the book.

1.2.2 The Nature of Reality

If asked to define reality, most contemporary scientists would probably find the question somewhat silly. After all, is not reality the state of the material universe around us? In philosophy, *ontology* is the study of the nature of reality, being, or existence. Since Aristotle (384–322 B.C.), the *empiricist* tradition of philosophy has held that material reality was indeed largely independent of human thought and best understood through experience. Science is still informed to a large degree through this empiricist perception. *Rationalism*, however, has an equally long tradition in philosophy. Rationalists such as Pythagoras (ca. 582–507 B.C.), Socrates (ca. 470–399 B.C.), and Plato (427/428–348 B.C.) argued that the ideal, grounded in reason, was in many ways more "real" than the material. From this perspective, the criterion for reality was not sensory experience, but instead was intellectual and deductive. Certain aspects of this perspective are still an integral part of modern science. For many contemporary philosophers, social scientists, and humanists, however, reality is ultimately a social construction (Berger and Luckmann 1966). That is, reality is to some degree contingent upon human perceptions and social interactions (Lincoln and Guba 1985; Jasinoff et al. 1995). While philosophers voiced arguments consistent with *social constructionism* as far back as the writing of Heraclitus (ca. 535–475 B.C.), this perspective toward the nature of being became well established during the mid-twentieth century.

Whatever the precise nature of reality, *knowledge* is society's accepted portrayal of it. Over the centuries, societies have – mistakenly or not – accessed knowledge through a variety of methods, including experience, astrology, experimentation, religion, science, and mysticism (Rosenberg 2000; Kitcher 2001). Because the quest for knowledge is fundamental to wildlife science, we now flesh out the permutations of knowing and knowledge acquisition.

1.2.3 Knowledge

What is knowledge, how is knowledge acquired, and what is it that we know? These are the questions central to *epistemology*, the branch of Western philosophy that studies the nature and scope of knowledge. The type of knowledge typically discussed in epistemology is propositional, or "knowing-that" as opposed to "knowing-how," knowledge. For example, in mathematics, one "knows that" 2 + 2 = 4, but "knows how" to add.

In Plato's dialogue *Theaetetus* (Plato [ca. 369 B.C.] 1973), Socrates concluded that *knowledge* was justified true belief. Under this definition, for a person to know a proposition, it must be true and he or she must simultaneously believe the proposition and be able to provide a sound justification for it. For example, if your friend said she knew that a tornado would level her house in exactly 365 days, and the destruction indeed occurred precisely as predicted, she still would not have known of the event 12 months in advance because she could not have provided a rational justification for her belief despite the fact that it turned out later to be true. On the other hand, if she said she knew a tornado would level her house sometime within the next 20 years, and showed you 150 years of records indicating that houses in her neighborhood were severely damaged by tornadoes approximately every 20 years, her statement would count as knowledge and tornado preparedness might be in order. This definition of knowledge survived without serious challenge by philosophers for thousands of years. It is also consistent with how most scientists perceive knowledge today.

Knowledge as justified true belief became a less adequate definition in the 1960s. First, Edmund L. Gettier (1963), in a remarkably brief paper (less than 3 pages), provided what he maintained were examples of beliefs that were both true and justified, but that should not be considered knowledge. In his and similar examples, the justified true belief depended on either false premises or justified false beliefs the protagonist was unaware of (see Box 1.1). Philosophers have been wrestling with the "Gettier problem" since then, and are yet to agree on a single definition of knowledge. A second problem with Plato's definition relates to ontology. If reality is to any degree socially constructed, then truth regarding this reality is to the same degree a social construct, and so society's accepted portrayal of a proposition – whether justified true belief or not – becomes a more relevant definition. At any rate, wildlife scientists attempt to acquire knowledge about wild animals, wildlife populations, and ecological systems of interest, and apply that knowledge to management and conservation, and so they must understand the nature of knowledge and knowledge acquisition.

Box 1.1 The Gettier Problem

Gettier (1963, pp. 122–123) provided the following two examples to illustrate the insufficiency of justified true belief as the definition of knowledge (see Plato [ca. 369 B.C.]1973).

Case I:

Suppose that Smith and Jones have applied for a certain job. And suppose that Smith has strong evidence for the following conjunctive proposition:

(d) Jones is the man who will get the job, and Jones has ten coins in his pocket.

Smith's evidence for (d) might be that the president of the company assured him that Jones would in the end be selected, and that he, Smith, had counted the coins in Jones's pocket ten minutes ago. Proposition (d) entails:

(e) The man who will g et the job has ten coins in his pocket.

Let us suppose that Smith sees the entailment from (d) to (e), and accepts (e) on grounds of (d), for which he has strong evidence. In this case, Smith is clearly justified in believing that (e) is true.

But imagine, further, that unknown to Smith, he himself, not Jones, will get the job. And, also, unknown to Smith, he himself has ten coins in his pocket. Proposition (e) is then true, though proposition (d), from which Smith inferred (e), is false. In our example, then, all of the following are true: (*i*) (e) is true, (*ii*) Smith believes that (e) is true, and (*iii*) Smith is justified in believing that (e) is true. But it is equally clear that Smith does not *know* that (e) is true; for (e) is true in virtue of the number of coins in Smith's pocket, while Smith does not know how many coins are in Smith's pocket, and bases his belief in (e) on a count of the coins in Jones's pocket, whom he falsely believes to be the man who will get the job.

(continued)

Box 1.1 (continued)

Case II:

Let us suppose that Smith has strong evidence for the following proposition:

(f) Jones owns a Ford.

Smith's evidence might be that Jones has at all times in the past within Smith's memory owned a car, and always a Ford, and that Jones has just offered Smith a ride while driving a Ford. Let us imagine, now, that Smith has another friend, Brown, of whose whereabouts he is totally ignorant. Smith selects three place names quite at random, and constructs the following three propositions:

(g) Either Jones owns a Ford, or Brown is in Boston;

(h) Either Jones owns a Ford, or Brown is in Barcelona;

(i) Either Jones owns a Ford, or Brown is in Brest-Litovsk.

Each of these propositions is entailed by (f). Imagine that Smith realizes the entailment of each of these propositions he has constructed by (f), and proceeds to accept (g), (h), and (i) on the basis of (f). Smith has correctly inferred (g), (h), and (i) from a proposition for which he has strong evidence. Smith is therefore completely justified in believing each of these three propositions. Smith, of course, has no idea where Brown is.

But imagine now that two further conditions hold. First, Jones does *not* own a Ford, but is at present driving a rented car. And secondly, by the sheerest coincidence, and entirely unknown to Smith, the place mentioned in proposition (h) happens really to be the place where Brown is. If these two conditions hold then Smith does *not* know that (h) is true, even though (*i*) (h) is true, (*ii*) Smith does believe that (h) is true, and (*iii*) Smith is justified in believing that (h) is true.

1.2.3.1 Knowledge Acquisition

Beginning with the Age of Enlightenment (seventeenth and eighteenth centuries), the *empiricist* tradition of inquiry exhibited new vigor. Important thinkers associated with the maturation of empiricism include Francis Bacon (1561–1626), John Locke (1632–1704), David Hume (1711–1776), and John Stuart Mill (1806–1873). From the empiricist perspective, we acquire knowledge only through experience, particularly as gained by observations of the natural world and carefully designed experiments. Thus, from a pure empiricist perspective, humans cannot know except by experience. Experience, however, can mean more than just counting or measuring things. For example, we can know by our senses that a fire is hot without measuring its precise temperature. Thus, physically sensing a phenomenon and employing metrics designed to quantify the magnitude of the phenomenon are both experiential.

Also during this period, philosophers informed by the *rationalist* tradition were busily honing their epistemological perspective. René Descartes (1596–1650), Baruch Spinoza (1632–1677), Gottfried Leibniz (1646–1716), and others are often associated with this epistemological tradition and were responsible for integrating mathematics into philosophy. For rationalists, reason takes precedence over experience for acquiring knowledge and, in principle, all knowledge can be acquired through reason alone. In practice, however, rationalists realized this was unlikely except in mathematics.

Philosophers during the Classical era probably would not have recognized any crisp distinction between empiricism and rationalism. The seventeenth century debate between Robert Boyle (1627–1691) and Thomas Hobbes (1588–1679) regarding Boyle's air pump experiments and the existence of vacuums fleshed out this division (Shapin and Schaffer 1985). Hobbes argued that only self-evident truths independent of the biophysical could form knowledge, while Boyle promoted experimental verification, where knowledge was reliably produced in a laboratory and independent of the researcher (Latour 1993). Even in the seventeenth century, many rationalists found empirical science important, and some empiricists were closer to Descartes methodologically and theoretically than were certain rationalists (e.g., Spinoza and Leibniz). Further, Immanuel Kant (1724–1804) began as rationalist, then studied Hume and developed an influential blend of rationalist and empiricist traditions. At least two important combinations of empiricism and certain aspects of rationalism followed.

One of these syntheses, *pragmatism*, remains the only major American philosophical movement. Pragmatism originated with Charles Saunders Peirce (1839–1914) in the early 1870s and was further developed and popularized by William James (1842–1910), John Dewey (1859–1952), and others. Peirce, James, and Dewey all were members of The Metaphysical Club in Cambridge, Massachusetts, during the 1870s and undoubtedly discussed pragmatism at length. Their perspectives on pragmatism were influenced by Kant, Mill, and Georg W.F. Hegel (1770–1831), respectively (Haack and Lane 2006, p. 10), although other thinkers such as Bacon and Hume were undoubtedly influential as well. James perceived pragmatism as a synthesis of what he termed the "tough-minded empiricist" (e.g., "materialistic, pessimistic, ... pluralistic, skeptical"), and "tender-minded rationalist" (e.g., "idealistic, optimistic, ... monistic, dogmatical") traditions of philosophy (1907, p. 12). Similarly, Dewey argued that pragmatism represented a marriage between the best of empiricism and rationalism (Haack 2006, pp. 33–40). James (1912, pp. 41–44) maintained the result of this conjunction was a "radical empiricism" that must be directly experienced. As he put it,

> To be radical, an empiricism must neither admit into its constructions any element that is not directly experienced, nor exclude from them any element that is directly experienced. ... a real place must be found for every kind of thing experienced, whether term or relation, in the final philosophic arrangement. (p. 42)

To these classical pragmatists, at least, the merits of even experimentation and observation were weighed by direct experience. Pragmatism is one of the most active fields of philosophy today. For this reason, there are several versions of neo-pragmatism that differ in substantive ways from the classical pragmatism of Peirce, James, Dewey, or George H. Mead (1863–1931). However, philosophers who consider themselves pragmatists generally hold that truth, knowledge, and theory are inexorably connected with practical consequences, or real effects.

The other important philosophical blend of empiricism and rationalism, *logical positivism* (later members of this movement called themselves *logical empiricists*), emerged during the 1920s and 1930s from the work of Moritz Schlick (1882–1936) and his Vienna Circle, and Hans Reichenbach (1891–1953) and his Berlin Circle (Rosenberg 2000). Logical positivists maintain that a statement is meaningful only if it is (1) analytical (e.g., mathematical equations) or (2) can reasonably be verified empirically. To logical positivists, *ethics* and *aesthetics*, for example, are *metaphysical* and thus scientifically meaningless because one cannot evaluate such arguments analytically or empirically. A common, often implicit assumption of those informed by logical positivism is that given sufficient ingenuity, technology, and time, scientists can ultimately come to understand material reality in all its complexity. Similarly, the notion that researchers should work down to the ultimate elements of the system of interest (to either natural or social scientists), and then build the causal relationships back to eventually develop a complete explanation of the universe in question, tends to characterize logical positivism as well. The recently completed mapping of the human genome and promised medical breakthroughs related to this genomic map characterizes this tendency.

The publication of Karl R. Popper's (1902–1994) *Logik der Forschung* by the Vienna Circle in 1934 (given a 1935 imprint) called into question the sufficiency of logical positivism. After the chaos of WWII, Popper translated the book into English and published it as *The Logic of Scientific Discovery* in 1959. Popper's (1962) perspectives were further developed in *Conjectures and Refutations: The Growth of Scientific Knowledge*. Unlike most positivists, Popper was not concerned with distinguishing meaningful from meaningless statements or verification, but rather distinguishing scientific from metaphysical statements using falsification. For him, metaphysical statements were unfalsifiable, while scientific statements could potentially be falsified. On this basis, scientists should ignore metaphysical contentions; instead, they should deductively derive tests for *hypotheses* that could lead to falsification. This approach often is called the *hypothetico–deductive model* of science. Popper argued that hypotheses that did not withstand a rigorous test should immediately be rejected and researchers should then move on to alternatives that were more productive. He acknowledged, however, that metaphysical statements in one era could become scientific later if they became falsifiable (e.g., due to changes in technology). Under Popper's model of science, while material reality probably exists, the best scientists can do is determine what it is not, by systematically falsifying hypotheses related to the topic of interest. Thus, for Popperians, knowledge regarding an issue is approximated by the explanatory hypothesis that has best survived substantive experimental challenges to date. From this perspective, often called *postpositivism*, knowledge ultimately is conjectural and can be modified based on further investigation.

Physicist Thomas Kuhn (1922–1996), in *The Structure of Scientific Revolutions* (1962), also argued that because science contains a social dimension it does not operate under the simple logical framework outlined by the logical positivists. His publication originally was part of the *International Encyclopedia of Unified Science* begun by the Vienna Circle. Kuhn's model of science includes "*normal science*," or periods where there is general consensus within a scientific community regarding theory, methods, terminology, and types of experiments likely to contribute useful insights. He argued that although advances occur during normal science, they are typically incremental in nature. Normal science at some point is interrupted by "*revolutionary science*," where a shift in *paradigm* occurs, followed by a new version of normal science, and eventually another paradigm shift, and so on. Kuhn argued that transition from an old to a new paradigm is neither rapid nor seamless, largely because the two paradigms are incommensurable. That is, a paradigm shift is not just about transformation of theory, but includes fundamental changes in terminology, how scientists perceive their field, and perhaps most importantly, what questions are deemed valid and what decision rules and methodological approaches are determined appropriate for evaluating scientific concepts. Thence new paradigms are not simply extensions of the old, but radically new worldviews, or as he put it, "*scientific revolutions*." Despite the importance of societal influences, Kuhn's model of science resonated with scientists because it provided a workable explanation for the obvious revolutionary changes observed historically in science (e.g., ecology from the Linnaean versus Darwinian perspective; Worster 1994).

Imre Lakatos (1922–1974) attempted to resolve the perceived conflict between Popper's falsification and Kuhn's revolutionary models of science in *Falsification and the Methodology of Scientific Research Programmes* (1970). He held that groups of scientists involved in a research program shielded the theoretical core of their efforts with a "protective belt" of hypotheses related to their central area of inquiry. Hypotheses within this belt could be found inadequate while the core theoretical construct remained protected from falsification. This approach protected the core ideas from premature rejection due to anomalies or other problems, something many viewed as a shortcoming of Popper's model of science. Under Lakatos' model, the question is not whether a given hypothesis is false, but whether a research program is progressive (marked by growth and discovery and potentially leading to a shift in paradigm) or degenerative (marked by lack of growth and novel facts and leading to oblivion).

Paul Feyerabend (1924–1994) took the cultural aspects of science further. In *Against Method* and *Science in a Free Society* (Feyerabend 1975, 1978, respectively), he argued that there was no single prescriptive scientific method, that such a method – if it existed – would seriously limit scientists and thus scientific progress, and that science would benefit from theoretical anarchism in large part because of its obsession with its own mythology. Feyerabend maintained that Lakatos' philosophy of research programs was actually "anarchism in disguise" (Feyerabend 1975, p. 14), because it essentially argued that there was no single, prescriptive scientific method. Feyerabend also challenged the notion that scientists or anyone else could objectively compare scientific theories. After all, as Kuhn had previously pointed out, scientific paradigms were incommensurable and so they were incomparable by definition. Feyerabend went on to argue that the condescending attitudes many scientists exhibited toward astrology, voodoo, folk magic, or alternative medicine had more to do with elitism and racism than to the superiority of science as an epistemological approach.

While Popper placed a small wedge in the door of natural science's near immunity to social criticism, Kuhn, Lakatos, Feyerabend, and other philosophers of science working in the 1960s and 1970s tore it from its hinges. Those interested in learning about these philosophies should read *Criticism and the Growth of Knowledge* (Lakatos and Musgrave 1970). This volume is based on a 1965 symposium, chaired by Popper, where the leading philosophers of science, including Popper, Lakatos, and Feyerabend, critiqued Kuhn's revolutionary model of science, and he responded to their criticisms. Similarly, Lakatos and Feyerabend's (1999) posthumous work, *For and Against Method*, further clarifies these authors' perspectives toward the philosophy of natural science from 1968 through 1974 (Lakatos died in February 1974). Whatever the merit of these philosophies, the juggernaut of critical studies of science, grounded in constructivist epistemology, had been unleashed.

Social constructivism is based on the philosophical perspective that all knowledge is ultimately a social construction regardless of whether material reality exists (Berger and Luckmann 1966; Lincoln and Guba 1985). After all, humans cannot escape being human; they know only through the lens of experience, perception, and social convention. For this reason, our individual and collective perspectives toward race, ethnicity, gender, sexuality, and *anthropocentrism*, to name only a few, form an important component of our knowledge on any topic, including science. Although the thinking of Hegel, Karl Marx (1818–1883), and Émile Durkheim (1858–1917) were important to the development of constructivism, Peter L. Berger and Thomas Luckmann's (1966) *The Social Construction of Reality* greatly enhanced the prominence of social constructionism, particularly in the United States. Constructionist critiques make use of *dialectic approaches*, or discussion and reasoning by logical dialogue, as the method of intellectual investigation. There are now an imposing array of subdisciplines and related academic journals in the humanities and social sciences informed by social constructionism that are dedicated to the critical study of science. These include ethnographic accounts of science, feminist studies of science, the rhetoric of science, and social studies of science. For those who take epistemology seriously, social constructivism has moved into a mainstream position from where it functions as an integrator for researchers working from empiricist, rationalist, pragmatist, and logical positivist perspectives.

1.2.3.2 Inductive, Deductive, and Retroductive Reasoning

Regardless of the epistemological approach one is informed by, logical thought remains an integral component of the process. During the Classical era, Aristotle and others developed important aspects of logical reasoning (Table 1.1). *Induction* con-

Table 1.1 The purpose, logical definition, and verbal description of inductive, deductive, and retroductive reasoning, given the preconditions α, postconditions β, and the rule R_1: $\alpha \rightarrow \beta$ (α therefore β; after Menzies 1996)

Method	Purpose	Definition[a,b]	Description
Induction	Determining R_1	$\alpha \rightarrow \beta \Rightarrow R_1$	Learning the rule (R_1) after numerous examples of α and β
Deduction	Determining β	$\alpha \wedge R_1 \Rightarrow \beta$	Using the rule (R_1) and its preconditions (α) to deterministically make a conclusion (β)
Retroduction	Determining α	$\beta \wedge R_1 \Rightarrow \alpha$	Using the postcondition (β) and the rule (R_1) to hypothesize the preconditions (α) that could best explain the observed postconditions (β)

[a] $\rightarrow$, $\wedge$, and $\Rightarrow$ signify "therefore," "and," and "logically implies," respectively
[b] Note that deduction and retroduction employ the same form of logical statement to determine either the post- or precondition, respectively

sists of forming general conclusions based on multiple instances, where a class of facts appears to entail another [e.g., each of thousands of common ravens (*Corvus corax*) observed were black, therefore all common ravens are black]. Stated differently, we believe the premises of the argument support the conclusion, but they cannot ensure it, and the strength of the induction depends in part on how large and representative the collection of facts is that we have to work with. *Deduction* consists of deriving a conclusion necessitated by general or universal premises, often in the form of a *syllogism* [e.g., all animals are mortal, northern bobwhites (*Colinus virginianus*) are animals, therefore northern bobwhites are mortal]. If the premises indeed are true, then the conclusion by definition must be true as well. Although philosophers of previous generations often perceived induction and deduction to be competing methods of reasoning, Peirce demonstrated that they actually were complementary (Haack 2006). Essentially, inductively derived general rules serve as the basis for deductions; similarly, should the deductive consequences turn out experimentally other than predicted, then the inductively derived general rule is called into question. Peirce also proposed a third type of logical reasoning he initially called *abduction*; he later referred to this concept as *retroduction* (retroduction hereafter; some philosophers argue that retroduction is a special case of induction and others argue that Peirce did not always use abduction and retroduction synonymously). Retroduction consists of developing a hypothesis that would, if true, best explain a particular set of observations (Table 1.1). Retroductive reasoning begins with a set of observations or facts, and then infers the most likely or best explanation to account for these facts (e.g., all the eggs in a northern bobwhite nest disappeared overnight and there were no shell fragments, animal tracks, or disturbance of leaf litter at the nest site, therefore a snake is the most likely predator).

All three forms of reasoning are important epistemologically. Retroductive reasoning is in many ways the most interesting because it is much more likely to result in novel explanations for puzzling phenomena than are induction or deduction. It is also much more likely to be wrong! Inductive reasoning is an effective way to derive important principles of association and is less likely to prove incorrect than retroduction. It has been the workhorse of science for centuries. Deductively derived conclusions are uninteresting in themselves; after all, they follow deterministically from the major premise. Instead, the value of deductive reasoning is that it allows us to devise ways to critically challenge and evaluate retroductively developed hypotheses or inductively derived rules of association.

1.2.4 Values and Science

Axiology is the study of value or quality. The nature, types, and criteria of values and value judgments are critical to science. At least three aspects of value are directly relevant to our discussion: (1) researcher ethics, (2) personal values researchers bring to science, and (3) how we determine the quality of research.

Ethics in science runs the gambit from humane and appropriate treatment of animal or human subjects to honesty in recording, evaluating, and reporting data. Plagiarism, fabrication and falsification of data, and misallocation of credit by scientists are all too often news headlines. While these ethical problems are rare, any fraud or deception by scientists undermines the entire scientific enterprise. Ethical concerns led the National Academy of Sciences (USA) to form the Committee on the Conduct of Science to provide guidelines primarily for students beginning careers in scientific research (Committee on the Conduct of Science 1989). All graduate students should read the updated and expanded version of this report (Committee on Science, Engineering, and Public Policy 1995). It also serves as a brief refresher for more seasoned scientists.

Perhaps two brief case studies will help put bookends around ethical issues and concerns in science. The first involves Hwang Woo-suk's meteoric rise to the pinnacle of fame as a stem-cell researcher, and his even more rapid fall from grace. He and his colleagues published two articles in *Science* reporting truly remarkable results in 2004 and 2005. These publications brought his laboratory, Seoul National University, and South Korea to the global forefront in stem cell research, and Professor Hwang became a national hero nearly overnight. The only problem was that Woo-suk and his coauthors fabricated data used in the two papers (Kennedy 2006). Additional ethical problems relating to sources of human embryos also soon surfaced. In less than a month (beginning on 23 December 2005), a governmental probe found the data were fabricated, Dr. Hwang admitted culpability and resigned his professorship in disgrace, and the editors of *Science* retracted the two articles with an apology to referees and those attempting to replicate the two studies. This episode was a severe disgrace for Professor Hwang, Seoul National University, the nation of South Korea, and the entire scientific community.

Although breaches of ethics similar to those in the previous example receive considerable media attention and near universal condemnation, ethical problems in science often are more insidious and thence less easily recognized and condemned. Wolff-Michael Roth and Michael Bowen (2001) described an excellent example of the latter. They used ethnographic approaches to explore the enculturation process of upper division undergraduate and entry level graduate student researchers beginning their careers in field ecology. These students typically had little or no direct supervision at their study areas and had to grapple independently with the myriad problems inherent to fieldwork. Although they had reproduced experiments as part of highly choreographed laboratory courses (e.g., chemistry), these exercises probably were more a hindrance than a help. In these choreographed exercises, the correct results were never in doubt, only the students' ability to reproduce them was in question. Roth and Bowen found that the desire to obtain the "right" or expected results carried over to fieldwork. Specifically, one student was to replicate a 17-year old study. He had a concise description of the layout, including maps. Unfortunately, he was unable to interpret the description and maps well enough to lay out transects identical to those used previously, despite the fact that most of the steel posts marking the original transects still were in place (he overlooked the effects of topographical variation and other issues). He knew the layout was incorrect, as older trees were not where he expected them to be. Instead of obtaining expert assistance and starting over, he bent "linear" transects to make things work out, assumed the previous researcher had incorrectly identified trees, and that published field guides contained major errors. " 'Creative solutions,' 'fibbing,' and differences that 'do not matter' characterized his work…" (p. 537). He also hid a major error out of concern for grades. As he put it

> I am programmed to save my ass. And saving my ass manifests itself in getting the best mark I can by compromising the scruples that others hold dear.…That's what I am made of. That is what life taught me. (p. 543)

Of course, his "replication" was not a replication at all, but this fact would not be obvious to anyone reading a final report. Roth and Bowen (2001) concluded that

> …the culture of university ecology may actually encourage students to produce 'creative solutions' to make discrepancies disappear. The pressures that arise from getting right answers encourage students to 'fib' and hide the errors that they know they have committed. (p. 552)

While this example of unethical behavior by a student researcher might not seem as egregious as the previous example, it actually is exactly the same ethical problem; both researchers produced data fraudulently so that their work would appear better than it actually was for purposes of self-aggrandizement.

Another important axiological area relates to the values researchers bring to science. For example, Thomas Chrowder Chamberlin (1890; 1843–1928) argued that scientists should make use of multiple working hypotheses to help protect themselves from their own biases and to ensure they did not develop tunnel vision. John R. Platt (1964) rediscovered Chamberlin's contention and presented it to a new generation of scientists (see Sect. 1.4.1 for details). That researchers' values impinge to some degree upon their science cannot be doubted. This is one of the reasons philosophers such as Kuhn, Lakatos, and Feyerabend maintained there were cultural aspects of science regardless of scientists' attempts to be "objective" and "unbiased" (see Sect. 1.2.3.1). Moreover, scientists' values are directly relevant to social constructionism and thence critical studies of science.

Finally, how do scientists determine whether knowledge they are developing, however they define such knowledge, matters? To whom does it matter (e.g., themselves, colleagues, some constituency, society)? How do scientists and the public determine whether science is of high quality? These also are axiological questions without clear answers. Even within the scientific community, researchers debate the answers (e.g., the merits of basic versus applied science). Such judgments hinge on one's values, and thus are axiological. To complicate matters further, these questions must be answered at multiple scales. One person might think that science that actually changes things on the ground to directly benefit wildlife conservation, for example, is the most valuable sort of scientific inquiry. If this person is an academician, however, he or she cannot safely ignore what colleagues working for funding agencies, peer reviewed journals, or tenure and promotion committees perceive to be valuable work. We could make the same sort of argument for scientists who maintain that theoretical breakthroughs are the ultimate metric of quality in science. Moreover, society is influenced by, and influences, these value judgments. Society ultimately controls the purse strings for governmental, industrial, and nongovernmental organizations and thus indirectly, scientific funding. In sum, the quality of scientific knowledge is important to scientists and nonscientists alike, and social influence on the scientific process – at whatever scale – is axiomatic.

1.3 Science and Method

In a general sense, science is a process used to learn how the world works. As discussed in Sect. 1.2, humans have used a variety of approaches for explaining the world around them, including mysticism, religion, sorcery, and astrology, as well as science. The scientific revolution propelled science to the forefront during the last few centuries, and despite its shortcomings, natural science (the physical and life sciences) has been remarkably effective in explaining the world around us (Haack 2003). What is it about the methods of natural science that has proven so successful? Here, we address this question for natural sciences in general and wildlife science in particular. We begin this task by discussing research studies designed to evaluate *research hypotheses* or *conceptual models*. We end this section by contextualizing how *impact assessment* and studies designed to *inventory* or *monitor* species of interest fit within the methods of natural science. The remainder of the book addresses specifics as they apply to wildlife science.

1.3.1 Natural Science Research

We avoided the temptation to label Sect. 1.3 "The Scientific Method." After all, as Sect. 1.2 amply illustrates, there is no single philosophy, let alone method, of science. As philosopher of science Susan Haack (2003, p. 95) put it, "Controlled

Table 1.2 Typical steps used in the process of conducting natural science

1	Observe the system of interest
2	Identify a broad research problem or general question of interest
3	Conduct a thorough review of the refereed literature
4	Identify general research objectives
5	In light of these objectives, theory, published research results, and possibly a pilot study, formulate specific research hypotheses and/or a conceptual model
6	Design (1) a manipulative experiment to test whether conclusions derived deductively from each research hypothesis are supported by data or (2) another type of study to evaluate one or more aspects of each hypothesis or the conceptual model
7	Obtain peer reviews of the research proposal and revise as needed.
8	Conduct a pilot study if needed to ensure the design is practicable. If necessary, circle back to steps 6 or 5
9	Conduct the study
10	Analyze the data
11	Evaluate and interpret the data in light of the hypotheses or model being evaluated. Draw conclusions based on data evaluation and interpretation as well as previously published literature
12	Publish results in refereed outlets and present results at scientific meetings
13	In light of the results and feedback from the scientific community, circle back and repeat the process beginning with steps 5, 4, or even steps 3, 2, or 1, as appropriate

experiments, for example – sometimes thought of as distinctive of the sciences – aren't used by all scientists, or only scientists; astronomers and evolutionary theorists don't use them, but auto mechanics, plumbers, and cooks do." The lack of a single, universal scientific method, however, does not imply that the natural sciences do not employ certain intellectual and methodological approaches in common. Here, we discuss general steps (Table 1.2) and the feedback process typically used during research in the natural sciences.

Readers should not take the precise number of steps we presented in Table 1.2 too literally. Others have suggested taxonomies for scientific research with as few as 4 and as many as 16 steps (e.g., Platt 1964; Ford 2000; Garton et al. 2005). These differences are typically matters of lumping or splitting to emphasize points the authors wished to make. Instead, readers should focus on (1) the importance of familiarity with the system of interest, the question being addressed, and the related scientific literature, (2) the role of research hypotheses and/or conceptual models and how they relate to theory and objectives, (3) appropriate study design, execution, and data analysis, (4) obtaining feedback from other scientists at various stages in the process, including through publication in referred outlets, and (5) the circular nature of science.

Step 1 (Table 1.2) is an obvious place to start because progress in science begins with researchers becoming familiar with the system of interest. This helps one to identify a research area of interest as well as develop important questions to be answered. One way to enhance this familiarity is to conduct a thorough review of the relevant scientific literature. This facilitates a better understanding of existing *theory* and previous research results relevant to the system and research objectives. By making numerous observations over time and studying similar systems in the scientific literature, one can inductively derive rules of association among classes of facts based on theory regarding how some aspect of the system works (see Guthery (2004) for a discussion of facts and science). Similarly, one can retroductively derive hypotheses that account for interesting phenomena observed (see Guthery et al. (2004) for a discussion of hypotheses in wildlife science). The development of research hypotheses and/or conceptual models that explain observed phenomena is a key attribute of the scientific process.

Step 6 (Table 1.2) is the principal topic of this book; we discuss the details in subsequent chapters. In general, one either designs a manipulative experiment to test whether conclusions derived deductively from one or more research hypotheses are supported by data, or designs another type of study to evaluate one or more aspects of each hypothesis or conceptual model. There are basic principles of design that are appropriate for any application, but researchers must customize the details to fit specific objectives, the scope of their study, and the system or subsystem being studied. It is critically important at this juncture to formally draft a research proposal and have it critically reviewed by knowledgeable peers. It is also important to consider how much effort will be required to achieve the study objectives. This is an exercise in approximation and requires consideration of how the researcher will analyze collected data, but can help identify cases where the effort required is beyond the capabilities and budget of the investigator, and perhaps thereby prevent wasted effort. *Pilot studies* can be critical here; they help researchers determine whether data collection methods are workable and appropriate, and also serve as sources of data for sample size calculations. We consider sample size further in Sect. 2.5.7.

Once the design is evaluated and revised, the researcher conducts the study and analyzes the resulting data (steps 9–10, Table 1.2). In subsequent chapters, we discuss practical tips and pitfalls in conducting wildlife field studies, in addition to general design considerations. We do not emphasize analytic methods because an adequate exposition of statistical methodology is beyond the scope of this book. Regardless, researchers must consider some aspects of statistical inference during the design stage. In fact, the investigator should think about the entire study process, including data analysis and even manuscript preparation (including table and figure layout), in as much detail as possible from the beginning. This will have implications for study design, especially sampling effort.

On the basis of the results of data analysis, predictions derived from the hypotheses or conceptual models are compared against the results, and interpretations are made and conclusions drawn (step 11, Table 1.2). The researcher then compares and contrasts these results and conclusions with those of similar work published in the refereed literature. Researchers then must present their results at professional meetings and publish them in refereed journals. A key aspect of science is obtaining feedback from other scientists. It is difficult to adequately accomplish this goal without publishing in scientific journals. Remember, if a research project was worth conducting in the first place, the results are worth publishing in a refereed outlet. We hasten to add that sometimes field research studies, particularly, do not work out as planned. This fact does not necessarily imply that the researcher did not learn something useful or that the effort was unscientific. In subsequent chapters, we discuss ways to salvage field studies that went awry. Similarly, some management-oriented studies do not lend themselves to publication in refereed outlets (see Sect. 1.3.2 for more details).

This brings us to possibly the single most important aspect of the scientific process, the feedback loop inherent to scientific thinking (step 13, Table 1.2). Once researchers complete a study and publish the results, they take advantage of what they learned and feedback from the scientific community. They then use this new perspective to circle back and repeat the process beginning with steps 5, 4, or possibly even steps 3, 2, or 1 (Table 1.2). In other words, researchers might need to begin by formulating new hypotheses or by modifying conceptual models addressing the same objectives used previously. In some cases, however, they might need to rethink the objectives or conduct additional literature reviews and descriptive studies. This reflexive and reflective thinking is the essence of science.

Although a broad research program typically uses all the steps outlined in Table 1.2 and discussed above, not all individual research projects or publications necessarily do so. Instead, different researchers often address different aspects of the same research program. For example, the landmark publications on the equilibrium theory of island biogeography by Robert H. MacArthur and Edward O. Wilson (1967) and MacArthur (1972) focused primarily on steps 1–5 (Table 1.2). They conducted thorough literature reviews and used the results of numerous observational studies to develop their theoretical perspective. From it, they deductively derived four major predictions. Experimental tests and other evaluations of these predictions were left primarily to others (e.g., Simberloff and Wilson 1969; Wilson and Simberloff 1969; Diamond 1972; Simberloff 1976a,b; Wilcox 1978; Williamson 1981). These and other publications provided feedback on equilibrium theory. At a more practical level, this continuously modified theoretical perspective toward the nature of islands still informs protected area design, linkage, and management, because wildlife refuges and other protected areas are increasingly becoming islands in seas of cultivation, urban sprawl, or other *anthropogenic* landscape changes (see Diamond 1975, 1976; Simberloff and Abele 1982; Whittaker et al. 2005). The point here is that not all useful research projects must employ all 13 of our steps (Table 1.2). Some might produce descriptive data that other researchers use to develop theoretical breakthroughs, while other researchers experimentally test or otherwise evaluate theoretically driven hypotheses, and still others could employ this information to produce important syntheses that close the feedback loop or support specific applications.

1.3.2 Impact Assessment, Inventorying, and Monitoring

Natural resource management agencies often implement field studies to collect data needed for management decision making (often required to do so by statute) rather than to test hypotheses or evaluate conceptual models. For example, agencies may need to determine which species of interest occur on a state wildlife management area or another tract of land (*inventory*). They also commonly need to monitor species of interest. After all, it is difficult to know whether management plans designed to increase abundance of an endangered species are effective without reliably *monitoring* the species' abundance over time. Similarly, state wildlife agencies must monitor intensely hunted elk (*Cervus elaphus*) populations if they are to regulate harvest safely and effectively. Further, state or federal management agencies or environmental consulting companies might need to determine the impact of proposed wind plants, highways, or other developments on wildlife or their habitat. Agencies also might want to evaluate the impact of an intense wildfire, a 100-year flood on a riparian area, or a proposed management treatment. Such *impact assessment* often cannot be conducted using replicated manipulative experiments with adequate controls; moreover, one rarely can assign treatments (e.g., floods, wind turbine locations) probabilistically. Despite the limitations of surveys (e.g., inventorying, monitoring) and impact assessment, these are among the most common types of wildlife studies and are important for natural resource management.

These management-oriented studies typically must employ more constrained study designs than those used for "ideal" replicated manipulative experiments. This does not imply, however, that wildlife scientists can safely ignore scientific methodology when designing these studies. Close attention to all details under the biologist's control is still critical. When study planning begins for inventorying, monitoring, and impact assessments, biologists typically already have much of the information listed for steps 1–5 in Table 1.2, although additional review of the literature probably is required. Appropriate study design, the importance of peer reviews of the proposed design, possibly a pilot study, data analysis, and data evaluation are just as important as with other sorts of wildlife research (see Sect. 1.3.1). Some impact assessments and extensive inventories lend themselves to publication in refereed outlets, and steps 12–13 (Table 1.2) follow as outlined in Sect. 1.3.1. In other cases, however, a single impact analysis, an inventory of a state wildlife management area, or the first few years of monitoring data are not suitable for publication in refereed outlets. This does not imply that these data were collected inappropriately or are unimportant. Instead, the purposes for data collection were different. In these cases, however, it is still critical for wildlife scientists to obtain feedback from peers not involved with these projects by presenting results at scientific meetings or via other approaches so that the feedback loop represented by steps 12–13 (Table 1.2) is completed.

Finally, the value of many impact assessments, inventories, and monitoring goes beyond immediate relevance to wildlife management, although this certainly is reason enough to conduct these studies. Researchers interested in complex ecological phenomena, for example, could conduct a metaanalysis (Arnqvist and Wooster 1995; Osenberg et al. 1999; Gurevitch and Hedges 2001; Johnson 2002) using numerous impact assessments that address the same sort of impacts. These studies also could serve as part of a metareplication (Johnson 2002). If researchers have access to raw data from multiple impact assessments or surveys, they can evaluate these data to address ecological and conservation questions beyond the scope of an individual field survey. Syntheses using multiple sets of data include some of the more influential ecology and conservation publications in recent years (e.g., Costanza et al. 1997; Vitousek et al. 1997; Myers et al. 2000; Jackson et al. 2001). Such analyses can be extraordinarily effective approaches epistemologically, and typically would not be possible without basic long-term survey data, impact assessments, and other studies that individually might have limited scope. At any rate, impact assessment, inventorying, and monitoring are so important to wildlife ecology and management that we deal with these topics to some extent in all subsequent chapters. Moreover, Chaps. 6 and 7 are devoted entirely to discussions of impact assessment and inventory and monitoring studies, respectively.

1.4 Wildlife Science, Method, and Knowledge

Thus far, we primarily have addressed natural science generally. Here we attempt to place wildlife science more specifically within the context of the philosophy of natural science. One way wildlife scientists have contextualized their discipline is by comparing what is actually done to what they consider to be ideal based on the philosophy of science. As we have seen in Sect. 1.2, however, the ideal was somewhat a moving target during the twentieth century. Additionally, the understandable tendency of wildlife scientists to cite one another's second, third, or fourth hand summaries of Popper or Kuhn's ideas, for example, rather than read these philosophers' writings themselves, further clouded this target. For this reason, many publications citing Popper or Kuhn do not accurately represent these authors' ideas. Here we discuss a few critiques of science by scientists that influenced how researchers conduct wildlife science. We then attempt to contextualize where wildlife science falls today within the philosophy of natural science.

1.4.1 Methodological Challenges

Critiques of scientific methodology written by natural scientists, as opposed to philosophers or social scientists, have greatly influenced how investigators conduct wildlife ecology and conservation research. One reason these publications were so influential is they were more accessible to wildlife scientists than philosophical tomes or social studies of science that some might argue were more agenda-driven deconstructions of science than constructive criticisms.

One of the most influential critiques of science by a scientist was "Strong Inference" by Platt (1964; originally titled "The New Baconians"). One reason Platt's essay in *Science* was so influential was that, directly or indirectly, it introduced wildlife scientists to Poppers' hypothetico–deductive method of science (1959, 1962), Kuhn's (1962) idea of normal versus revolutionary science, and Chamberlin's (1890) call for multiple working hypotheses. Briefly, Platt (1964) argued that The New Baconians, exemplified by leading researchers in molecular biology and high-energy physics, made much more rapid scientific progress and significant breakthroughs than did those working in other natural sciences because they utilized an approach he called strong inference. Platt maintained that strong inference was nothing more than an updated version of Bacon's method of inductive inference. Specifically, he argued, researchers should (1) inductively develop multiple alternative hypotheses (after Chamberlin 1890), (2) deduce from these a critical series of outcomes for each hypothesis, then devise a crucial experiment or series of experiments that could lead to the elimination of one or more of the hypotheses (after Popper 1959, 1962), (3) obtain decisive results through experimentation, and (4) recycle the procedure to eliminate subsidiary hypotheses. He also argued that these New Baconians used logic trees to work out what sort of hypotheses and questions they should address next. He provided numerous examples of extraordinarily productive scientists whom he felt had used this approach. As Platt concluded (1964, p. 352)

> The man to watch, the man to put your money on, is not the man who wants to make "a survey" or a "more detailed study" but the man with the notebook, the man with the alternative hypotheses and the crucial experiments, the man who knows how to answer your Question of disproof and is already working on it.

Rowland H. Davis (2006) maintained that while Platt's (1964) essay was influential in an array of natural and social sciences, it probably had its greatest impact in ecology. One reason was that in 1983, the *American Naturalist* prepared a dedicated issue titled "A Round Table on Research in Ecology and Evolutionary Biology" that included some of the most highly cited theoretical papers in ecology till that date. Some of these authors directly suggested that researchers use Platt's method of strong inference to address their theoretical questions (Quinn and Dunham 1983; Simberloff 1983) and others made similar suggestions somewhat less directly (Roughgarden 1983; Salt 1983; Strong 1983). Several other essays invoking aspects of Platt's approach also appeared in ecology and evolutionary biology outlets during the 1980s (e.g., Romesburg 1981; Atkinson 1985; Loehle 1987; Wenner 1989). There is little doubt that wildlife ecology and conservation researchers were inspired directly or indirectly to improve the sophistication of their study designs by Platt's essay.

Strong inference (Platt 1964) was not without problems, however, including some that were quite serious. Only one year after its publication, a physicist and a historian (Hafner and Presswood 1965) demonstrated in *Science* that historical evidence did not support the contention that strong inference had been used in the high-energy physics examples that Platt provided. Instead, they maintained "... that strong inference is an idealized scheme to which scientific developments seldom conform" (p. 503). More recently, two psychologists concluded that (1) Platt failed to demonstrate that strong inference was used more frequently in rapidly versus slowly progressing sciences, (2) Platt's historiography was fatally flawed, and (3) numerous other scientific approaches had been used as or more successfully than strong inference (O'Donohue and Buchanan 2001). Davis (2006, p. 247) concluded that "... the strongest critiques of his [Platt's] recommendations were entirely justified."

One might logically ask why Platt's essay was so influential, given its many shortcomings. The answer, as Davis (2006, p. 238) put it, is "that the article was more an inspirational tract than the development of a formal scientific methodology." It was effective because it "imparted to many natural and social scientists an ambition to test hypotheses rather than to prove them" (p. 244). Davis concluded that the value of "Strong Inference" was that it "encouraged better ideas, better choices of research problems, better model systems, and thus better science overall, even in the fields relatively resistant to the rigors of strong inference" (p. 248). This undoubtedly was true for wildlife science.

Numerous influential essays more directly targeting how wildlife scientists should conduct research also appeared during the last few decades. For example, H. Charles Romesburg (1981) pointed out that wildlife scientists had used induction to generate numerous rules of association among classes of facts, and had retroductively developed many intriguing hypotheses. Unfortunately, he argued, these "research hypotheses either are forgotten, or they gain credence and the status of laws through rhetoric, taste, authority, and verbal repetition" (p. 295). He recommended that wildlife science attempt to falsify retroductively derived research hypotheses more often using the hypothetico–deductive approach to science championed by Popper (1959, 1962) and discussed by Platt (1964). Similarly, Stuart H. Hurlbert (1984) maintained that far too many ecological researchers, when attempting to implement the hypothetico–deductive method using replicated field experiments, actually employed pseudoreplicated designs (see Sect. 2.2 for details). Because of these design flaws, he argued, researchers were much more likely to find differences between treatments and controls than actually occurred.

One of the difficulties faced by wildlife science and ecology is that ecological systems typically involve middle-number systems, or systems made up of too many parts for a complete individual accounting (*census*), but too few parts for these parts to be substituted for by averages (an approach successfully used by high-energy physics) without yielding fuzzy results (Allen and Starr 1982; O'Neill et al. 1986). For this reason, wildlife scientists often rely on statistical approaches or modeling to make sense of these problematic data. Thence, the plethora of criticisms regarding how wildlife scientists evaluate data should come as no surprise. For example, Romesburg (1981) argued that wildlife scientists had a "fixation on statistical methods" and noted that "scientific studies that lacked thought ... were dressed in quantitative trappings as compensation" (307). Robert K. Swihart and Norman A. Slade (1985) argued that sequential locations of radiotelemetered animals often lacked statistical independence and that many researchers evaluated such data inappropriately. Douglas H. Johnson (1995) maintained that ecologists were too easily swayed by the allure of nonparametric statistics and used these tools when others were more appropriate. Patrick D. Gerard and others (1998) held that wildlife scientists should not use retrospective power analysis in the manner *The Journal of Wildlife Management* editors had insisted they should (The Wildlife Society 1995). Steve Cherry (1998), Johnson (1999), and David R. Anderson and others (2000) maintained that null hypothesis significance testing was typically used inappropriately in wildlife science and related fields, resulting in far too many p-values in refereed journals (See Sect. 2.5.2 for details). Anderson (2001, 2003) also made a compelling argument that wildlife field studies relied far too much on (1) convenience sampling and (2) index values. Since one leads to the other and neither are based on probabilistic sampling designs, there is no valid way to make inference to the population of interest or assess the precision of these parameter estimates. Finally, Fred S. Guthery and others (2001, 2005) argued that wildlife scientists still ritualistically applied statistical methods and that this tended to transmute means (statistical tools) into ends. They also echoed the view of previous critics (e.g., Romesburg 1981; Johnson 1999) that wildlife scientists should give *scientific hypotheses* and *research hypotheses* a much more prominent place in their research programs, while deemphasizing *statistical hypotheses* and other statistical tools because they are just that – tools. These and similar critiques will be dealt with in more detail in subsequent chapters. The take home message is that wildlife science is still struggling to figure out how best to conduct its science, and where to position itself within the firmament of the natural sciences.

1.4.2 The Puzzle of Scientific Evidence

Even if we ignore the serious deficiencies Kuhn, Lakatos, Feyerabend, and other philosophers found in Popper's model of science (see Sect. 1.2.3.1), and argue that hypothesis falsification defines science, there is still a major disconnect between this ideal and what respected wildlife researchers actually do. For example, although most wildlife scientists extol the hypothetico–deductive model of science, Fig. 1.1 represents common study designs actually employed by wildlife researchers.

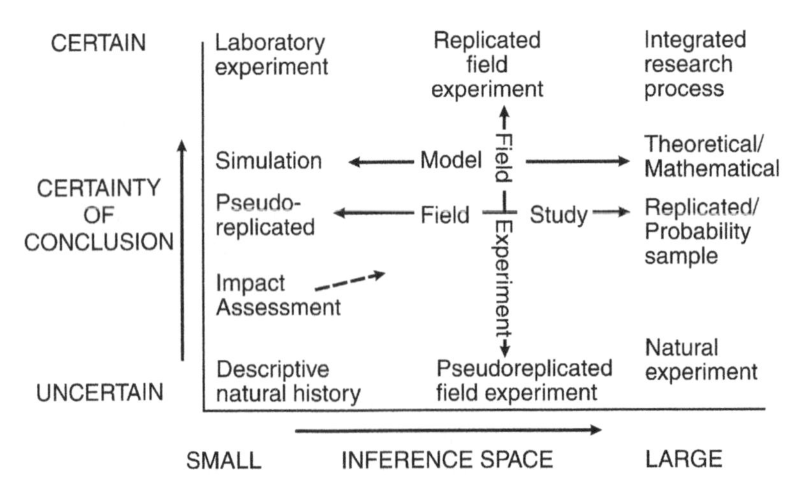

Fig. 1.1 The potential for various wildlife study designs to produce conclusions with high certainty (few plausible alternative hypotheses) and widespread applicability (diversity of populations where inferences apply). Reproduced from Garton et al. (2005), with kind permission from The Wildlife Society

Only a few of these can be construed as clearly Popperian. This does not imply the remaining designs are not useful. In fact, much of the remaining chapters deal with how to implement these and related designs. Instead, although Popper's postpositivist model of science is sometimes useful, it is often insufficient for the scope of wildlife science.

Is there a philosophical model of science that better encompasses what wildlife researchers do? Yes, there probably are several. For example, Lakatos' (1970) attempt to reconcile Popper (1959, 1962) and Kuhn's (1962) representations of science resulted in what we expect many wildlife scientists assume was Popper's model of science. Lakatos' formulation still gives falsification its due, but also makes a place for historically obvious paradigm shifts and addressed other potential deficiencies in Popper's model (see Sect. 1.2.3.1 for details). Lakatos' model, however, still cannot encompass the array of designs represented in Fig. 1.1 (and discussed in subsequent chapters). Although Feyerabend's (1975, 1978) "anything goes" approach to science certainly can cover any contingency, it offers wildlife scientists little philosophical guidance.

Haack (2003) developed a model of scientific evidence that offers a unified philosophical foundation for natural science. Further, Fig. 1.1 makes perfect sense in light of this model. Essentially, she argues that, from an epistemological perspective (see Sect. 1.2.3.1), natural science is a pragmatic undertaking. Her retro-classical version of American pragmatism places science firmly within the empiricist sphere of epistemology as well, due to the criticality of experience. She developed an apt analogy, beginning in the early 1990s (Haack 1990, 1993), which should help contextualize her model. Haack (2003) maintained that natural science research programs are conducted in much the way one completes a crossword puzzle, with warranted scientific claims anchored by experiential evidence (analogous to clues) and enmeshed in reasons (analogous to the matrix of completed entries). As she put it

> How reasonable a crossword entry is depends not only on how well it fits with the clue and any already-completed intersecting entries, but also on how plausible those other entries are, independent of the entry in question, and how much of the crossword has been completed. Analogously, the degree of warrant of a [scientific] claim for a person at a time depends not only on how supportive his evidence is, but also on how comprehensive it is, and how secure his reasons are, independent of the claim itself. (p. 67)

Following the crossword analogy, a group of researchers, each with 20 years of experience working with a system of interest, should be able to solve a scientific puzzle more easily than a first semester graduate student. While some writers find the social nature of science problematic (see Sect. 1.2.3.1), Haack (2003) maintained it is beneficial. After all, "scientific work… is much like carrying a heavy log, which can be lifted by several people but not by one. It is complex, intricate, multi-faceted – yes! – like working on a vast crossword puzzle" (p. 106). Different investigators employing different study designs and different methodologies might solve different portions of the puzzle. Because many researchers work on the puzzle simultaneously, there also must be "ways of discriminating the nut and the incompetent from the competent inquirer – credentials, peer review – so as to ensure that what the journals make available is not rubbish but worthwhile work"

(p. 107). Further, just as someone completing a crossword puzzle might make inappropriate entries, and be forced to rethink their approach, scientists are fallible as well. In fact, learning from mistaken results, concepts, or theories, and having to begin certain aspects of a research program repeatedly, seems to characterize natural science (Hafner and Presswood 1965; Haack 2003).

We hasten to point out that others noted the puzzle-like nature of natural science prior to Haack (1990, 1993, 2003). For example, Albert Einstein (1879–1955; 1936, pp. 353–354) wrote that

> The liberty of choice [of scientific concepts and theories], however, is of a special kind; it is not in any way similar to the liberty of a writer of fiction. Rather, it is similar to that of a man engaged in solving a well-designed word puzzle. He may, it is true, propose any word as the solution; but, there is only *one* word which really solves the puzzle in all it forms. It is an outcome of faith that nature – as she is perceptible to our five senses – takes the character of such a well-formulated puzzle. The successes reaped up to now by science do, it is true, give a certain encouragement to this faith.

Haack (2003) added that "scientific inquiry is a highly sophisticated, complex, subtle, and socially organized extension of our everyday reliance on experience and reasoning" (pp. 124–125). She also clarified that "there is a real world knowable to some extent by creatures with sensory and intellectual powers such as ours" (p. 125), despite the fact that our understanding of this world is to some degree a social construction (see Sect. 1.2.3.1). Despite the fact that scientists sometimes blunder about while attempting to solve scientific puzzles, there is a world outside of us we can come to know to some degree, and natural science is one of the more effective ways to acquire this knowledge. In fact, "unless theories in mature science were at least approximately true, their predictive power would be miraculous" (p. 145). This should give us hope, if nothing else.

In sum, Haack (2003) provides wildlife science a pragmatic model for knowledge acquisition (see Sect. 1.2.3.1). It does more than explain why wildlife scientists commonly employ study designs incongruent with Popper's (1959, 1962) falsification approach (e.g., Fig. 1.1). Her pragmatic model of science allows for material reality on Earth before (and possibly after) human existence, despite the contentions of radical social constructionists. It allows the social aspects of science to be explicitly included within the enterprise. It also permits any study design that can provide reliable solutions to the scientific puzzle, including various types of descriptive research, impact assessment, information–theoretic approaches using model selection, replicated manipulative experiments attempting to falsify retroductively derived research hypotheses, and qualitative designs to name just a few examples. She did not argue that each of these study designs was equally likely to provide reliable information in all circumstances. Instead, researchers must determine the best approach for each individual study, given specific constraints. There is no rote checklist for effective wildlife research. Finally, for Haack's pragmatic epistemology, truth, knowledge, and theory are inexorably connected with practical consequences, or real effects. This should resonate with wildlife scientists for whom practical conservation consequences are the ultimate metric of success.

1.5 What is it We Study?

If the objective of a wildlife study is to make inference, it is important to ask the following question: "To what biological entity do I wish to make inference"? Researchers must define this entity specifically. Is it a biological population, a species, the set of animals (of possibly different species) that serve as prey for a predator of interest, the trees in a patch of forest? The entity that is defined will be the entity that you will try to measure in the study, and the extent to which you access it will determine how well you can make inference to it from the results. Defining and accessing the entity of interest requires tools of both biology and sampling.

In designing wildlife studies, we are faced with two sets of definitions related to *populations*, one biological, and the other statistical. We start with statistical definitions, as they underpin all inference in wildlife studies. We then move on to biological definitions, the notion of significance, and whether one's focus is on wildlife or wildlife habitat.

1.5.1 Statistical Definitions

A *target population* is the collection of all *sampling* or *experimental units* about which one would like to make an inference. With respect to wildlife studies, this could be all the individuals in a biological population, subspecies, or species, all individuals or species in a community, or their habitat. The target population is just that, a target. If you had the ability and desire to measure every element in the target population, that would be a *census*. This is rarely the case in ecological systems.

In many cases, there is a subset of the target population not accessible using chosen field methods. In this case, the subset of the target population that is accessible is the *sampled population*. Because a census of even the sampled population is rarely feasible, researchers take a representative sample. A *sample* is the collection of experimental or *sampling units* from the sampled population that are actually measured. If researchers choose the sample appropriately, then they can make statistical inferences about the sampled population. However, to extend inference to the target population, they must argue that the sampled population is representative of the target population. For example, suppose you are studying the recruitment of wood ducks (*Aix sponsa*) in the Mississippi Alluvial Valley over time and your annual measure is the occupancy rate of nest boxes. To draw inference for the entire Mississippi Alluvial Valley, the target population would be all potential wood duck nesting sites in the valley. The sampled population is already smaller than the target population because the study is restricted to nesting boxes, thus ignoring natural cavities in trees. If, in addition, you only had access to the wood duck boxes found on government-owned land, such as state wildlife management areas, the sampled population would be further restricted to all wood duck boxes on government-owned land. Therefore, even with sophisticated methods of sampling design, the only resulting statistical inference strictly justified by the study design would be to recruitment in nest boxes on government-owned land. Extension of that inference to the entire Mississippi Alluvial Valley would require an argument, based on subjective expertise or previous studies where both nest boxes and natural cavities on both public and private lands were included, that trends in recruitment should be equivalent between public and private lands, and between nest boxes and natural cavities.

1.5.2 Biological Definitions

The target population of a wildlife study could include a broad array of biological entities. It is important to be clear and specific in defining what that entity is. It is just as important to identify this before the study begins as when it is explained in a manuscript or report at the end, because the sampled population and thus the sample stem directly from the target population. If some part of the target population is ignored when setting up the study, then there will be no chance of sampling that portion, and therefore drawing statistical inference to the entire population of interest cannot be done appropriately, and any inference to the target population is strictly a matter of professional judgment.

If a target population is well defined, and the desirable situation where the sampled population matches the target population is achieved, then the statistical inference will be valid, regardless of whether the target matches an orthodox definition of a biological grouping in wildlife science. Nevertheless, we believe that reviewing general definitions of biological groupings will assist the reader in thinking about the target population he or she would like to study.

In ecology, a *population* is a group of individuals of one species in an area at a given time (Begon et al. 2006, p. 94). We assume these individuals have the potential to breed with one another, implying there is some chance they will encounter one another. Dale R. McCullough (1996, p. 1) describes the distinction between *panmictic populations*, where the interactions between individuals (including potential mating opportunities) are relatively continuous throughout the space occupied by the population, and metapopulations. A *metapopulation* (Levins 1969, 1970) is a population subdivided into segments occupying patches of habitat in a fragmented landscape. An environment hostile to the species of interest separates these patches. Movement, and presumably gene flow, between patches is inhibited, but still occurs. Metapopulations provide a good example of where the design of a population study could go awry. Suppose a metapopulation consists of sources and sinks (Pulliam 1988), where the species remains on all patches and the metapopulation is stable, but those that are sinks have low productivity and must rely on dispersal from the sources to avoid local extinction. If an investigator considers individuals on just one patch to constitute the entire population, then a demographic study of this subpopulation could be misleading, as it could not address subpopulations on other patches. By considering only natality and survival of this subpopulation, the investigator might conclude that the population will either grow exponentially (if a source) or decline to extinction (if a sink).

This example illustrates the importance of including all elements of population dynamics when studying populations, metapopulations, or subpopulations. Births, deaths, immigration from other areas, or emigration to other areas defines the state of the population. Emigration can be permanent, as in dispersal of young to find new territories, or temporary. We must consider all these population parameters, in addition to other measures such as age structure and age at first reproduction, to study population dynamics properly. The population picture becomes more complicated for migratory populations, where animals that breed in distinct breeding populations often mix in staging or wintering areas. These additional dimensions must be taken into account to understand their dynamics.

The *biotic community* is "an assemblage of species populations that occur together in space and time" (Begon et al. 2006, p. 469). Sometimes the usage is more specific, such as a plant community or a small-mammal community. There are concepts of community dynamics that parallel those of population dynamics. *Species richness* is the number of species in the community at a given time, and *species diversity* refers to indices of community diversity that take into account both species richness and the relative abundance of species (Begon et al. 2006, pp. 470–471). *Local extinction probability* is the probability that a species currently present will not be in the community by the next time period. The *number of colonizing species* is the number of species currently in the community that were absent during the previous time period.

Biodiversity is one of the most commonly used ecological terms in both the scientific literature and the popular media today. Unfortunately, it rarely appears with an unambiguous definition. In its most general sense, biodiversity refers to all aspects of variety in the living world (Begon et al. 2006, p. 602). More specifically, the term is used to describe the number of species (species richness), the amount of genetic variation, or the number of community types present in an area of interest.

Habitat also has many definitions in the literature. For our purposes, it is "the physical space within which an organism lives, and the abiotic and biotic entities (e.g., resources) it uses and selects in that space" (Morrison et al. 2006, p. 448). Further, because habitat is organism-specific, "it relates the presence of a species, population, or individual (animal or plant) to an area's physical and biological characteristics" (Hall et al. 1997, p. 175).

1.5.3 Biological vs. Statistical vs. Social Significance

John Macnab (1985) argued that wildlife science was plagued with "slippery shibboleths," or code words having different meanings for individuals or subgroups within the field. "Significance" is as slippery as any shibboleth in wildlife science. We typically use this term in one of three ways: biological, statistical, or social significance. All too often, authors either do not specify what they mean when they say something is significant, or appear to imply that simply because results are (or are not) statistically significant, they must also be (or not be) significant biologically and/or socially.

When wildlife scientists say that something is biologically significant, they mean that it matters biologically. Because one of wildlife sciences' primary objectives is to determine what is biologically important, this is not a trivial matter. In fact, the reason we use inferential statistics at all, and sometimes compute statistical significance in the process, is to learn what is biologically important. The problem is that, based on a particular study, not all statistically significant differences matter biologically, and just because we cannot find statistically significant differences does not imply that biological differences do not indeed exist in the system being studied (Cherry 1998; Johnson 1999). Further, as Johnson (1999, p. 767) maintained, "the hypotheses usually tested by wildlife ecologists... are statistical hypotheses [see glossary].... Unlike scientific hypotheses [see glossary], the truth of which is truly in question, most statistical hypotheses are known a priori to be false." For example, successful hunter–gathers in North America since the Pleistocene have known that white-tailed deer (*Odocoileus virginianus*) do not use habitat at random, so designing a study to determine whether deer use habitat in proportion to availability is silly; it is also silly to consider this question for any species known well by humans. The more relevant question is "how much time are animals spending in available habitats" (Cherry 1998, p. 948), and what important life requisites do each of these cover types provide. Much of the time, wildlife scientists are actually attempting to find the magnitude of some effect rather than determine whether the effect actually exists – we already know that answer.

Another complication is that just because wildlife scientists find something to be biologically significant does not imply that society will reach the same conclusion. Moreover, society might well find something to be extraordinarily important that wildlife scientists do not think matters much biologically. For contentious environmental issues, various segments of society will undoubtedly disagree with one another as well. As case studies amply illustrate, differences in the moral cultures of various segments of society, and disagreement regarding what is or is not socially or biologically significant, contribute greatly to wildlife-related environmental conflicts (Wondolleck and Yaffee 2000; Peterson et al. 2002, 2004, 2006a). These differences also form one of the primary challenges to public participation processes designed to work through such environmental conflicts (Daniels and Walker 2001; Depoe et al. 2004; Peterson and Franks 2005; Peterson et al. 2005, 2006b). Because the majority of wildlife scientists work for regulatory agencies at the state or federal level, for nongovernmental organizations, or for environmental consulting firms – or train those who do – what various publics and related interest groups perceive to be significant, and why they reach these conclusions, are questions central to wildlife science.

1.5.4 Focus on Wildlife vs. Focus on Wildlife Habitat

We have defined the statistical sampling concepts of target population, sampled population, and sample, as well as the biological concepts of population, metapopulation, community, and habitat. The statistical concepts will be applied to the biological ones (i.e., the set of experimental or sampling units will be identified), based on the objectives of the study. We can divide wildlife studies into those whose objectives focus on groupings of animals and those whose objectives focus on the habitat of the animals.

We can further divide studies of animals into those that focus on measuring something about the individual animal (e.g., sex, mass, breeding status) and those that focus on how many animals are there. Consider a study of a population of cotton rats (*Sigmodon hispidus*) in an old field where there are two measures of interest: the size of the population and its sex ratio. The sampling units would be individual rats and the target population would include all the rats in the field (assume the field is isolated enough that this is not part of a metapopulation). If capture probabilities of each sex are the same (perhaps a big assumption), then by placing a set of traps throughout the field one could trap a representative sample and estimate the sex ratio. If the traps are distributed probabilistically, the sampled population would match the target population (and in this case the target population would coincide with a biological population) and therefore the estimated sex ratio should be representative of the population sex ratio.

The estimation of abundance is an atypical sampling problem. Instead of measuring something about the sampling units, the objective is to estimate the total number of units in the target population. Without a census, multiple samples and capture–recapture statistical methodology are required to achieve an unbiased estimate of the population size (see Sect. 2.5.4.). If traps are left in the same location for each sample, it is important that there be enough traps so that each rat has some chance of being captured during each trapping interval.

Estimates of abundance are not limited to the number of individual animals in a population. The estimation of species richness involves the same design considerations. Again, in the absence of a census of the species in a community (i.e., probability of detecting at least one individual of each species is 1.0), designs that allow the use of capture–recapture statistical methodologies might be most appropriate (see reviews by Nichols and Conroy 1996; Nichols et al. 1998a,b; Williams et al. 2002). In this case, the target population is the set of all the species in a community. We discuss accounting for detectability more fully in Sect. 2.4.1.

If wildlife is of ultimate interest, but the proximal source of interest is something associated with the ecosystem of which wildlife is a part, then the target population could be vegetation or some other aspect of the animals' habitat (e.g., Morrison et al. 2006). For example, if the objective of the study is to measure the impact of deer browsing on a given plant in a national park, the target population is not the deer, but the collection of certain plants within the park. The researcher could separate the range of the plant into experimental units consisting of plots; some plots could be left alone but monitored, whereas exclosures could be built around others to prevent deer from browsing. In this way, the researcher could determine the impact of the deer on this food plant by comparing plant measurements on plots with exclosures versus plots without exclosures.

1.6 Summary

Because wildlife scientists conduct research in the pursuit of knowledge, they must understand what knowledge is and how it is acquired. We began Sect. 1.2 using "the science wars" to highlight how different ontological, epistemological, and axiological perspectives can lead to clashes grounded in fundamentally different philosophical perspectives. This example also illustrates practical reasons why wildlife scientists should become familiar with philosophy as it relates to natural science. Differing perspectives on the nature of reality (*ontology*) explain part of this clash of ideas. Most scientists, grounded in the empiricist tradition, hold that reality independent of human thought and culture indeed exists. Conversely, many social scientists and humanists argue that reality ultimately is socially constructed because it is to some degree contingent upon human percepts and social interactions. Several major perspectives toward the nature and scope of knowledge (*epistemology*) have developed in Western philosophy. Influential approaches to knowledge acquisition include empiricism, rationalism, pragmatism, logical positivism, postpositivism, and social constructionism. Regardless of the epistemological perspective one employs, however, logical thought, including inductive, deductive, and retroductive reasoning (Table 1.1), remains an integral component of knowledge acquisition. At least three aspects of value or quality (*axiology*) influence natural science. Ethical behavior by scientists supports the integrity of the scientific enterprise, researchers bring their own values into the scientific process, and both scientists and society must determine the value and quality of scientific research.

As Sect. 1.2 illustrates, there is no single philosophy of science, and so there can be no single method of science either. Regardless, natural science serves as a model of human ingenuity. In Sect. 1.3, we addressed why natural science has proven such a successful enterprise. Much of the reason relates to general steps commonly employed (Table 1.2). These include (1) becoming familiar with the system of interest, the question being addressed, and the related scientific literature, (2) constructing meaningful research hypotheses and/or conceptual models relating to theory and objectives, (3) developing an appropriate study design and executing the design and analyzing the data appropriately, (4) obtaining feedback from other scientists at various stages in the process, such as through publication in referred outlets, and (5) closing the circle of science by going back to steps 3, 2, or 1 as needed. Often, because of the complex nature of scientific research, multiple researchers using a variety of methods address different aspects of the same general research program. Impact assessment, inventorying, and monitoring studies provide important data for decision making by natural resource policy makers and managers. The results of well-designed impact and survey studies often are suitable for publication in refereed outlets, and other researchers can use these data in conjunction with data collected during similar studies to address questions beyond the scope of a single study.

In Sect. 1.4, we discussed how wildlife scientists have honed their approaches to research by studying influential critiques written by other natural scientists (e.g.,

Platt 1964; Romesburg 1981; Hurlbert 1984). Because ecological systems contain too many parts for a complete individual accounting (*census*), but too few parts for these parts to be substituted for by averages, wildlife scientists typically rely on statistical approaches or modeling to make sense of data. For this reason, numerous critiques specifically addressing how wildlife scientists handle and mishandle data analysis were published in recent decades. These publications continue to shape and reshape how studies are designed, data analyzed, and publications written.

As Fig. 1.1 illustrates, wildlife science commonly employs a number of study designs that do not follow Popper's (1959, 1962) falsification approach to science. Epistemologically, wildlife science probably is better described by Haack's (2003) pragmatic model of natural science, where research programs are conducted in much the same way one completes a crossword puzzle, with warranted scientific claims anchored by experiential evidence (analogous to clues) and enmeshed in reasons (analogous to the matrix of completed entries). This pragmatic model permits any study design that can provide reliable solutions to the scientific puzzle, including various types of descriptive research, impact assessment, information–theoretic approaches using model selection, replicated manipulative experiments attempting to falsify retroductively derived research hypotheses, and qualitative designs to name just a few. Under this pragmatic epistemology, truth, knowledge, and theory are inexorably connected with practical consequences, or real effects.

We ended the chapter by clarifying what it is that wildlife scientists study (Sect. 1.5). We did so by defining a number of statistical, biological, and social terms. This is important as the same English word can describe different entities in each of these three domains (e.g., significance). We hope that these common definitions will make it easier for readers to navigate among chapters. Similarly, this chapter serves as a primer on the philosophy and nature of natural science that should help contextualize the more technical chapters that follow.

References

Allen, T. F. H., and T. B. Starr. 1982. Hierarchy: Perspectives for Ecological Complexity. University of Chicago Press, Chicago, IL.

Anderson, D. R. 2001. The need to get the basics right in wildlife field studies. Wildl. Soc. Bull. 29: 1294–1297.

Anderson, D. R. 2003. Response to Engeman: index values rarely constitute reliable information. Wildl. Soc. Bull. 31: 288–291.

Anderson, D. R., K. P. Burnham, and W. L. Thompson. 2000. Null hypothesis testing: problems, prevalence, and an alternative. J. Wildl. Manag. 64: 912–923.

Arnqvist, G., and D. Wooster. 1995. Meta-analysis: synthesizing research findings in ecology and evolution. Trends Ecol. Evol. 10: 236–240.

Ashman, K. M., and P. S. Barringer, Eds. 2001. After the Science Wars. Routledge, London.

Atkinson, J. W. 1985. Models and myths of science: views of the elephant. Am. Zool. 25: 727–736.

Begon, M., C. R. Townsend, and J. L. Harper. 2006. Ecology: From Individuals to Ecosystems, 4th Edition. Blackwell, Malden, MA.

Berger, P. L., and T. Luckmann. 1966. The Social Construction of Reality: A Treatise in the Sociology of Knowledge. Doubleday, Garden City, NY.
Chamberlin, T. C. 1890. The method of multiple working hypotheses. Science 15: 92–96.
Cherry, S. 1998. Statistical tests in publications of The Wildlife Society. Wildl. Soc. Bull. 26: 947–953.
Committee on Science, Engineering, and Public Policy. 1995. On Being a Scientist: Responsible Conduct in Research, 2nd Edition. National Academy Press, Washington, D.C.
Committee on the Conduct of Science. 1989. On being a scientist. Proc. Natl. Acad. Sci. USA 86: 9053–9074.
Costanza, R., R. d'Arge, R. de Groot, S. Farber, M. Grasso, B. Hannon, K. Limburg, S. Naeem, R. V. Oneill, J. Paruelo, R. G. Raskin, P. Sutton, and M. van den Belt. 1997. The value of the world's ecosystem services and natural capital. Nature 387: 253–260.
Daniels, S. E., and G. B. Walker. 2001. Working Through Environmental Conflict: The Collaborative Learning Approach. Praeger, Westport, CT.
Davis, R. H. 2006. Strong inference: rationale or inspiration? Perspect. Biol. Med. 49: 238–249.
Denzin, N. K., and Y. S. Lincoln, Eds. 2005. The Sage Handbook of Qualitative Research, 3rd Edition. Sage Publications, Thousand Oaks, CA.
Depoe, S. P., J. W. Delicath, and M.-F. A. Elsenbeer, Eds. 2004. Communication and Public Participation in Environmental Decision Making. State University of New York Press, Albany, NY.
Diamond, J. M. 1972. Biogeographic kinetics: estimation of relaxation times for avifaunas of southwest Pacific islands. Proc. Natl. Acad. Sci. USA 69: 3199–3203.
Diamond, J. M. 1975. The island dilemma: lessons of modern biogeographic studies for the design of nature reserves. Biol. Conserv. 7: 129–146.
Diamond, J. M. 1976. Island biogeography and conservation: strategy and limitations. Science 193: 1027–1029.
Dillman, D. A. 2007. Mail and Internet Surveys: The Tailored Design Method, 2nd Edition. Wiley, Hoboken, NJ.
Einstein, A. 1936. Physics and reality. J. Franklin Inst. 221: 349–382.
Feyerabend, P. 1975. Against Method: Outline of an Anarchistic Theory of Knowledge. NLB, London.
Feyerabend, P. 1978. Science in a Free Society. NLB, London.
Ford, E. D. 2000. Scientific Method for Ecological Research. Cambridge University Press, Cambridge.
Garton, E. O., J. T. Ratti, and J. H. Giudice. 2005. Research and experimental design, in C. E. Braun, Ed. Techniques for Wildlife Investigations and Management, 6th Edition, pp. 43–71. The Wildlife Society, Bethesda, MD.
Gerard, P. D., D. R. Smith, and G. Weerakkody. 1998. Limits of retrospective power analysis. J. Wildl. Manage. 62: 801–807.
Gettier, E. L. 1963. Is justified true belief knowledge? Analysis 23: 121–123.
Gross, P. R., and N. Levitt. 1994. Higher Superstition: The Academic Left and its Quarrels With Science. Johns Hopkins University Press, Baltimore, MD.
Gross, P. R., N. Levitt, and M. W. Lewis, Eds. 1997. The Flight From Science and Reason. New York Academy of Sciences, New York, NY.
Gurevitch, J. A., and L. V. Hedges. 2001. Meta-analysis: combining the results of independent experiments, in S. M. Scheiner, and J. A. Gurevitch, Eds. Design and Analysis of Ecological Experiments, 2nd edition, pp. 347–369. Oxford University Press, Oxford.
Guthery, F. S. 2004. The flavors and colors of facts in wildlife science. Wildl. Soc. Bull. 32: 288–297.
Guthery, F. S., J. J. Lusk, and M. J. Peterson. 2001. The fall of the null hypothesis: liabilities and opportunities. J. Wildl. Manag. 65: 379–384.
Guthery, F. S., J. J. Lusk, and M. J. Peterson. 2004. Hypotheses in wildlife science. Wildl. Soc. Bull. 32: 1325–1332.

Guthery, F. S., L. A. Brennan, M. J. Peterson, and J. J. Lusk. 2005. Information theory in wildlife science: critique and viewpoint. J. Wildl. Manag. 69: 457–465.
Haack, S. 1990. Rebuilding the ship while sailing on the water, in R. B. Gibson, and R. F. Barrett, Eds. Perspectives on Quine, pp. 111–128. Blackwell, Oxford.
Haack, S. 1993. Evidence and Inquiry: Towards Reconstruction in Epistemology. Blackwell, Oxford.
Haack, S. 2003. Defending Science – Within Reason: Between Scientism and Cynicism. Prometheus Books, Amherst, NY.
Haack, S. 2006. Introduction: pragmatism, old and new, in S. Haack, and R. Lane, Eds. Pragmatism, Old and New: Selected Writings, pp. 15–67. Prometheus Books, Amherst, NY.
Haack, S., and R. Lane, Eds. 2006. Pragmatism, Old and New: Selected Writings. Prometheus Books, Amherst, NY.
Hacking, I. 1999. The Social Construction of What? Harvard University Press, Cambridge, MA.
Hafner, E. M., and S. Presswood. 1965. Strong inference and weak interactions. Science 149: 503–510.
Hall, L. S., P. R. Krausman, and M. L. Morrison. 1997. The habitat concept and a plea for standard terminology. Wildl. Soc. Bull. 25: 173–182.
Hurlbert, S. H. 1984. Pseudoreplication and the design of ecological field experiments. Ecol. Monogr. 54: 187–211.
Jackson, J. B. C., M. X. Kirby, W. H. Berger, K. A. Bjorndal, L. W. Botsford, B. J. Bourque, R. H. Bradbury, R. Cooke, J. Erlandson, J. A. Estes, T. P. Hughes, S. Kidwell, C. B. Lange, H. S. Lenihan, J. M. Pandolfi, C. H. Peterson, R. S. Steneck, M. J. Tegner, and R. R. Warner. 2001. Historical overfishing and the recent collapse of coastal ecosystems. Science 293: 629–638.
James, W. 1907. Pragmatism, a new name for some old ways of thinking: popular lectures on philosophy. Longmans, Green, New York, NY.
James, W. 1912. Essays in Radical Empiricism. Longmans, Green, New York, NY.
Jasinoff, S., G. E. Markle, J. C. Petersen, and T. Pinch, Eds. 1995. Handbook of Science and Technology Studies. Sage Publications, Thousand Oaks, CA.
Johnson, D. H. 1995. Statistical sirens: the allure of nonparametrics. Ecology 76: 1998–2000.
Johnson, D. H. 1999. The insignificance of statistical significance testing. J. Wildl. Manag. 63: 763–772.
Johnson, D. H. 2002. The importance of replication in wildlife research. J. Wildl. Manag. 66: 919–932.
Kennedy, D. 2006. Editorial retraction. Science 311: 335.
Kitcher, P. 2001. Science, truth, and democracy. Oxford University Press, New York.
Koertge, N., Ed. 1998. A House Built on Sand: Exposing Postmodernist Myths About Science. Oxford University Press, New York, NY.
Kuhn, T. S. 1962. The Structure of Scientific Revolutions. University of Chicago Press, Chicago, IL.
Lakatos, I. 1970. Falsification and the methodology of scientific research programmes, in I. Lakatos, and A. Musgrave, Eds. Criticism and the Growth of Knowledge, pp. 91–196. Cambridge University Press, Cambridge.
Lakatos, I., and P. Feyerabend. 1999. For and Against Method: Including Lakatos's Lectures on Scientific Method and the Lakatos–Feyerabend Correspondence. M. Motterlini, Ed. University of Chicago Press, Chicago, IL.
Lakatos, I., and A. Musgrave, Eds. 1970. Criticism and the Growth of Knowledge. Cambridge University Press, London.
Latour, B. 1993. We Have Never Been Modern. C. Porter, translator. Harvard University Press, Cambridge, MA.
Levins, R. 1969. Some demographic and genetic consequences of environmental heterogeneity for biological control. Bull. Entomol. Soc. Am. 15: 237–240.
Levins, R. 1970. Extinction, in M. Gerstenhaber, Ed. Some Mathematical Questions in Biology, pp. 77–107. American Mathematical Society, Providence, RI.

Lincoln, Y. S., and E. G. Guba. 1985. Naturalistic Inquiry. Sage Publications, Newbury Park, CA.

Loehle, C. 1987. Hypothesis testing in ecology: psychological aspects and the importance of theory maturation. Q Rev Biol 62: 397–409.

MacArthur, R. H. 1972. Geographical Ecology: Patterns in the Distribution of Species. Harper and Row, New York, NY.

MacArthur, R. H., and E. O. Wilson. 1967. The Theory of Island Biogeography. Princeton University Press, Princeton, NJ.

Macnab, J. 1985. Carrying capacity and related slippery shibboleths. Wildl. Soc. Bull. 13: 403–410.

McCullough, D. R. 1996. Introduction, in D. R. McCullough, Ed. Metapopulations and Wildlife Conservation, pp. 1–10. Island Press, Washington, D.C.

Menzies, T. 1996. Applications of abduction: knowledge-level modelling. Int. J. Hum. Comput. Stud. 45: 305–335.

Morrison, M. L., B. G. Marcot, and R. W. Mannan. 2006. Wildlife–habitat relationships: concepts and applications, 3rd Edition. Island Press, Washington, D.C.

Myers, N., R. A. Mittermeier, C. G. Mittermeier, G. A. B. da Fonseca, and J. Kent. 2000. Biodiversity hotspots for conservation priorities. Nature 403: 853–858.

Nichols, J. D., and M. J. Conroy. 1996. Estimation of species richness. in D. E. Wilson, F. R. Cole, J. D. Nichols, R. Rudran, and M. Foster, Eds. Measuring and Monitoring Biological Diversity: Standard Methods for Mammals, pp. 226–234. Smithsonian Institution Press, Washington, D.C.

Nichols, J. D., T. Boulinier, J. E. Hines, K. H. Pollock, and J. R. Sauer. 1998a. Estimating rates of local species extinction, colonization, and turnover in animal communities. Ecol. Appl. 8: 1213–1225.

Nichols, J. D., T. Boulinier, J. E. Hines, K. H. Pollock, and J. R. Sauer. 1998b. Inference methods for spatial variation in species richness and community composition when not all species are detected. Conserv. Biol. 12: 1390–1398.

O'Donohue, W., and J. A. Buchanan. 2001. The weaknesses of strong inference. Behav. Philos. 29: 1–20.

O'Neill, R. V., D. L. DeAngelis, J. B. Waide, and T. F. H. Allen. 1986. A Hierarchical Concept of Ecosystems. Princeton University Press, Princeton, NJ.

Osenberg, C. W., O. Sarnelle, and D. E. Goldberg. 1999. Meta-analysis in ecology: concepts, statistics, and applications. Ecology 80: 1103–1104.

Peterson, M. N., T. R. Peterson, M. J. Peterson, R. R. Lopez, and N. J. Silvy. 2002. Cultural conflict and the endangered Florida Key deer. J. Wildl. Manag. 66: 947–968.

Peterson, M. N., S. A. Allison, M. J. Peterson, T. R. Peterson, and R. R. Lopez. 2004. A tale of two species: habitat conservation plans as bounded conflict. J. Wildl. Manag. 68: 743–761.

Peterson, M. N., M. J. Peterson, and T. R. Peterson. 2005. Conservation and the myth of consensus. Conserv. Biol. 19: 762–767.

Peterson, M. N., M. J. Peterson, and T. R. Peterson. 2006a. Why conservation needs dissent. Conserv. Biol. 20: 576–578.

Peterson, T. R., and R. R. Franks. 2006. Environmental conflict communication, in J. Oetzel, and S. Ting-Toomey, Eds. The Sage Handbook of Conflict Communication: Integrating Theory, Research, and Practice, pp. 419–445. Sage Publications, Thousand Oaks, CA.

Peterson, T. R., M. N. Peterson, M. J. Peterson, S. A. Allison, and D. Gore. 2006b. To play the fool: can environmental conservation and democracy survive social capital? Commun. Crit. / Cult. Stud. 3: 116–140.

Plato. [ca. 369 B.C.] 1973. Theaetetus. J. McDowell, translator. Clarendon Press, Oxford.

Platt, J. R. 1964. Strong inference: certain systematic methods of scientific thinking may produce much more rapid progress than others. Science 146: 347–353.

Popper, K. R. 1935. Logik der forschung: zur erkenntnistheorie der modernen naturwissenschaft. Springer, Wien, Österreich.

Popper, K. R. 1959. The Logic of Scientific Discovery. Hutchinson, London.

Popper, K. R. 1962. Conjectures and Refutations: The Growth of Scientific Knowledge. Basic Books, New York, NY.
Pulliam, H. R. 1988. Sources, sinks, and population regulation. Am. Nat. 132: 652–661.
Quinn, J. F., and A. E. Dunham. 1983. On hypothesis testing in ecology and evolution. Am. Nat. 122: 602–617.
Romesburg, H. C. 1981. Wildlife science: gaining reliable knowledge. J. Wildl. Manag. 45: 293–313.
Rosenberg, A. 2000. Philosophy of science: a contemporary introduction. Routledge, London.
Roth, W.-M., and G. M. Bowen. 2001. 'Creative solutions' and 'fibbing results': enculturation in field ecology. Soc. Stud. Sci. 31: 533–556.
Roughgarden, J. 1983. Competition and theory in community ecology. Am. Nat. 122: 583–601.
Salt, G. W. 1983. Roles: their limits and responsibilities in ecological and evolutionary research. Am. Nat. 122: 697–705.
Shapin, S., and S. Schaffer. 1985. Leviathan and the air-pump: Hobbes, Boyle, and the experimental life. Princeton University Press, Princeton, NJ.
Simberloff, D. 1976a. Experimental zoogeography of islands: effects of island size. Ecology 57: 629–648.
Simberloff, D. 1976b. Species turnover and equilibrium island biogeography. Science 194: 572–578.
Simberloff, D. 1983. Competition theory, hypothesis-testing, and other community ecological buzzwords. Am. Nat. 122: 626–635.
Simberloff, D., and L. G. Abele. 1982. Refuge design and island biogeographic theory: effects of fragmentation. Am. Nat. 120: 41–50.
Simberloff, D. S., and E. O. Wilson. 1969. Experimental zoogeography of island: the colonization of empty islands. Ecology 50: 278–296.
Sokal, A. D. 1996a. A physicist experiments with cultural studies. Lingua Franca 6(4): 62–64.
Sokal, A. D. 1996b. Transgressing the boundaries: toward a transformative hermeneutics of quantum gravity. Soc. Text 46/47: 217–252.
Sokal, A. D., and J. Bricmont. 1998. Fashionable Nonsense: Postmodern Intellectuals' Abuse of Science. Picador, New York, NY.
Strong Jr., D. R., 1983. Natural variability and the manifold mechanisms of ecological communities. Am. Nat. 122: 636–660.
Swihart, R. K., and N. A. Slade. 1985. Testing for independence of observations in animal movements. Ecology 66: 1176–1184.
Theocharis, T., and M. Psimopoulos. 1987. Where science has gone wrong. Nature 329: 595–598.
Vitousek, P. M., H. A. Mooney, J. Lubchenco, and J. M. Melillo. 1997. Human domination of Earth's ecosystems. Science 277: 494–499.
Wenner, A. M. 1989. Concept-centered versus organism-centered biology. Am. Zool. 29: 1177–1197.
Whittaker, R. J., M. B. Araujo, J. Paul, R. J. Ladle, J. E. M. Watson, and K. J. Willis. 2005. Conservation biogeography: assessment and prospect. Divers. Distrib. 11: 3–23.
Wilcox, B. A. 1978. Supersaturated island faunas: a species–age relationship for lizards on post-Pleistocene land-bridge islands. Science 199: 996–998.
The Wildlife Society. 1995. Journal news. J. Wildl. Manag. 59: 196–198.
Williams, B. K., J. D. Nichols, and M. J. Conroy. 2002. Analysis and management of animal populations. Academic Press, San Diego, CA.
Williamson, M. H. 1981. Island populations. Oxford University Press, Oxford.
Wilson, E. O., and D. S. Simberloff. 1969. Experimental zoogeography of islands: defaunation and monitoring techniques. Ecology 50: 267–278.
Wondolleck, J. M., and S. L. Yaffee. 2000. Making Collaboration Work: Lessons From Innovation in Natural Resource Management. Island Press, Washington, D.C.
Worster, D. 1994. Nature's Economy: A History of Ecological Ideas, 2nd edition. Cambridge University Press, Cambridge.

Chapter 2
Concepts for Wildlife Science: Design Application

2.1 Introduction

In this chapter, we turn our attention to the concept of basic study design. We begin by discussing variable classification, focusing on the types of variables: explanatory, disturbing, controlling, and randomized. We then discuss how each of these variable types is integral to wildlife study design. We then detail the necessity of randomization and replication in wildlife study design, and relate these topics to variable selection.

We outline the three major types of designs in decreasing order of rigor (i.e., manipulative experiments, quasi-experiments, and observational studies) with respect to controls, replication, and randomization, which we further elaborate in Chap. 3. We provide a general summary on adaptive management and we briefly touch on survey sampling designs for ecological studies, with a discussion on accounting for detectability, but leave detailed discussion of sampling design until Chap. 4.

We discuss the place of statistical inference in wildlife study design, focusing on *parameter* estimation, *hypothesis* testing, and *model* selection. We do not delve into specific aspects and applications of statistical models (e.g., generalized linear models or correlation analysis) as these are inferential, rather than design techniques. We discuss the relationships between statistical inference and *sampling distributions*, covering the topics of statistical *accuracy*, *precision*, and *bias*. We provide an outline for evaluating Type I and II errors as well as sample size determination. We end this chapter with a discussion on integrating project goals with study design and those factors influencing the design type used, and conclude with data storage techniques and methods, programs for statistical data analysis, and approaches for presenting results from research studies.

2.2 Variable Classification

There are many things to be considered when designing a wildlife field study and many pitfalls to be avoided. Pitfalls usually arise from unsuccessfully separating sources of variation and relationships of interest from those that are extraneous or nuisances. Nomenclature in study design is not standardized and can be confusing. Kish (1987) presented a classification scheme for these variables that we find useful. He calls variables of interest, including those that predict and those that are predicted, *explanatory* variables, and those that are extraneous *disturbing* variables, *controlled* variables, or *randomized* variables.

2.2.1 Explanatory Variables

Explanatory variables are the focus of most scientific studies. They include response variables, dependent variables, or *Y* variables, and are the variables of interest whose behavior we wish to predict on the basis of our research hypotheses. Predictor variables are those variables that are purported by the hypothesis to cause the behavior of the response variable. Predictors can be discrete or continuous, ordered or unordered. When a predictor is continuous, such as in studies where some type of regression analysis is used, it is often called a covariate, an independent variable, an *X* variable, and sometimes an explanatory variable (adding confusion in this case). When a predictor is discrete, it is often called a factor, as in *analysis of variance*, or class variables as in some statistical programs (SAS Institute Inc. 2000). Regardless of nomenclature, the goal of all studies is to identify the relationship between predictors and response variables in an unbiased fashion, with maximal precision. This is difficult to do in ecology, where the system complexity includes many other extraneous sources of variation that are difficult to remove or measure on meaningful spatial or temporal scales.

For most wildlife studies, the goal is to estimate the trend, or change in population size, over time. Therefore, for example, the response variable could be population size for each species, with the predictor being time in years. The resulting analysis is, in essence, measuring the effect that time has on populations.

2.2.2 Disturbing Variables

Extraneous variables, if not dealt with properly either through control or randomization, can *bias* the results of a study. Such variables potentially affect the behavior of the response variable, but are more of a nuisance than of interest to the scientist or manager. For example, consider that in some surveys, individuals at a survey point are not counted (i.e., probability of detection is <1.0). Figure 2.1 illustrates this point, using simulated data. Based on the raw count of 20 birds of a species in years 1 and 2, an obvious but biased estimate for trend would be a 0% increase. However, the actual abundance decreased by 20% from 50 to 40. What is the reason for this bias? The probability of detection was 0.4 in year 1 and 0.6 in year 2. What causes variation in detection probability? There are many possibilities, including changing observers, change of skill for a given observer, variation in weather or

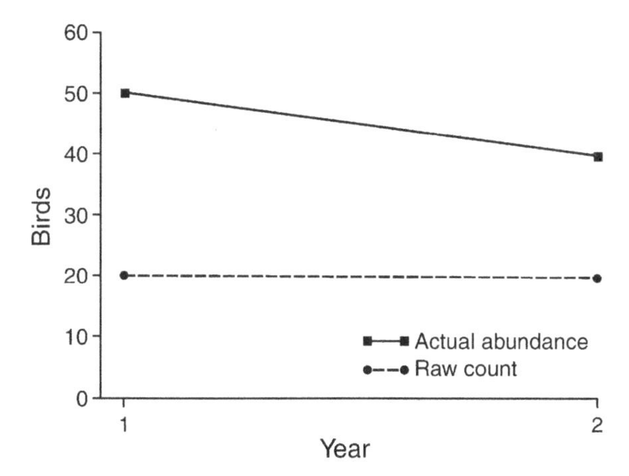

Fig. 2.1 Comparison of a hypothetical actual trend across 2 years with an estimated trend based on raw counts. The difference in trend is based on difference detection probabilities in year 1 (0.40) and year 2 (0.60). Reproduced from Morrison et al. (2001), with kind permission from Springer Science + Business Media

noise levels, changes in habitat characteristics, survey timing, behavioral changes by the birds, and numerous other potential disturbing variables. The choices an investigator has in dealing with individual disturbing variables is to provide control for them (Sect. 2.1.4), randomize to remove or minimize their effect (Sects. 2.1.5 and 2.3), or ignore them. For example, the United States Geological Survey (USGS) Patuxent Wildlife Research Center, which coordinates the Breeding Bird Survey (BBS), controls for disturbing factors that have been consistently measured and shown to be systematic over time in the BBS (Link and Sauer 1998), including observer differences (Sauer et al. 1994) and novice observer effects (Kendall et al. 1996).

2.2.3 Controlling Variables

The best way to deal with disturbing variables, if feasible, is to make them controlled variables, thus removing the potential bias and increasing precision. In many design and analysis of variance books, we use controlling (blocking) in the design, but we can potentially block in the statistical analysis as well (Kuehl 2000). By sampling the same locations each year, we can control for differences in locations. We could control for observer differences by ensuring that the same observers conduct the surveys at a given location each year. However, using the same observers at the same locations is not practical for many long-term surveys and sampling at the same location ignores spatial variability; also, sites wear out. Thus, during our analysis for a given location, we compute the average of the trend estimates over all the observers who surveyed at that location. This should remove or negate the bias due to observer differences (Verner and Milne 1990). However, the more frequently observers change within a given location, the more variability will be introduced. When the number of observers is equal to the number of years a location has been surveyed, there is no basis for computing a trend for a given observer, as (1) the change in observer is confounded with change over time and (2) the trend may be confounded with changing observer ability.

2.2.4 *Randomized Variables*

Some potentially disturbing variables cannot be controlled for through either design or a posteriori analysis. This can occur in at least one of three ways:

1. The variable might simply be unrecognized, given that there are numerous potential sources of variation in nature.
2. The investigator might recognize a potential problem conceptually, but either does not know how to measure it or finds it impractical to do so.
3. Due to sample size, there is a limit to the number of disturbing variables that can be accounted for in an analysis.

To minimize the effect of these remaining disturbing variables, we convert them to randomized variables. Using random selection avoids bias due to any systematic pattern in trend over space. We discuss randomization further in Sect. 2.3.

2.3 Randomization and Replication

It is important that researchers be as objective as possible when conducting wildlife field studies, and that the resulting inference be representative of the *target population*. Randomization and replication are important tools for maintaining the integrity of a study. *Randomization* consists of two facets: (1) choosing study (sampling or experimental) units randomly from the target population of interest and (2) if the study is a true experiment, assigning various treatments of the experiment randomly to the experimental units (Fisher 1935). If we cannot choose study sites at random, the investigator cannot automatically claim that the results apply to the entire *target population*. Instead, he or she must provide a reasonable argument to justify that the study units used are representative of the target population. In some cases, randomization will produce a collection of study units that one could argue is not representative of the target population, especially with a small sample size (Hurlbert 1984). Generally, however, randomization is still advantageous because it results in a probabilistic design. If the various treatments are not assigned randomly, then the study is vulnerable to disturbing variables, as described earlier in Sect. 2.1.3.

Suppose you are interested in promoting the use of water impoundments (moist soil management units) on national wildlife refuges of the northeastern United

States by migrating shorebirds. Migrating birds use stopovers to build up energy reserves for the remainder of the migration to the breeding grounds (Lehnen and Krementz 2005). Therefore, part of the management for migrating shorebirds is to manage for their food, which includes invertebrates in shallow water. Now suppose you wish to determine which of the two methods of drawing down the water in impoundments is more beneficial for producing invertebrates during the spring migration. You choose a sample of impoundments from the target population – the impoundments of national wildlife refuges of the northeast that could conceivably support shorebirds. Each impoundment should be independent of the next with respect to invertebrate production and the ability to draw water down. First, consider choosing two impoundments, one for a quick drawdown and one for a slow drawdown. If you choose these two randomly from all the impoundments in the region, then you can say that each of the impoundments had the same probability of selection and therefore the sample is representative of the impoundments in the region (ignoring for now the problems associated with small sample size). Similarly, you should assign the two treatments to the two impoundments randomly. If not, you run the risk of biasing the study with your own preconceived notions or desires, perhaps by assigning the treatment you think will produce the best results to the impoundment that has had a lot of shorebird activity in the past. Attempting to keep your prejudices out of the process might cause you to bias the process in the other direction. The point is that although it is desirable to have experimental units that are homogeneous with respect to everything except the treatment factor of interest (e.g., drawdown method), this is often not the case in nature, due to disturbing variables. Therefore, the process of assigning treatments should be objective, preferably by randomly assigning treatments to experimental units.

Replication, where the study includes more than one experimental unit for each *treatment*, is another crucial element of a study. Generally, replicates incorporate experimental units that are physically separated (either temporally or spatially) and allow for independent application of treatments. Suppose the difference in invertebrate density is large between two impoundments when comparing the two drawdown methods. You would not know whether this was due to the treatments or just part of the inherent variation among plots (i.e., no two plots are identical). However, by replicating, even adding one more impoundment (randomly) for each treatment, you would get a measure of the variability in density of invertebrates among impoundments within a treatment. Then, you could compare variability within a treatment too and the variability between treatments, that is, the difference in average density of invertebrates between those impoundments under a quick drawdown versus those impoundments under a slow drawdown. The comparison of variability among treatments with variability within treatments is the essence of analysis of variance (Kuehl 2000), taught in most introductory statistics classes.

Including at least two experimental units (e.g., impoundments) per treatment is crucial to being able to test for a treatment effect and for extrapolating results to the population of impoundments. However, the more units assigned per treatment the better the ability to detect a treatment effect and make inferences about the population being studied. The variability among treatments mentioned earlier is not dependent on sample size but the variability of the average for a given treatment is reduced as sample size increases, which makes the test for treatment effect more powerful. The appropriate sample size for the study should be determined at the design stage. We discuss sample size estimation in Sect. 2.5.8.

It is important to avoid confusing replication with pseudoreplication (Hurlbert 1984; Stewart-Oaten et al. 1986). In the impoundment example, suppose that you measure density of invertebrates on a given impoundment at 20 different locations. This is not a problem if these 20 samples are in reference to the impoundment. Pseudoreplication here would consist of treating these 20 measurements as being taken from 20 independent experimental units for the purpose of evaluating the effect of the treatment (and thus the variation therein as experimental error), when in fact it is a sample of 20 from one experimental unit for the purpose of estimating the effect of a particular drawdown treatment (the variation therein is sampling error). The effect of pseudoreplication is to underestimate experimental error and increase the probability of detecting a treatment effect that does not really exist.

2.4 Major Types of Study Designs

In this section, we introduce general classes of study designs, in decreasing order of rigor: manipulative experiments, quasi-experiments, and observational studies. We briefly compare them with respect to the notions of control and randomization discussed earlier, and devote more attention to them in later chapters. We also briefly discuss adaptive resource management (ARM) as it relates to wildlife studies.

2.4.1 Manipulative Experiments

We define a manipulative experiment as the observation of an ecological system of interest under specific, controllable circumstances in an effort to evaluate system response. Fundamentally, manipulative or comparative experiments (Hurlbert 1984) require (1) random allocation of treatments (including controls) to experimental units (entity subjected or not subjected to the treatment independent of other units) from the population under study and (2) replication (independent replicate of the experiment) of each treatment over several experimental units (Fisher 1925). Controls are unmanipulated experimental units. Controls are also treatments, as controls are the benchmarks used to evaluate impacts of treatments (e.g., water drawdown from our previous example) on system response. In many cases, the experimental unit differs from the observational unit in that observational units can be samples taken from the experimental unit. For example, the experimental unit could be a randomly selected pasture on which a treatment (e.g., prescribed burn, herbicide application) was independently applied, whereas the observational units (units on which measurements were taken) would be the plants within that pasture. Ideally, through randomization and replication, researchers control confounding factors through design, hence providing unbiased results. Additionally, researchers use randomization and replication to estimate experimental error and to separate natural variation from treatment effect (Kuehl 2000; Williams et al. 2002).

When wildlife ecologists are interested in identifying the specific *mechanisms* that drive the wildlife system of interest, rather than conducting general monitoring and assessment studies, a manipulative experiment is recommended. The basic concept behind most ecological experiments is that researchers have the ability to define a set of unique biological conditions that are equally applicable across differing ecological situations and species. However, in many cases, it is difficult to use strict experimental designs for the system under study. Difficulties in replication due to the spatial scale under study or limitations on treatment randomization due to limited study sites all compromise the experiment's integrity. When developing an experimental study, the number of replicates influences parameter precision and the ability of the researcher to evaluate biological hypotheses. However, replication frequently requires a trade-off among number of replicates (cost and logistical feasible), parameter precision, extrapolation, and the reproducibility of results. Randomization ensures that assignments of experimental treatments are independent of researcher bias and that treatment effects are independent, hopefully ensuring representative results. Replication provides the information to estimate experimental error; thus, without replication the researcher loses the ability to test research hypotheses of interest.

Figure 2.2 illustrates the impoundment example. To qualify as an experiment, we chose 20 experimental units randomly from all impoundments in the northeast, to increase the likelihood that they are representative of the target population. Then we randomly assigned two treatments of quick drawdown and slow drawdown to the experimental units. Because we randomly selected impoundments and treatments, we should be able to make statistical inferences about the population, with estimated bounds of uncertainty. Nevertheless, if the investigator feels that there are factors that could have an effect, the investigator could control for these to increase the efficiency of the study. For example, one could block on the five refuges that contributed to the study, four impoundments per refuge (see Fig. 2.2) and two replications per treatment per block. The outstanding deficiency in this design is that although it is an experiment, formal (i.e., statistical) inference from the results is limited to national wildlife refuges of the northeast. To make inferences beyond these, such as to all wetlands in the northeast, one would have to argue that they are similar, based on subjective arguments or past studies involving both types.

2.4.2 *Quasi-Experiments*

In many settings in wildlife science, a replicated manipulative experiment is not practical, due to infeasibility or lack of budget. In these cases, we frequently sacrifice randomization, in which case we use the term quasi-experiment to characterize the study. Quasi-experiments occur when the assignment to treatment or control groups is not random. When this occurs, we compromise our ability to make statistical inference. The extent of this compromise might be large or small, depending on the goal

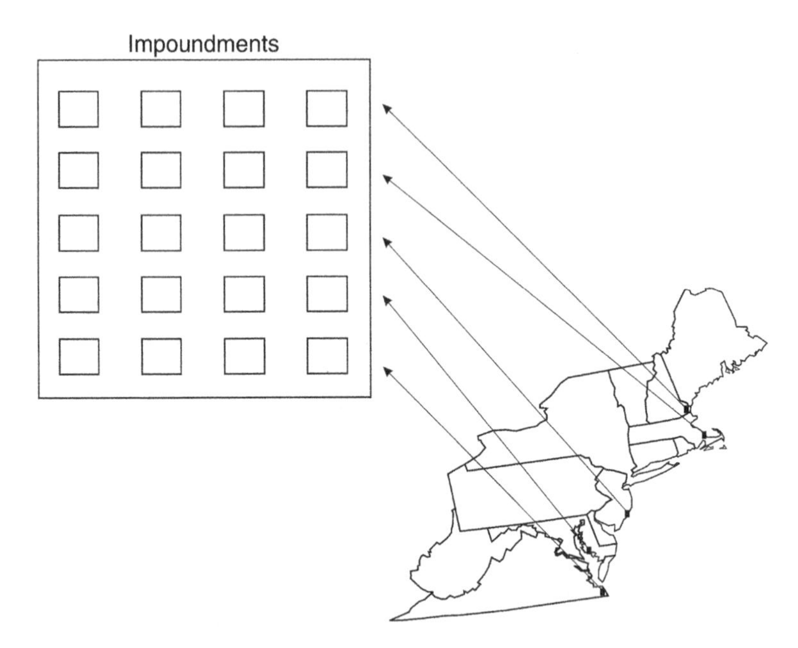

Fig. 2.2 Randomly drawn 20 impoundments from five national wildlife refuges of the northeast U.S. for a hypothetical experiment to study the effect of drawdown speed on spring-migrating shorebirds. They could be drawn completely randomly, or one could block on refuge by randomly choosing four impoundments from each refuge. Reproduced from Morrison et al. (2001), with kind permission from Springer Science + Business Media

of the study. Even where the compromise is great, inference might still be required (e.g., legal requirements to evaluate impacts). Figure 2.3 (Skalski and Robson 1992) presents a general classification of studies based on the presence/absence of randomization and replication.

Stewart-Oaten et al. (1986) described a quasi-experiment that falls in the category of impact assessment (see Fig. 2.3), where they infer impact of a power plant on the abundance of a polychaete, with minimal compromise of rigor. We develop the field of impact assessment in detail in Chap. 6, but must briefly introduce the topic here to ensure continuity with other material in the present chapter. They called this a BACI (before–after/control–impact) design, which is equivalent to Green's (1979) "optimal impact study design." We use the Stewart-Oaten et al. (1986) example here to illustrate the principles behind this commonly used design, but treat impact assessment in more detail in Chap. 6.

2.4.3 Mensurative Studies

Mensurative studies (Hurlbert 1984) represent the class of observational studies for which the researcher suspects certain conditions apply, but where it is not practical to conduct a manipulative or quasi-experiment. Typical in wildlife research, obser-

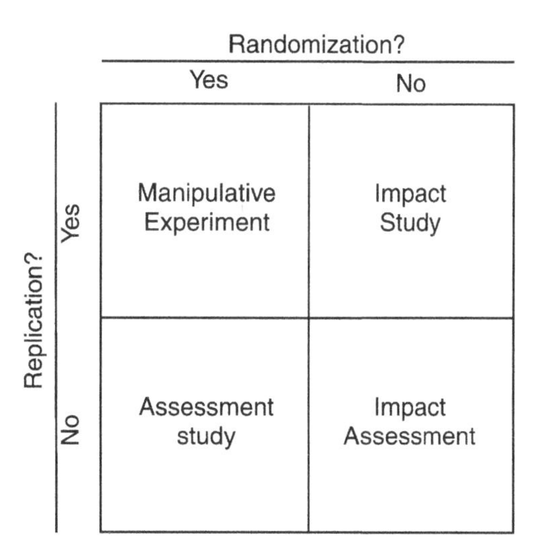

Fig. 2.3 Relationship between design principles of randomization and replication, and the nature of environmental field studies. Reproduced from Skalski and Robson (1992:12), with kind permission from Elsevier

vational studies are conducted in lieu of experimentation and consist of comparing natural processes and populations over time or between locations through estimation of various attributes. Given an appropriate sampling design, mensurative studies can be quite useful for evaluating patterns and processes within the population of interest. In many cases, the intent of mensurative studies is to go beyond description and draw subjective inference to causal factors related to the measurement of interest, primarily with correlative analysis. For example, the relationship between density and survival or recruitment of a population is of common interest in wildlife ecology. We may conduct annual monitoring to measure both, and use regression to develop a predictive model for recruitment as a function of density. However, note that in this type of mensurative study, we make no effort to manipulate the system of interest with the intent to test research hypotheses. Additionally, we take no action to reduce the impact of confounding variables through randomization. Finally, our judgment of the relationship of density and survival is subjective and based on the properties of the model and not from inference based on a controlled experiment.

Recording ancillary information can benefit many experiments. If we suspect a factor influences the response and is not controlled for in the experimental design, such as organism age, recording its value allows for the possibility of controlling forthat variable in the analysis after the fact. This process is especially important in an observational study, and indeed provides the only basis for its usefulness beyond simple description. Relationships between the response of interest and concomitant variables can also serve as a basis for designing experiments to evaluate and establish causation.

2.4.4 Descriptive Studies

Descriptive studies (Hurlbert 1984) are a class of studies providing description of a system of interest. Descriptive studies cannot provide detailed answers to "how" or "why" questions (Gavin 1991), but they can increase our knowledge. However, understanding of causation under descriptive studies is limited or nonexistent, thus our inferences tend to be weaker than those derived from experimental manipulation. To illustrate, consider Sinclair's (1991) hypothetical study of black-tailed deer (*Odocoileus hemionus*) diets during the winter in British Columbia. In this study, the objective is to document diets of black-tailed deer in a specific location at a specific time. Studies such as this do not allow us to make broader inferences to all black-tailed deer in British Columbia because (1) we have not replicated across different landscapes, (2) we have not randomized or applied multiple treatments (e.g., different forage availability), and (3) we have not considered any other populations in winter in British Columbia. Other descriptive studies include, for example, evaluation of changes in habitat parameters over time based on GIS mapping techniques. However, these mapping approaches are descriptive as there is little ability to infer specific causation for the changes in habitat structure other than time as no treatments are applied and areas are not randomly selected, although plots within an area could be randomly selected if area-specific inferences are warranted.

Descriptive studies, however, can and do provide a wealth of information and have been the foundation for ecological research for many years. Perhaps the most useful studies of organisms have been descriptive work, as these studies provide the foundation for future evaluation of hypotheses regarding mechanisms causing changes in populations over time.

2.4.5 Adaptive Resource Management

ARM (Holling 1978; Walters 1986; Williams 1996), sometimes called management by experiment, is an approach to management that emphasizes continued learning about a system in order to improve management in the future. Adaptive management helps scientists learn about ecological systems by monitoring the results from a suite of management programs (Gregory et al. 2006a). There are two basic types of adaptive management, passive management, or management where historical data and expert opinions combine into a best guess format focused on a singular hypothesis, and active management, or management where systems are deliberately perturbed in several ways and managers define competing hypotheses about the impact of these perturbations (Walters and Holling 1990; Gregory et al. 2006a,b).

Although the process of adaptive management is often termed learning by doing, it is important to recognize that the management process can be broken into several elements (Johnson et al. 1993):

- Objective function
- Set of management options
- Set of competing system models
- Model weights
- Monitoring program

An objective function is the mechanism by which stakeholders evaluate results of a management action. Simple examples include maximizing total harvest over an infinite time horizon, to minimize the probability that a species goes extinct over a 100-year period or to maintain a population at a target level.

The set of management options includes possible actions that a manager might choose to achieve objectives. This could include an array of harvest regulations; the timing and speed of a drawdown for an impounded wetland; and the size, distribution, and frequency of clear-cuts in a forest.

Under active adaptive management, we posit a set of models, each model representing a competing hypothesis about how we expect the system to respond to management actions. This set of predictive models should account for various types of uncertainty. These include partial controllability, which acknowledges that the relationship between the chosen management action and the mechanism that affects the system of interest according to the model is not perfectly precise. For example, a fixed set of hunting regulations does not result in the same kill rate every year in every location. Environmental variation includes sources of variability that remain unexplained by a model and are included as random noise. Structural uncertainty is reflected in the number and variety of models used, acknowledging that the structural form of the processes that drive the system dynamics are not completely known. Of course, any model is a simplification of the real process it approximates, but some models reflect the truth better than others.

A monitoring and research program is a crucial aspect of ARM (Chap. 7) or any informed approach to management. We use monitoring and research programs to evaluate the impact of the management actions against the objectives. We evaluate comparisons between competing models and system response. Study design is a crucial part of this element, both to produce an unbiased estimate of the relevant state of the system and to minimize the remaining source of uncertainty in management or partial observability (i.e., *sampling variation*).

Given the elements of ARM, the approach to finding the optimal decision can fall into any of three categories. In passively adaptive management, initial assessment of alternatives is conducted, and the management action deemed best is designed and implemented. Results of management actions are monitored and compared against predictions under various hypotheses, leading to adjustments in the management actions. If we forego traditional management objectives to learn about the system as quickly as possible, the result is true experimentation. Actively, adaptive management is a hybrid of the two, focusing on management objectives while pursuing learning to the extent that it promotes those management objectives.

ARM is difficult to conduct from a scientific point of view. Adaptive management requires that we conduct the evaluation in a realistic setting so that we can make inferences based on the management action. That is, we must conduct manipulations at the same scale that management is conducted (Walters and Holling 1990). If management is conducted on large scales, it is difficult to identify spatial replicates, and therefore we must use temporal replication. Additionally, on large scales, it is difficult or impossible to control potential disturbing variables. Despite these difficulties, we must be able to make decisions for managed systems. ARM brings design principles to bear to manage systems in the face of uncertainty while allowing for learning more about the system to improve future management.

With the possible exception of the inclusion of multiple models, we might view ARM as simply reflecting what an astute manager might be doing regularly through subjective assessment and reevaluation of his or her decisions. Unfortunately, many managers subscribe to this incorrect view of adaptive management. Nevertheless, as the scientific method promotes rigor and objectivity in studies of natural history, ARM promotes rigor and objectivity in the management of natural resources. This rigor becomes more useful as the size and complexity of the managed system grows, and the number of stakeholders increases.

2.5 Sampling

Sampling in wildlife field studies is most often associated with observational studies where there is no control of the system under study. We sample when the ecological unit of interest cannot be censused. Thus, we use sampling when it is impractical to measure every element of interest within an ecological unit of interest. Typically, sampling consists of selecting (based on some probabilistic scheme) a subset of a population, allowing one to estimate something about the entire target population (Thompson 2002a). In classical manipulative experimental design applications, experimental units are frequently small enough for us to measure the entire experimental unit. For example, when applying different fertilizer types to small plots of ground, we can collect the entire biomass of the plot to measure biomass differences between fertilizer types.

With ecological manipulative experiments, however, the scale of treatment (or observation) could be too large to conduct a census of the target population. For example, if we subject a pine (*Pinus* spp.) plantation of 40 ha to a controlled burn to estimate the effect on subsequent understory growth within that plantation, we must evaluate the number of new shoots and their subsequent growth and survival for multiple years. However, enumeration of all new shoots in just the first year across the 40-ha plantation would be logistically infeasible (and probably unnecessary); thus, we must take a sample of new shoots, perhaps among ten 5 m × 5 m vegetation plots randomly placed throughout the plantation.

In probability sampling, each element (e.g., animal or plot of vegetation) in the unit has some nonzero probability of selection. We will measure those selected units for the variable of interest (e.g., number of new shoots). Each selected element is measured and summaries (e.g., means and variances) of these measurements serve as the measurement for the sampling unit of interest. Using the summarized measurements, we extrapolate to the ecological unit of interest.

Given that a complete count of each element is only rarely achieved in wildlife studies, the purpose of sampling is to estimate the parameter (survival, abundance, or recruitment) of interest while accounting for (1) spatial, (2) temporal, and (3) sampling variations as well as accounting for imperfect *detectability* (Sect. 2.4.1).

We frequently combine spatial and temporal variation into process variation (Thompson et al. 1998) as they are process based (e.g., not related to the sampling procedure). Differences within a population due to heterogeneity between study areas is spatial variation while temporal variation is influenced by changes in the number or location of individuals in response to factors that are changing through time (e.g., habitat, management regime, precipitation). Sampling variation consists of among-unit variation and enumeration variation (Williams et al. 2002). Among-unit variation occurs because only a subset of the available plots are sampled, thus repeated sampling of the same number of randomly selected plots could give different results. Enumeration variation, or variation due to incomplete counts, is a function of the number of sampled plots and detection probabilities (Sect. 2.4.1).

In survey sampling, the population of interest consists of a number of elements (N) whose values are fixed. Random variation is introduced through the selection methods used to choose elements for a sample. For example, if we divided a 40-ha pine plantation into 100-m^2 plots, and drew a random sample of those plots, we could enumerate the number of new shoots (assuming 100% detection of all shoots within the plot). If we surveyed all the sample plots at the same approximate time, the sample variance would summarize spatial variation of new shoot production in the burned study area. If we replicated the sampling process in the same area later, we could estimate temporal variation in shoot production. In Sect. 2.5, we discuss how to make inferences regarding the target population, based on the sampling results.

The sampling design described earlier is *simple random sampling*. This design produces unbiased estimates of the population total, mean, and variance (Thompson 2002a). However, the underlying distribution of the population (e.g., clumped or rare) can inflate variance estimates under simple random sampling designs. Thus, we recommend that sampling designs be based on some knowledge of the life history of the species under study as this will allow for researchers to reduce variation in estimates of population parameter. For example, some potentially more efficient sampling designs include stratified sampling, cluster and systematic sampling, multistage and adaptive sampling, network or double sampling, and various combinations of the earlier sampling approaches (Cochran 1977; Thompson and Seber 1996; Thompson 2002a). We discuss some of these topics in more detail in Chaps. 4 and 5.

The designs mentioned earlier lead to design-based inference, where the design itself (the selection method for the sample *elements*/sampling units) justifies the inferential results. In some cases, additional inference may be desirable; for example, instead of being interested in a simple expression of variability over space, one might be interested in the way in which things vary spatially (e.g., a common assumption is that the closer organisms are together the more similar they are). Another example includes the belief that a measurement on a sample unit (e.g., animal or plot) will covary with other ancillary information (e.g., new shoot growth varies with precipitation or seed density). In this case, the investigator conceptualizes these relationships and attempts to explain them using models; hence, it is a model-based inference. Model-based inference, while common in wildlife studies, requires more assumptions and therefore requires good tests of those assumptions.

The properties of a sample are a function of the design, and therefore the anticipated data analysis should have a bearing on the design. We suggest that researchers use pilot studies to (1) evaluate the data collection methodologies, (2) determine necessary sample sizes to obtain estimates with some accepted level of variation and minimal bias, and (3) allow for optimal allocation of sampling effort over space and time.

2.5.1 Accounting for Detectability

As a first step in understanding the structure and dynamics of wildlife populations, managers usually require estimates of population size (Seber 1982). A majority of effort expended by resource managers when studying terrestrial population of plants and animals has been on estimating abundance (Seber 1982; Nichols et al. 2000; Rosenstock et al. 2002), and a wide variety of approaches are available to assist managers with abundance estimation (Otis et al. 1978; Seber 1982; Williams et al. 2002). When tracking population dynamics of terrestrial mammals, fish, birds, or plants through time (e.g., survival, recruitment, local colonization, extinction) or space (movements, dispersal, immigration, emigration), it is unreasonable to assume that you detected all organisms of interest in the sample. That is, if you do not observe a species in an area (or during a formal count), is it because the species was not present or because you were unable to detect its presence? Additionally, when conducting auditory bird surveys, is it likely that you will be able to hear a species in a forest adjacent to a running stream with as much confidence as you can hear the same species in a forest away from the stream? Population indexes based on uncorrected counts rely on the unrealistic assumption that the organisms under study are all detected equally across multiple habitat types, observers, or time frames (Anderson 2001). In any survey of animals (we discuss animals for the rest of this section), we must assume that within a population some animals will not be detected during survey efforts, and the probability of detecting individuals will vary as the result of a number of interacting factors. Thus, in the case of imperfect detection, researchers must directly estimate detection or use modeling methods that account for varying detectability of the target organism.

If the objective of sampling is to measure an attribute about animals in a population, such as mass or length, then one approach to dealing with nondetection might be to assume that the animals detected are representative of those not detected, or that detection is not influenced by the attribute of interest. This is also true in design-based approaches, but in that case, we assure representativeness because the researcher has equal access to each element, at least in theory. Thus, use of average weight or length from the sample to estimate the average weight and length for the target population is justified. However, when the attribute of interest is the attribute totaled over all animals, or the number of animals, or their dynamics, the issue becomes more difficult, and detection probability becomes another potential disturbing variable within the study.

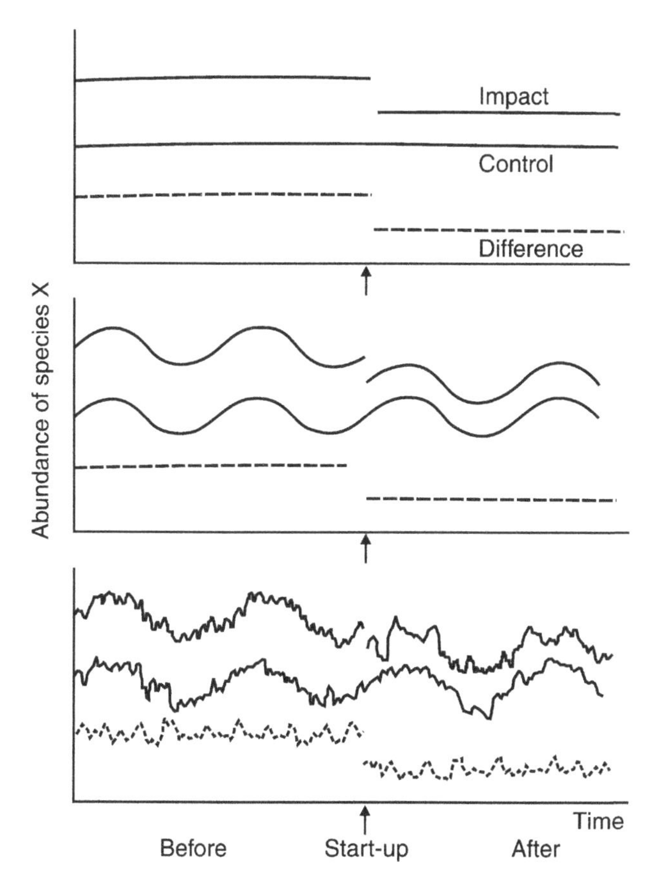

Fig. 2.4 The abundance of "Species X" at the Impact and Control stations, and the difference of the abundances, as functions of time, in three versions of impact assessment. (**a**) In the most naive view, each station's abundance is constant except for a drop in the Impact station's abundance when the power plant starts up. (**b**) In a more plausible but still naive view, the abundance fluctuates (e.g., seasonally), but the difference still remains constant except at start-up of the power plant. (**c**) In a more realistic view, the abundance fluctuates partly in synchrony and partly separately; the former fluctuations disappear in the differences but the latter remain, and the power plant effect must be distinguished from them. Reproduced from Stewart-Oaten et al. (1986), with kind permission from Springer Science + Business Media

Consider the historical approaches to evaluating population changes over time. Probably the most well-known historical survey technique in wildlife literature is the BBS for landbirds (Link and Sauer 1998; Robbins et al. 1986) and road-spotlight counts for white-tailed deer (Farfarman and DeYoung 1986; Mitchell 1986; Collier et al. 2007). Historical survey approaches have relied on raw counts (C) (e.g., number of birds/deer seen/heard at each survey point or along a transect) to be an *index* of abundance, such that the index is assumed to be proportionally related to the true population size (Anderson 2001, 2003). Thus, we have often used changes in the index value between temporally separated samples to indicate changes in population status (Thompson 2002b). However, the above approach assumes that detection rates remain constant among all survey sites, observers, weather conditions, species, and time periods, a seemingly "absurd" assumption (Anderson 2001, p. 1295).

Consider the case where N_1 and N_2 are the population sizes in years 1 and 2, and the change in population size N_2/N_1 is of interest. A common approach is to estimate this ratio with the ratio of an index to population size for the 2 years. If the index is a count (C) from some standardized protocol (e.g., number of deer seen along a road transect), it is appropriate to characterize this count C_t at time t as $C_t = N_t p_t$, where p_t is the detection probability. The ratio of the counts would thus be $C_2/C_1 = N_2 p_2/N_1 p_1$. This estimator for the ratio N_2/N_1 is therefore unbiased if and only if (*iff*) $p_1 = p_2$.

Given the above example, assume that a deer transect survey was used to evaluate a deer population. Assume an average detection probability of 0.74 for the observers during the initial survey and 0.44 for the observers during second survey (Thompson 2002b; Anderson 2003). In addition, assume that the population remains closed between survey periods (e.g., no births, death, immigration, or emigration) at 250 deer. The count during the first transect survey (C_1) would be 185 (0.74 × 250), whereas the count during the second survey (C_2) would be 110 (0.44 × 250). Using the above ratio estimator (N_2/N_1), these values would indicate a decline (110/185) of approximately 41%, when actually no decline had occurred. Additionally, a mechanism for evaluating which disturbing variables (e.g., multiple sampling occasions or interobserver variability) contributed to the different detection probabilities is unknown in this simple example.

This example illustrates the problems that can result in this type of sampling. We deal with detection probability in at least one of three ways.

(1) Assume that detection probability varies randomly across time, space, treatments, observers, and other factors of interest, and therefore, on average, detection probabilities will be the same. This is the most common, but often questioned, approach to the issue of detection probability and we do not recommended this approach (Anderson 2001, 2003; Rosenstock et al. 2002; Thompson 2002b).
(2) Identify the disturbing variables that cause detection probability to vary and model them as predictors of the counts. This is the approach discussed earlier (Sect. 2.2).
(3) Estimate the detection probabilities and those factors influencing variation in detection directly (Williams et al. 2002), which is the most desirable option because it relies on weaker assumptions of the attribute of interest (population size), but typically requires substantially more effort than the other approaches.

Option 1 is the most naive approach, but it is also the simplest and cheapest option. Option 3 is the only reasonable choice if aspects of population dynamics are of interest to the researchers.

The most commonly used methods that account for detectability fall into two general categories: capture–recapture methods and distance sampling methods

(Buckland et al. 2001; Williams et al. 2002). Capture–mark–recapture (CMR) methods entail monitoring marked individuals (animals or plants) across samples taken over time or space. Methods using CMR usually require physically capturing, marking, and recapturing or resighting live individuals or recovering dead individuals (Otis et al. 1978; White 1996). Additionally, CMR techniques to estimate detection also include techniques using multiple observers (Nichols et al. 2000), double sampling (Bart and Earnst 2002), occupancy estimation (MacKenzie et al. 2006), and other model-based approaches (Royle 2003). Under these approaches, we estimate detectability from the pattern of each individual's presence or absence in each sampling occasion (e.g., if every marked individual was detected in every sample, then the estimated detectability would be 1.0). In the field of wildlife ecology and management, there are numerous peer-reviewed CMR studies (e.g., Otis et al. 1978; Brownie et al. 1985; Burnham et al. 1987; Pollock et al. 1990; Lebreton et al. 1992) as well as several excellent books covering a host of CMR topics (Borchers et al. 2002; Williams et al. 2002; Amstrup et al. 2005), which we suggest interested readers consider. In addition, Skalski and Robson (1992) provide a detailed explanation of putting CMR studies within the larger frameworks of experimental, quasi-experimental, and observational studies.

How might we account for detectability directly in the earlier examples? First, let us look at bird point counts, in which a single observer listens for a set amount of time and counts the total number of birds seen/heard, such as the BBS (Peterjohn et al. 1996). Nichols et al. (2000) experimented with using two observers on each survey route, alternating in roles as primary and secondary observers at each point count. In each case, the primary observer would identify every bird he or she could see or hear, while the secondary observer would record these and separately record those birds missed by the primary observer. Nichols et al. (2000) then used the methods of Cook and Jacobsen (1979) for aerial surveys to estimate abundance. They extracted the number of birds missed by both observers from the ratio of birds counted by the primary observer to those counted only by the secondary observer.

Estimation of population density using distance sampling involves running a probabilistically selected transect or conducting a point count and recording those individuals (or groups of individuals) counted from the transect or point, as well as the perpendicular distance of the observed individuals to the transect line or to the point. An estimate of density is then derived under the assumption that the detection probability decreases with increasing distance from the transect line of observation point (Buckland et al. 2001).

Inferences in CMR and distance sampling methods are dependent upon the descriptive model, which we use to estimate the parameters (e.g., detectability; see Chap. 4 for a detailed discussion of both topics). We evaluate model validity by examining how well a chosen model fits the data contingent upon the set of models evaluated. Model-based inference is not trivial, and model evaluation is never easy (Ripley 1996); however, model-based inference has received much attention lately in wildlife and statistical sciences (Burnham and Anderson 2002). When nuisance parameters such as detectability are an issue, a model-based approach to inference is required.

2.6 Statistical Inference

Wildlife field research requires that scientists ask challenging questions before, during, and after the design process. In addition to the common question of "What is my objective?" researchers must also ask questions like "How am I going to collect data for this research?" and "What analyses will I do with the data I have collected?" Too many researchers treat not only statistical design, but statistical inference as afterthoughts, only to discover that they cannot evaluate the research question of interest because either data collection methodology was flawed or sample sizes were inadequate. We maintain that scientists can do a much better job if they think out the entire study design thoroughly, from formulation of the general scientific question to specific research hypotheses, through designs most likely to allow strong inferences.

Investigators should consider several aspects of statistical inference during the design stage to achieve their goals. These topics are covered in undergraduate courses in statistics. Thus, each is an important topic and is worth reviewing here.

2.6.1 Hypothesis Testing and Estimation

There are two primary areas of statistical inference: testing of hypotheses and estimation of parameters. We will limit our focus in this section to the classical approach to statistical analysis (frequentist) rather than the approaches based on Bayesian theory, acknowledging that both approaches have a place in wildlife research. To illustrate the comparison of the two approaches, consider a study comparing the average thickness of the eggshells of osprey (*Pandion haliaetus*) treated with the insecticide DDT (dichlorodiphenyl-trichloroethane). Suppose the overlying biological hypothesis is that DDT reduces the productivity of osprey, and more specifically, that it thins the eggshell, thus making eggs more fragile. Under a hypothesis-testing approach, the null statistical hypothesis might be H_0: There is no difference in average eggshell thickness between lakes with and without DDT residues. The alternative hypothesis would likely be H_A: Average eggshell thickness is lower where there is DDT residue. The researcher then collected a sample of eggs from lakes both with and without DDT exposure, computed sample means and variances, and performed a statistical test on the difference between the sample means (e.g., a two-sample *t*-test) to determine if the difference was statistically significant. An estimation approach to address the same question would be to estimate the difference in average eggshell thickness for samples of eggs from lakes with and without DDT exposure, construct a confidence interval around this difference, and determine if the confidence interval includes 0. If the interval does not include 0, then the difference would be statistically significant. Obviously, there is an intrinsic link between hypothesis testing and estimation, and frequently the two approaches give similar, if not identical, results.

There is a philosophical distinction between testing and estimation, and we urge the reader to remember that these two topics, although closely related, are independent. Several authors have put forth detailed discussions regarding the usefulness of classical hypothesis testing in wildlife sciences (Cherry 1998; Johnson 1999), given the nature of the hypotheses under question and the nature of the alternative hypothesis. Consider our osprey example: when measuring osprey eggs near any two lakes, regardless of the presence or absence of DDT, it is almost impossible that the average eggshell thickness would be identical. Thus, the alternative hypothesis that they are different is given, although the alternative that those exposed to DDT are thinner is more meaningful. The point that these authors make is that the magnitude of the difference between the means is most important. In our example, if there is a difference in the shell thickness between the ospreys exposed and those not exposed to DDT, but that difference is not "large" enough to produce a meaningful (e.g., biologically relevant) change in the rate of production (hatching rate or fledgling health), then there is likely no impact on population productivity, which is our scientific question of interest.

Regardless of whether an estimation or hypothesis-testing approach to inference is used, the primary focus should be on evaluating the magnitude of effect. If the estimated magnitude is not biologically meaningful, then that is the conclusion. Typically, if researchers determine that the magnitude is biologically meaningful, they then assess its statistical significance. However, we concur with Burnham and Anderson (2002) and Guthery et al. (2001, 2005) in that ecological systems can have biologically, but not statistically, significant effects. Therefore, rather than evaluating statistical significance after biological importance has been determined, we suggest that researchers should focus their effort on a priori specifying what constitutes a "biologically significant difference" and use that measure for determining the effects of a treatment, perhaps by conducting power analysis (Cohen 1988).

Biological significance is a value which must be defined by the researchers, based upon the reason for the study and should be quantifiable and influenced through management (see Sect. 1.5.3 for further discussion). Scientists must be able to quantify a biologically significant change for measurement of biological significance to be useful. Additionally, for management recommendations, scientists must be able to manipulate those factors that cause the biologically significant value to change. For example, knowledge that nest survival is most influenced by the age of the nest (Dinsmore et al. 2002), although interesting, is irrelevant to management as managers cannot manipulate nest age.

Suppose that in the earlier osprey example, average eggshell thickness is lower where DDT is present. The direction of this difference is consistent with that predicted by the investigators' research hypotheses. However, to determine biological significance, the investigator must look at the implications of the difference in eggshell thickness on the population trajectory over time. Thus, the question perhaps becomes: How does reduction in eggshell thickness translate into hatching rate and fledgling health, then into reductions in productivity, and ultimately to a decrease in the population (e.g., a 10% decline in eggshell thickness causes a 25% decline in population productivity)? For our example, there is no specified level of eggshell thickness that would be considered biologically significant, as biological significance must be based on the researcher's understanding of the system and life history of the species under study.

2.6.2 Hypothesis Testing and Model Selection

Model selection and inference procedures have become increasingly common in the field of wildlife ecology since the early 1990s (Lebreton et al. 1992; Burnham et al. 1995; Burnham and Anderson 2002), primarily as an alternative to statistical null hypothesis significance testing (Anderson et al. 2000). Estimation procedures in wildlife ecology have slowly shifted toward evaluating "nuisance" parameters (detectability, capture rates) as well as parameters of interest like survival and abundance (Lebreton et al. 1992), creating a clearly defined break with hypothesis testing (Sect. 2.5.1). Model-based inference has become more important in ecological studies, with its focus being primarily on the analysis of data collected from capture–recapture studies (Lebreton et al. 1992; Burnham et al. 1995) as most programs used for population parameter estimation use model selection criteria (Sect. 2.6.2; White and Burnham 1999; Arnason and Schwarz 1999; Buckland et al. 2001). Design-based inferences, which are the foundation of sampling literature, are more uncommon in wildlife sciences, likely due to the logistical difficulties with replication and randomization. Although design-based inferences are the most statistically powerful, and in many cases can justify the use of hypothesis-testing approaches, additional inference is necessary if stochastic processes determine the distribution, detectability, or characteristics of a population of interest (Buckland et al. 2000).

Currently, there are four groups with respect to evaluation of hypotheses and use of model selection in wildlife sciences. We suggest that the first two groups include scientists who are uninterested in analytical or statistical ecology (probably the largest group) and those scientists who are interested in a specific analytical cookbook that suits their specific needs (probably the other largest group). The other two groups, in our opinion, represent a minority, although highly vocal, (relative to all scientists), who focus on the development and evaluation of different statistical procedures. Neither group disagrees with the basic fact that the "Immediate issue is how to present useful and sensible results from field studies" (Eberhardt 2003, p. 241), and both seem to agree that exorbitant usage of silly null hypotheses and *p* values are unnecessary (Cherry 1998; Johnson 1999). One group suggests that model selection is superior to other analytic methods (Anderson et al. 2000; Burnham and Anderson 2002; Lukacs et al. 2007), whereas the other group suggests that wildlife ecologists might simply be substituting one rote statistical technique (model selection) for another (null hypothesis significance testing), while losing track of the more fundamental biological questions and related research hypotheses (Guthery et al. 2001, 2005; Stephens et al. 2007a,b).

One of the primary criticisms of hypothesis testing is that scientists take the results from a single, unreplicated study and make wide-ranging management suggestions based on estimates of statistical significance (Johnson 1999). This differs considerably from Fisher's belief that hypotheses (and thus hypothesis tests) were only valid across a series of experiments, as they would confirm the size and direction with replication (Fisher 1929). However, this same issue holds true with respect to statistical inference for model selection approaches to inference, in that studies are frequently not replicated; thus, although the inference engine has changed, the validity of the results should still be questioned until adequate meta-replication (replication of the entire study) has been conducted (Johnson 2002b).

Information-theoretic approaches suggest a priori (e.g., before data collection, preferably during study design) specification of candidate models (Burnham and Anderson 2002). We agree with this general approach to science (thinking before you act) as it forces scientists to evaluate and justify data collection needs. Indeed, this is the underlying motivation for this book in that wildlife studies should be conceived beforehand rather than as an afterthought. Careful planning upfront keeps scientists from using a shotgun approach to hypothesis creation (e.g., testing all possible relationships); we suspect that it is only rarely accomplished in observational studies. For example, one author of this book published a set of models he posited before study implementation. He was instructed to evaluate more than five additional models before his work could be published based on reviewer comments about the data he presented. Thus, we suggest that although critical thinking before study implementation is extremely important, most sets of candidate models should be posited after preliminary data collection and evaluation using graphical displays (Anscombe 1973), summary statistics, or some other supplementary method (Eberhardt 2003) or after initial evaluation of an a priori set (Norman et al. 2004). Either approach should reduce the frequency of vacuous candidate models in wildlife studies (Guthery et al. 2005). However, we do not endorse detailed exploratory data analysis or data mining, where a researcher looks for relationships between the data without considering biological plausibility.

Although there is a multitude of research extolling or deprecating many statistical approaches to wildlife ecology, there seems to be a little gray area in this discussion, with some treating model selection as "...the alternative to null hypothesis testing" (Franklin et al. 2001) while others question the usefulness of information-theoretic approaches as a replacement for all other ecological statistics (Guthery et al. 2001, 2005; Steidl 2006). Statistical hypothesis testing has several limitations in observation studies, but under Fisher's (1929) model of multiple experimentation can provide useful results. Model selection is a useful statistical tool for biologists to use in observational studies for estimation and prediction, but does not substitute for replicated experiments. There are numerous statistical tools available to wildlife scientists, and we suggest that the use of many tools can assist with furthering our understanding of ecological systems.

2.6.3 Sampling Distributions

Scientific progress is often credited to experimentation where data are collected, and we draw conclusions based on repeatability of results. This process of inference allows us to extend results from the specific to the general (Mood et al. 1974). As discussed in Chap. 1, the purpose of a research study is to make valid inference to the target population about some set of *parameters* that describe attributes of the target population. In wildlife field studies, sampling from a population and then drawing inference to the population based on the sample collected usually accomplishes this. Consider our osprey example: the parameter of interest was the average, or arithmetic mean, thickness of the shells of all osprey eggs found in nests around a particular lake. However, it is often impractical to measure every egg within a nest because the measurement would be invasive (i.e., the egg must be broken unless samples were taken after hatch); thus the osprey population would be negatively affected for the duration of the study because of the sampling protocols. Therefore, we take a sample of size n from the N eggs in the population. We then summarize those data collected from the sample (n) into one or more *statistics* (e. g., mean, variance, range). We make inferences about populations under study using one of these statistics. An *estimator* is a statistic that serves to approximate a parameter of interest. Because our interest is in the mean thickness (μ) of the osprey eggs in that population, a logical estimator is the mean thickness ($\bar{x}$) of the sample eggs. However, we cannot assume that $\bar{x} = \mu$; but we hope that it is close. Assuming repeated samples from a probability distribution (see below) that has mean (μ) and variance (σ^2), our expectation would be that the *expected value* of $\bar{x}$ (e.g., $E[\bar{x}] = \mu$), or that the average $\bar{x}$ is equal to the μ that we are interested in estimating (Mood et al. 1974). Thus, from our osprey example, we cannot assume that $\bar{x} = \mu$, nor can we assume that another random sample of size n from the osprey eggs in the same location will have $\bar{x} = \mu$. This is because the thickness of eggshells varies across the population, and thus so does the eggshell mean thickness ($\bar{x}$) from samples of eggs drawn from that population. The probability distribution that formed from the variation in an estimator across multiple samples is called a sampling distribution. This distribution has a mean and variance parameter (measures of central tendency and variation).

The objective of statistical inference is to identify the sampling distribution of the estimator in relation to the parameters of interest. Properties of the sampling distribution define the properties of the estimator in that we would assume that any measurements taken from wildlife (e.g., offspring number, size, or weight) would exhibit considerable variation from population to population and within a population because of differences in age, sex, or reproductive status. A researcher could determine the most likely sampling distribution by taking n samples and building a frequency diagram of the results. However, it is likely that the effort that would be necessary to fully specify this distribution using n samples would be more than it would take to census the population (eggshell widths), and would be disruptive to the population (osprey) under study.

The classical approach in ecological field studies is to assume a form for the sampling distribution (e.g., Normal distribution), and then use the data collected during the study to estimate the parameters that specify the distribution. There is considerable literature on the case where the parameter of interest is an arithmetic mean μ and the estimator is the sample mean $\bar{x}$, as in our osprey example. If the sample is drawn by simple random sampling (sampling from an finite population), then we know that the sample mean is an unbiased estimator for the population mean, and that the standard deviation of the mean $\sigma_{\bar{x}}$ is $\sigma/\sqrt{n}$, where σ is the standard deviation of the egg population and n is the sample size. If the distribution of the egg thickness (x) among all eggs in the population is normally distributed (e.g., bell shaped), then from statistical theory we know that the sampling distribution of $\bar{x}$ will also be normally distributed. Additionally, the *central limit theorem* tells us that even if the distribution of x is not normal (Fig. 2.5a), as sample size increases a normal distribution becomes a better approximation for the sampling distribution of $\bar{x}$ (Fig. 2.5b).

Once we identify and justify the form of a sampling distribution and estimate the parameters of interest, we can evaluate our statistical hypotheses, construct confidence intervals, and implement estimation procedures to develop general inferences from the sample to the population. Although we have outlined the normal distribution here given its frequency of use in wildlife studies, other distributions such as the Poisson, negative binomial, beta, and Weibull have also been used successfully in wildlife studies.

In some cases, we cannot assume or ascertain the distributional form of the data. In these cases, nonparametric methods are available for inference, although nonparametric approaches are not assumption free (Johnson 1995). Recent advances in computing power have also opened up opportunities for techniques such as simulation, randomization, and Markov Chain Monte Carlo modeling (Manly 1991; Link et al. 2002) for inference in wildlife sciences. Each of these provides alternative methods for hypothesis testing, estimation, and inference with varying assumptions regarding the sampling distribution for the data at hand.

2.6.4 Bias, Precision, and Accuracy

In Sect. 2.6, we outlined several factors used for making valid inferences in wildlife study design. The necessity for this should be apparent because the intention in wildlife field studies is to make inferences from the sample collected to the population. Thus, we require that our hypothesis tests or parameter estimates adequately represent the population as a whole (Thompson et al. 1998). When using the results from a sample to make population inferences, an estimator $\hat{\theta}$ is beneficial if it provides a good approximation of the population parameter θ. A good approximation depends upon the amount of error, which is associated with $\hat{\theta}$. There are two basic desirable properties for an estimator; *bias* and *precision* (note: accuracy is a combination of bias and precision). The first (bias) is that the estimator mean be as close to the parameter as possible and the second (precision) is that the estimator does not vary considerably over multiple samples.

Both *bias* and *precision* are well-defined properties of estimators. A useful statistical concept for defining these terms is the expected value, which is nothing more than the average of values parameter x can take, weighted by the frequency

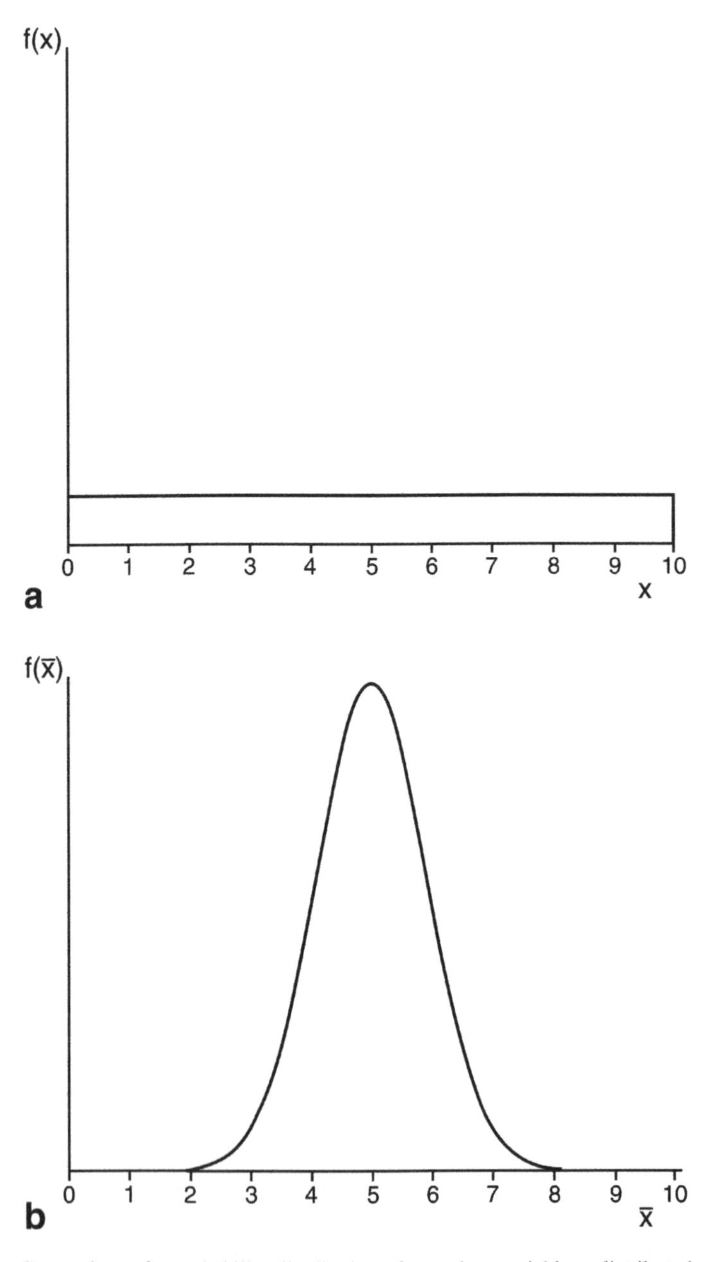

Fig. 2.5 Comparison of a probability distribution of a random variable x, distributed uniformly from 0 to 10 (**a**) with the sampling distribution for the mean, $\bar{X}$, of a sample of size 8 drawn from that distribution of x (**b**). Reproduced from Morrison et al. (2001), with kind permission from Springer Science + Business Media

of occurrence of those values (Mood et al. 1974; Williams et al. 2002). For example, the expected value of $\bar{x}$, designated as E($\bar{x}$), is equal to the population mean μ. The expected value of a sample is simply the arithmetic average of the values of x, where each value has equal weight. Thus, E($\bar{x}$) is the long-term limiting average from independent repeated experiments, thus an average of an average.

The bias of an estimator $\hat{\theta}$ is the difference between its expected value and the parameter of interest θ:

$$\text{Bias}(\hat{\theta}) = E[\hat{\theta}] - \theta$$

If $E[\hat{\theta}] - \theta = 0$, the estimator $\hat{\theta}$ is unbiased. Precision is "the variability among parameter estimates from repeated samples" (Thompson et al. 1998; Williams et al. 2002). Thus, precision is the variability in the estimator around its expected value, as represented by the variance:

$$\text{Var}(\hat{\theta}) = E[\hat{\theta} - E(\hat{\theta})]$$

The square root of the variance ($\sqrt{\text{Var}\ \hat{\theta}}$) of any random variable (e.g., osprey eggshell thickness) is called the standard deviation. However, when the random variable is an estimator for a parameter (e.g., if the random variable is a mean ($\bar{x}$)), the standard deviation is more commonly called a standard error. Wildlife science students commonly confuse standard deviation and standard error, and this confusion has carried through to professional wildlife biologists. From a given sampling distribution (estimated or assumed), the standard error is the standard deviation from this distribution. Confusion regarding estimation of standard error is the most relevant when considering population means as the parameter of interest. Consider our osprey eggshell thickness example. A sample of n eggs is randomly collected and we measured shell thickness (x) for each egg, and computed the sample mean $\bar{x}$ and sample variance s^2:

$$\bar{x} = \Sigma_{i=1}^{n} x_i / n$$

and

$$s^2 = \Sigma_{i=1}^{n} (x_i - \bar{x})^2 / (n - 1)$$

We are interested in the sampling distribution of $\bar{x}$, and statistical theory shows that the standard error of $\bar{x}$ is $\sigma_{\bar{x}} = \sigma/\sqrt{n}$, where σ is the standard deviation for the population of egg thickness. Thus, a reasonable estimator for this parameter is $s_{\bar{x}} = s/\sqrt{n}$, where s is an estimator for the standard deviation of the thickness of individual eggshells in the population and $s_{\bar{x}} = s/\sqrt{n}$ is an estimator for the standard deviation, or standard error, of the mean thickness of a sample of n eggshells randomly chosen from the population under study.

Here, we show the oft-seen bulls-eye graphic illustrating varying levels of bias and precision (Fig. 2.6). Figure 2.6a indicates a precise estimate, but biased;

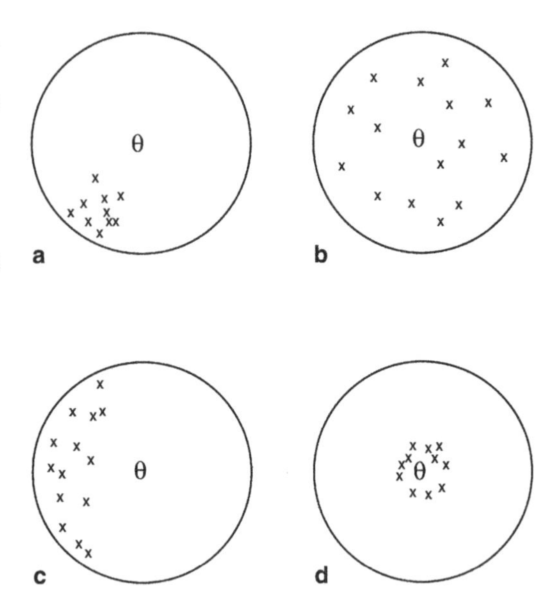

Fig. 2.6 Examples of estimator accuracy, precision, and bias. Estimates of $\hat{\theta}$ (denoted as x) are compared with the true value θ. (a) A precise estimate, but biased. (b) An unbiased estimate with low precision. (c) A estimate that is both imprecise and biased. (d) The optimal case where the estimate is both unbiased and precise. Reproduced from Williams et al. (2002), with kind permission from Elsevier Academic Press

Fig. 2.6b indicates an unbiased estimate with low precision. Figure 2.6c indicates an estimate that is both imprecise and biased, whereas Fig. 2.6d is the optimal case where the estimate is both unbiased and precise. Note that in those cases where bias is shown (Fig. 2.6a and c), there is a systematic difference between the replicated parameter $\hat{\theta}$ that causes it to differ from the population parameter θ (Williams et al. 2002). The most accurate estimator is usually one that minimizes bias and maximizes precision, and typically is a balance between the two. Estimator accuracy is commonly measured using mean squared error (MSE), which is a variation of the estimator $\hat{\theta}$ around the parameter θ:

$$\mathrm{mse}(\hat{\theta}) = \mathrm{E}[(\hat{\theta}_i - \theta)^2]$$

In general, mse can be written as $\mathrm{mse}(\hat{\theta}) = \mathrm{var}(\hat{\theta}) + [\mathrm{bias}(\hat{\theta})]^2$. Note that if $\hat{\theta}$ is an unbiased estimator for θ, $\mathrm{mse}(\hat{\theta}) = \mathrm{var}(\hat{\theta})$ because $\mathrm{E}(\hat{\theta}) = \hat{\theta}$.

2.6.5 Assumptions

We have alluded to assumptions required for statistical inference throughout the first two chapters of this book. Any inference that relies on statistics requires assumptions. The quantity and rigor of these assumptions depend upon the specific study, statistical approaches taken, and whether inferences will be design or model based. Most statistical methods require that the sample be selected using some probabilistic scheme from the target population. Design-based inference necessitates random selection of sample units, rather than characteristics specific to the variable of interest. Many methods require assumptions regarding the form of the sampling distribution (e.g., normally distributed). However, in some cases, the central limit theorem allows us to relax those assumptions as long as sample sizes are large. Model-based inferences (e.g., CMR studies) frequently require more stringent adherence to certain assumptions about the relationship between the predictands and predictors. Nonparametric methods still require assumptions, contrary to the application by many in the scientific community (Johnson 1995). Resampling (e.g., randomization, bootstrapping; Manly 1991), where the shape of the sampling distribution is derived empirically from repeated sampling of the data that were collected, has requirements such as initial random samples or observations that are exchangeable under the null hypothesis with the consequence that test of difference in location requires equal variance.

Although assumptions are ubiquitous in study design and statistical inference, many methods are robust to moderate violations of assumptions. For example, some methods requiring normality are robust to deviations from normality when distributions are symmetric. Additionally, model-based inferences for capture–recapture methods are in some cases robust to violations of the population closure assumption (Kendall 1999). Thus, researchers should not lose heart because there are numerous methods one can choose from, with varying degrees of assumption complexity. We recommend that investigators use available analytical tests and graphical evaluations to verify whether violations of assumptions have occurred.

At the design and inference stages, we recommend that the investigator identifies the assumptions necessary for the suggested approach and then ask the following questions: (1) Are there any assumptions that are likely to be severely violated? (2) For assumptions that will be difficult to achieve, is there anything I can do to meet those assumptions more closely? (3) Is the analytical method I will be employing robust to violations of the assumptions I am likely to violate? (4) If analytical problems such as bias are likely to be an issue, are there alternative design- or model-based approaches I can implement, which would provide me with results that are more robust? Critical thinking about the question under investigation and the study design at hand will greatly increase the probability that the study will produce biologically and statistically meaningful results.

2.6.6 Type I vs. Type II Error

Under a classical (frequentist) design for statistical inference using hypothesis testing where there is a specific null hypothesis, an omnibus alternative hypothesis, and a specified level of significance for that test, we can have two types of errors. A *Type I* error, rejecting the null hypothesis when it is true, occurs with probability α, which typically is set by the investigator (i.e., the α level of the test, often $\alpha = 0.05$; Cherry 1998). The p value is a related concept. Historically, scientists have viewed p values as a measure of how much evidence we have against the null hypothesis we are evaluating. However, we prefer the definition given by Anderson et al.

(2000) in that a p value is the "...probability of obtaining a test statistic at least as extreme as the one observed, conditional on the null hypothesis being true" (Anderson et al. 2000, p. 914). If the p value is then less than the fixed α-level, then the null hypothesis is rejected. A more detailed discussion on the use of p values in wildlife ecology can be found in Cherry (1998) and Johnson (1999, 2002a).

We make *Type II* errors by failing to reject the null hypothesis when it is false. The probability of a Type II error occurring is denoted as β, and depends upon the α-level of the test, as well as the hypothesized and actual sampling distributions of the estimator. The probability we reject the null hypothesis when it is false is the power of the test ($1 - \beta$). Power is a commonly abused concept in ecological research. Power can be used to estimate required sample sizes prior to study implementation. All too often, however, researchers incorrectly use power to evaluate confidence in nonsignificant null hypothesis tests after the fact (Hayes and Steidl 1997; Steidl et al. 1997; Gerard et al. 1998).

The arbitrary nature of α and β has contributed to reduced null hypothesis significance testing in wildlife science (Johnson 1999; Anderson et al. 2000; Burnham and Anderson 2002). An argument for using estimation rather than hypothesis testing is that in estimation, there is no notion of Type I or Type II error. Although we agree that estimation procedures are preferable to null hypothesis significance testing, there is a statistical subtlety in this argument, because estimation usually requires a confidence interval to evaluate statistical and possibly biological significance. We specify confidence levels before constructing a confidence interval around the estimate. Thus, the confidence level is the estimation counterpart to the α-level in a hypothesis-testing scenario, and is actually $1 - \alpha$. Therefore, one must still consider Type I and Type II errors, but the arbitrary nature of the confidence interval probably is less important than α levels during null hypothesis significance testing due to the descriptive nature of confidence intervals.

2.6.7 Sample Size Estimation

Whether we use null hypothesis significance testing or estimation approaches for scientific inference, more often than not the initial question asked during study planning and design is "What sample size should be used?" (Thompson 2002). Sample size determination is an important aspect underlying research achieving study objectives, as sample size influences the precision of our parameter estimates and the cost of the study. Under a hypothesis-testing framework, knowledge of sample size is required to achieve a desired level of power to detect an effect of specified magnitude, given the α-level of the test. Under an estimation framework, the criterion for adequate sample size could be a specified coefficient of variation (CV; i.e., proportional standard error, or the standard error of the estimate divided by the estimate) or a confidence interval with a specified width and confidence level. For each of these situations adequate sample size depends upon the sampling distribution. For some sampling distributions (e.g., normal, binomial), formulas used for sample size estimation are straightforward and can be found in most introductory statistical texts or statistical software. For others, especially when the estimator of interest is a function of other random variables, sample size is more easily determined numerically through simulation.

The assumption that an estimator follows some specified distribution is often only approximate. Thus, some investigators feel that a priori computation of sample size is not necessary. Rather, "getting as large a sample as you can" becomes the prevailing philosophy. Although this approach can be advantageous, as increasing sample size increases the likelihood that a statistical test will be significant (Johnson 1995), it is not good practice. First, eventually assumptions will be required to analyze the data. Second, although computed precision or power for a given sample size is never exactly achieved, a rough estimate of sample size is useful for planning. For example, if the required sample size under modest assumptions indicates that a sample size of 100 is necessary to meet study objectives, but the current logistical plans allow for the collection of only 10 samples, then the process of sample size determination was useful. Third, most of us do not have the luxury of limitless budgets. As a result, we need to be as efficient as possible in conducting research. This efficiency is possible when you define the sample sizes needed and design your study accordingly.

2.7 Project Goals, Design, and Data Collection, Analysis, and Presentation

The underlying reason for scientific inquiry in wildlife science is conservation and maintenance of species, communities, and biodiversity over time and space. Therefore, all wildlife research revolves around the development of methods to assist with studying populations and evaluating those factors that influence population trajectories. The first step in developing any research study, regardless of the topic, is to clearly define the project's goals (Thompson et al. 1998). Questions should be worthwhile (of some conservation or management importance; MacKenzie et al. 2006) and should be attainable (Sutherland 2006). Establishing general study goals and framework is critical for determining information needs, the necessary data, the time period of study, and the use of the data. In this section, we discuss how to link project goals, study design, data collection, data interpretation, and data presentation into a package that will result in meaningful conclusions.

2.7.1 Linking Goals to Design to Data

Ecological research projects require well thought-out questions (Chap. 1), adequate sampling and experimental designs (Chaps. 3 and 4 and this chapter), which ensure the target population is identifiable. As an example, consider the question we defined to evaluate abundance of prairie voles (*Microtus ochrogaster*) in reclaimed and nonreclaimed agricultural fields in east-central Illinois over 4 years. We would be interested in determining the impact reclamation has on vole population trajectory. Note that, based on our questions, we would only include reclaimed and nonreclaimed agricultural fields in our sampling frame, removing upland hardwood forests from our potential sampling frame as our interest is in agricultural fields, and voles are infrequently found in upland forests. Once we identify the target population, our sampling frame should include all potential reclaimed and nonreclaimed agricultural fields in east-central Illinois, from which we would then draw a random sample of fields on which data collection would be conducted.

Once our sample of fields to be surveyed is drawn, we must choose an enumeration method (Chap. 4). In this example, our question of interest is abundance, thus perhaps a CMR study using a trapping grid (Nichols et al. 1984) or using a distance-based estimator based on a trapping web (Buckland et al. 2001) is appropriate. However, as part of our enumeration, we must decide whether we should consider the population demographically closed, which will influence when and how we conduct our trapping. Several environmental factors could influence vole populations over time, such as precipitation and predation, and we must decide which, if any, of these values we should consider while collecting additional data. Obviously, there are considerable factors that influence study design and data collection once project goals are determined. Please see Chap. 7 for a more detailed evaluation of sample survey design and data collection, Thompson et al. (1998) for a dichotomous key to enumeration methods for fish, birds, mammals, amphibians, and reptiles, and Williams et al. (2002) for a broad outline of monitoring methods.

2.7.2 Tools to Analyze What You Collect

The emphasis of this book is on the design of ecological field studies; however, design and inference are intimately related (Williams et al. 2002). Since in most studies we are observing only a portion of the population, our usual interest is estimation of population parameters (abundance, survival, recruitment, and movement), which we hypothesize, based on our design, are characteristic of the entire population (Thompson 2002).

Statistical methods available for analysis of ecological data are extensive; current approaches include among classical frequentist and estimation methods, information-theoretic, and Bayesian approaches (Burnham and Anderson 2002; Link et al. 2002; Ellison 2004; Steidl 2006). However, because these approaches are tools, we should treat them as tools, or means, rather than ends. The list of potential statistical methodology used in wildlife sciences is considerable, and the choice of approach depends on the species under study, questions of interest, study design, and type of data collected. Thus, we will refrain from discussing the intricacies of specific analytical methods (e.g., AIC for linear regression) and instead focus on a general discussion of analytical systems under which most wildlife scientists conduct analyses.

First, and we quote, "Lets not kid ourselves: the most widely used piece of software for statistics is Excel" (Ripley 2002). We use spreadsheets such as Microsoft Excel for four primary purposes in wildlife studies: (1) data entry and storage, (2) data manipulation, (3) statistical analysis, and (4) graphic creation (see Sect. 2.4.7). In fact, Excel has become the "de rigueur" initial location where most data analyses are conducted and graphics developed in the wildlife sciences. This is likely due to Excel's availability and simplicity. This simplicity, however, comes at a price – considerable mathematical inaccuracies. Errors associated with statistical computations in Excel are common (McCullough and Wilson 1999, 2002, 2005), although the ecological community is slow to recognize the limitations of Excel for anything other than data entry, storage, and manipulation.

Databases are an alternative for data storage. Databases are collections of records linked through a data model and provide a description of how we represent and manipulate data. They come in many different forms, ranging from simple table models or two-dimensional arrays of data, in which columns indicate similar values and rows indicate individuals or groups, to hierarchical and relational models (Codd 1970; Date 2003). Data in hierarchical models are organized into a tree-like structure that allows for repeating information using a parent–child relationship. For example, a study site could be the parent, and the birds radiomarked on the site the children. Relational databases are databases that use a set of relations to order and manipulate data. A well-designed relational database helps ensure data are entered in the correct format, takes up considerably less disk space, and is much less likely to be corrupted by user errors as compared with a spreadsheet containing the same data. We use databases widely in wildlife sciences primarily for data storage and manipulation, but databases are probably underutilized, given their great flexibility and range of applications, including analysis and data reporting.

Next, after data collection and transfer to some data storage format, wildlife ecologists typically want to summarize and interpret the data using one or several statistical procedures. Luckily, there exist a number of statistical programs for analysis of ecological data. Nevertheless, these programs vary in functionality, ease of use, and accuracy. Summary statistics (means, variances, and medians) are estimated in nearly any program, and as such will not be discussed. Additionally, approaches to link data in these formats with each of the below statistics programs are readily available, although certain programs require specific data formats not outlined (e.g., .inp files in MARK).

Some of the more common statistical environments used in wildlife science include (this list is not comprehensive): Jump, SPSS Inc. (1999), SAS (SAS Institute Inc. 2000), SYSTAT (SYSTAT 2002), MINITAB (Minitab 2003), STATISTICA (StatSoft 2003), STATA (StataCorp 2005), and GENSTAT (Payne et al. 2006). Each of these programs has advantages and disadvantages. For example, SAS efficiently conducts batch processing, simplifying data manipulation and analysis for large datasets; GENSTAT, SPSS, and STATISTICA all have excellent GUIs (graphical user interfaces). SPSS is taught as the primary undergraduate and graduate statistics package in many universities across the United States while SAS has a considerable presence in both the academic research and business worlds. The downside to most of these, however, is cost, as most are expensive and some require annual licensing, although student versions are inexpensive. Scientific programming and statistical computing environments also include programming languages like S (Venables and Ripley 2002) and SPlus (Chambers 1998), and programming environments such as MATLAB (2005) and R (R Development Core Team 2006). These systems have been at the forefront of nearly all statistical computing for the last decade and have a wide group of active users involved with development and testing. Because each of these four environments allows command line, high-level programming, they provide more flexibility with modeling and figure development. R is open-source freeware while S, SPlus, and MATLAB are available for purchase.

Statistical programs designed to estimate population parameters are widely available and have seen a dramatic increase in use by wildlife scientists since the advent of powerful personal computers in the 1990s (Schwarz and Seber 1999; Amstrup et al. 2005). The most recognizable, Program MARK (White and Burnham 1999), is used for estimation of parameters from "marked" individuals (hence the name). MARK has become the standard engine for >100 different modeling approaches ranging from survival estimation using telemetry data to abundance estimation in closed systems from CMR data. However, other programs exist for population parameter estimation, including RMARK (R-based system invoking MARK for parameter estimation; Laake 2007), POPAN (open population mark–recapture/resight models; Arnason and Schwarz 1999), and abundance estimation using Distance (Buckland et al. 2001) and NOREMARK (White 1996) and occupancy estimation using Presence (MacKenzie et al. 2006). Regularly updated, as new methods become available, these programs represent state-of-the-art methods for population parameter estimation. Users should note, however, that nearly all of these "wildlife-specific" programs rely on information-theoretic approaches to model selection and inference (Burnham and Anderson 2002), which require considerable statistical background to ensure that resulting inferences are appropriately developed, applied, and interpreted.

2.7.3 How to Present What You Collect

Wildlife research is primarily descriptive; all that varies is the choice of methods (e.g., summary statistics, hypothesis tests, estimation procedures, and model selection) used to describe the system of interest. Statistical methods used in wildlife science range from simple data description to complex predictive models (Williams et al. 2002). As shown in the previous section, statistical applications have become increasingly important in the examination and interpretation of ecological data to the extent that entire programs have been developed for estimation of specific population parameters. Approaches to presenting data are unlimited and dependent upon the context (e.g., oral presentation, peer-reviewed article), so we will limit our discussion to a few key points. Note that Anderson et al. (2001) provided general suggestions regarding (1) clarification of test statistic interpretation, (2) presentation of summary statistics for information-theoretic approaches, (3) discussion of methods for Bayesian explanation, and (4) general suggestions regarding description of sample sizes and summary descriptive statistics (e.g., means).

Tables should be used to present numerical data that support specific conclusions. Tables have the following general characteristics: they should present relevant data efficiently in an unambiguous manner and each table should be readily interpretable without reference to textual discussion. Tables tend to outline specific cases (e.g., number of mortalities due to harvest) while graphics (see below) are used to describe relationships between parameters (Sutherland 2006). Table headings, row labels, and footnotes should precisely define what the data in each row–column intersection mean. Tables are amenable to a wide variety of data types, from absolute frequency data on captured individuals to summary parameter estimates (Figs. 2.7 and 2.8, respectively).

Graphics also are important for interpreting data as they allow the reader to visually inspect ecological parameter estimates. Graphics should display ecological data efficiently and accurately, and there is a wide range of graphical options available to researchers (Cleveland 1993; Maindonald and Braun 2003). Tufte (1983) suggests, "Excellence in statistical graphics consists of complex ideas communicated with clarity, precision, and efficiency. Good graphs should (from Tufte 2001):

- Illustrate the data
- Induce the viewer to think about the substance rather than methodology, design, or technology of graphic construction

Period	Species	Age	Sex	No. banded	No. recovered
1959-1966	Lesser scaup	AHY	Female	3,572	138
	Lesser scaup	AHY	Male	26,882	1,015
	Canvasback	AHY	Female	645	53
	Canvasback	AHY	Male	1,504	126
	Barrow's goldeneye	AHY	Female	1,537	27
	Barrow's goldeneye	AHY	Male	2,645	59
	American wigeon	AHY	Male	4,688	579
	Northern pintail	AHY	Female	787	24
	Northern pintail	AHY	Male	876	28
1989-2000	Northern pintail	AHY	Female	3,609	92
	Northern pintail	AHY	Male	919	74
	Northern pintail	HY	Female	4,319	142
	Northern pintail	HY	Male	2,738	203
	Green-winged teal	AHY	Female	1,265	52
	Green-winged teal	AHY	Male	1,579	110
	Green-winged teal	HY	Female	1,295	39
	Green-winged teal	HY	Male	1,229	70
	Mallard	AHY	Female	1,001	53
	Mallard	AHY	Male	1,317	114
	Mallard	HY	Female and male	985	68

Fig. 2.7 Example table showing summary data enumerating the number of individuals captured during a research study. Reproduced from Lake et al. (2006), with kind permission from The Wildlife Society

Parameter	Mean	SE
Cub litter size	1.7	0.15
Litter production rate		
4-yr-olds	0.0	0.0
5-yr-olds	0.11	0.11
6-yr-olds	0.29	0.47
>6-yr-olds	0.93	0.33
Proportion male cubs	0.55	0.06

Fig. 2.8 Example table showing summary parameter estimates for all individuals captured during a research study. Reproduced from Taylor et al. (2006), with kind permission from the Wildlife Society

- Present large datasets coherently and in a small area
- Reveal data at several levels from broad overview to finer structure
- Serve a clear purpose (e.g., description, tabulation)
- Be closely related to the statistical and verbal descriptions of the data"

The most widely used piece of software for ecological graphics is Excel. Although spreadsheet programs seem to provide a wealth of graphical options, few of these canned figures meet any of the criteria suggested by Tufte (1983, 2001) for use as graphs. For example, almost every "point and click" graph available in spreadsheets provides completely useless graphical options, such as false third dimensions, complicated and unnecessary gridlines, and moire effects (Tufte 2001). Thus, we caution ecologists to take care when constructing graphics using spreadsheets as most are deficient or misleading.

We suggest that scientists give considerable thought regarding graphical displays before data collection (Sutherland 2006). Graphics should display ecological data efficiently and accurately, and researchers should consider the range of graphical options available rather than relying on canned figure development (Tufte 1983; Cleveland 1993; Maindonald and Braun 2003). Additionally, we suggest that graphics be limited to datasets exceeding 20 values; otherwise, tables provide a more representative presentation (Tufte 2001). We highly recommend that scientists refer to works by Tufte (2001), Chambers et al. (1983), and Cleveland (1993, 1994) for detailed discussions on graph construction.

2.8 Summary

The emphasis of this book is on the design of ecological research with this chapter focusing on the relationship between study design and statistical inference. In Sect. 2.2, we discussed the different types of variables common to wildlife studies:

explanatory, disturbing, controlling, and randomized variables. We outlined and discussed how each of these variables can influence both study design and inferences from a field study. Next, in Sect. 2.3, we outlined the necessity of randomization and replication for strong inferences in wildlife studies. In this section, we covered the concept of pseudoreplication (Hurlbert 1984), its impacts on our inferences, and methods to avoid pseudoreplication.

We have outlined that a strong inference from ecological studies relies not only on replication and randomization, but also on data collection methods and analytical techniques for evaluating parameters of interest in the fact of nuisance information. In Sect. 2.4, we discussed major types of designs: manipulative, quasi-experiments, mensurative, and descriptive. For each design type, we outlined the basics of those designs, focusing on how *controls*, *replication*, and *randomization* are used in each. In Sect. 2.5, we outlined the importance of the survey sampling theory in the broader theme of wildlife study design. In that section, we discussed the impact that detectability has on a strong inference and outlined the several methods used to correct for imperfect detection. We considered the place of statistical inference in wildlife study design in Sect. 2.6 and provided a general discussion on approaches to inference, including relationships between statistical inference and sampling distributions, statistical accuracy, precision, and bias. In this section, we also provided general discussion on sample size estimation and statistical power. Finally, in Sect. 2.7, we discussed how we link project goals with design, data collection, analysis, and presentations, the factors influencing the design type used, and the methods wildlife biologists can use to analyze the data they collected, and provided an outline on methods for data interpretation and presentation.

References

Amstrup, S. C., T. L. McDonald, and B. F. J. Manly. 2005. Handbook of Capture–Recapture Analysis. Princeton University, New Jersey.

Anderson, D. R. 2001. The need to get the basics right in wildlife field studies. Wildl. Soc. Bull. 29: 1294–1297.

Anderson, D. R. 2003. Response to Engeman: index values rarely constitute reliable information. Wildl. Soc. Bull. 31: 288–291.

Anderson, D. R., K. P. Burnham, and W. L. Thompson. 2000. Null hypothesis testing: problems, prevalence, and an alternative. J. Wildl. Manage. 64: 912–923.

Anderson, D. R., W. A. Link, D. H. Johnson, and K. P. Burnham. 2001. Suggestions for presenting the results of data analyses. J. Wildl. Manage. 65: 373–378.

Anscombe, F. J. 1973. Graphs in statistical analysis. Am. Stat. 27: 17–21.

Arnason, A. N., and C. J. Schwarz. 1999. Using POPAN-5 to analyse banding data. Bird Study 46(Suppl.): 157–168.

Bart, J., and S. Earnst. 2002. Double sampling to estimate density and population trends in birds. Auk 119: 36–45.

Borchers, D. L., S. T. Buckland, and W. Zucchini. 2002. Estimating animal abundance-closed populations. Springer-Verlag, London.

Brownie, C., D. R. Anderson, K. P. Burnham, and D. R. Robson. 1985. Statistical inference from band recovery data – a handbook, 2nd Edition. U.S. Fish and Wildlife Service Resource Publication 156.

Buckland, S. T., I. B. J. Goudie, and D. L. Borchers. 2000. Wildlife population assessment: past developments and future directions. Biometrics 56: 1–12.
Buckland, S. T., D. R. Anderson, K. P. Burnham, J. L. Laake, D. L. Borchers, and L. Thomas. 2001. Introduction to Distance Sampling. Oxford University, Oxford.
Burnham, K. P. and D. R. Anderson. 2002. Model selection and multimodel inference: a practical information-theoretic approach, 2nd Edition. Springer-Verlag, New York.
Burnham, K. P., D. R. Anderson, G. C. White, C. Brownie, and K. P. Pollock. 1987. Design and analysis of methods for fish survival experiments based on release–recapture. Am. Fish. Soc. Monogr. 5: 1–437.
Burnham, K. P., G. C. White, and D. R. Anderson. 1995. Model selection strategy in the analysis of capture–recapture data. Biometrics 51: 888–898.
Chambers, J. M., W. S. Cleveland, B. Kleinez, and P. A. Turkey. 1983. Graphical methods for data analysis. Wadsworth International Group, Belmont, CA, USA.
Chambers, J. M. 1998. Programming with data. A guide to the S language. Springer-Verlag, New York.
Cherry S. 1998. Statistical tests in publications of The Wildlife Society. Wildl. Soc. Bull. 26: 947–953.
Cleveland, W. S. 1993. Visualizing Data. Hobart, Summit, NJ.
Cleveland, W. S. 1994. The Elements of Graphing Data. Hobart, Summit, NJ.
Cochran W. G. 1977. Sampling Techniques, 3rd Edition. John Wiley and Sons, New York.
Codd, E. F. 1970. A relational model of data for large shared data banks. Commun. ACM 13: 377–387.
Cohen, J. 1988. Statistical power analysis for the behavioral sciences, 2nd Edition. Lawrence Erlbaum Associates, Inc., Mahwah, NJ.
Collier, B. A., S. S. Ditchkoff, J. B. Raglin, and J. M. Smith. 2007. Detection probability and sources of variation in white-tailed deer spotlight surveys. J. Wildl. Manage. 71: 277–281.
Cook, R. D., and J. O. Jacobsen. 1979. A design for estimating visibility bias in aerial surveys. Biometrics 35: 735–742.
Date, C. J. 2003. An Introduction to Database Systems, 8th Edition. Addison Wesley, Boston, MA.
Dinsmore, S. J., G. C. White, and F. L. Knopf. 2002. Advanced techniques for modeling avian nest survival. Ecology 83: 3476–3488.
Eberhardt, L. L. 2003. What should we do about hypothesis testing? J. Wildl. Manage. 67: 241–247.
Ellison, A. M. 2004. Bayesian inference in ecology. Ecol. Lett. 7: 509–520.
Farfarman, K. R., and C. A. DeYoung. 1986. Evaluation of spotlight counts of deer in south Texas. Wildl. Soc. Bull. 14: 180–185.
Fisher, R. A. 1925. Statistical Methods for Research Workers. Oliver and Boyd, London.
Fisher, R. A. 1929. The statistical method in psychical research. Proc. Soc. Psychical Res. 39: 189–192.
Fisher, R. A. 1935. The Design of Experiments. Reprinted 1971 by Hafner, New York.
Franklin, A. B., T. M. Shenk, D. R. Anderson, and K. P. Burnham. 2001. in T. M. Shenk and A. B. Franklin, Eds. Statistical model selection: the alternative to null hypothesis testing, pp. 75–90. Island, Washington, DC.
Gavin, T. A. 1991. Why ask "Why": the importance of evolutionary biology in wildlife science. J. Wildl. Manage. 55: 760–766.
Gerard, P. D., D. R. Smith, and G. Weerakkody. 1998. Limits of retrospective power analysis. J. Wildl. Manage. 62: 801–807.
Green, R. H. 1979. Sampling Design and Statistical Methods for Environmental Biologists. Wiley, New York.
Gregory, R., D. Ohlson, and J. Arvai. 2006a. Deconstructing adaptive management: criteria for applications in environmental management. Ecol. Appl. 16: 2411–2425.
Gregory, R., L. Failing, and P. Higgins. 2006b. Adaptive management and environmental decision making: a case study application to water use planning. Ecol. Econ. 58: 434–447.

Guthery, F. S., J. J. Lusk, and M. J. Peterson. 2001. The fall of the null hypothesis: liabilities and opportunities. J. Wildl. Manage. 65: 379–384.
Guthery, F. S., L. A. Brennan, M. J. Peterson, and J. J. Lusk. 2005. Information theory in wildlife science: critique and viewpoint. J. Wildl. Manage. 69: 457–465.
Hayes, J. P., and R. J. Steidl. 1997. Statistical power analysis and amphibian population trends. Conserv. Biol. 11: 273–275.
Holling, C. S. (ed.) 1978. Adaptive Environmental Assessment and Management. Wiley, London.
Hurlbert, S. H. 1984. Pseudoreplication and the design of ecological field experiments. Ecol. Monogr. 54: 187–211.
Johnson, D. H. 1995. Statistical sirens: the allure of nonparametrics. Ecology 76: 1998–2000.
Johnson, D. H. 1999. The insignificance of statistical significance testing. J. Wildl. Manage. 63: 763–772.
Johnson, D. H. 2002a. The role of hypothesis testing in wildlife science. J. Wildl. Manage. 66: 272–276.
Johnson, D. H. 2002b. The importance of replication in wildlife research. J. Wildl. Manage. 66: 919–932.
Johnson, F. A., B. K. Williams, J. D. Nichols, J. E. Hines, W. L. Kendall, G. W. Smith, and D. F. Caithamer. 1993. Developing an adaptive management strategy for harvesting waterfowl in North America. In Transactions of the North American Wildlife and Natural Resources Conference, pp. 565–583. Wildlife Management Institute, Washington, DC.
Kendall, W. L. 1999. Robustness of closed capture–recapture methods to violations of the closure assumption. Ecology 80: 2517–2525.
Kendall, W. L., B. G. Peterjohn, and J. R. Sauer. 1996. First-time observer effects in the North American Breeding Bird Survey. Auk 113: 823–829.
Kish, L. 1987. Statistical Design for Research. Wiley, New York.
Kuehl, R. O. 2000. Design of Experiments: Statistical Principles of Research Design and Analysis, 2nd Edition. Brooks/Cole, Pacific Grove, California.
Laake, J. L. 2007. RMark, version 1.6.1. R package. http://nmml.afsc.noaa.gov/Software/marc/marc.stm
Lebreton, J.-D., K. P. Burnham, J. Clobert, and D. R. Anderson. 1992. Modeling survival and testing biological hypotheses using marked animals: a unified approach with case studies. Ecol. Monogr. 62: 67–118.
Lehnen, S. E., and D. G. Krementz. 2005. Turnover rates of fall-migrating pectoral sandpipers in the lower Mississippi Alluvial Valley. J. Wildl. Manage. 69: 671–680.
Link, W. A., and J. R. Sauer. 1998. Estimating population change from count data. Application to the North American Breeding Bird Survey. Ecol. Appl. 8: 258–268.
Link, W. A., E. Cam, J. D. Nichols, and E. G. Cooch. 2002. Of BUGS and birds: Markov Chain Monte Carlo for hierarchical modeling in wildlife research. J. Wildl. Manage. 66: 227–291.
Lukacs, P. M., W. L. Thompson, W. L. Kendall, W. R. Gould, P. F. Doherty Jr., K. P. Burnham, and D. R. Anderson. 2007. Concerns regarding a call for pluralism of information theory and hypothesis testing. J. Appl. Ecol. 44: 456–460.
MacKenzie, D. I., J. D. Nichols, J. A. Royle, K. H. Pollock, L. L. Bailey, and J. E. Hines. 2006. Occupancy Estimation and Modeling. Academic, Burlington, MA.
Maindonald, J H., and J. Braun. 2003. Data Analysis and Graphics Using R. Cambridge University, United Kingdom.
MATLAB. 2005. Learning MATLAB. The MathWorks, Inc., Natick, MA.
Manly, B. F. J. 1991. Randomization and Monte Carlo Methods in Biology. Chapman and Hall, New York.
McCullough, B. D., and B. Wilson. 1999. On the accuracy of statistical procedures in Microsoft Excel 97. Comput. Stat. Data Anal. 31: 27–37.
McCullough, B. D., and B. Wilson. 2002. On the accuracy of statistical procedures in Microsoft Excel 2000 and Excel XP. Comput. Stat. Data Anal. 40: 713–721.
McCullough, B. D., and B. Wilson. 2005. On the accuracy of statistical procedures in Microsoft Excel 2003 Comput. Stat. Data Anal. 49: 1224–1252.

Minitab. 2003. MINITAB Statistical Software, Release 14 for Windows. State College, Pennsylvania.
Mitchell, W. A. 1986. Deer spotlight census: Section 6.4.3, U.S. Army Corp of Engineers Wildlife Resources Management Manual. Technical Report EL-86–53, U.S. Army Engineer Waterways Experiment Station, Vicksburg, MS.
Mood, A. M., F. A. Graybill, and D. C. Boes. 1974. Introduction to the Theory of Statistics, 3rd Edition, McGraw-Hill, Boston, MA.
Nichols, J. D., J. E. Hines, and K. H. Pollock. 1984. The use of a robust capture–recapture design in small mammal population studies: a field example with *Microtus pennsylvanicus*. Acta Therilogica 29: 357–365.
Nichols, J. D., J. E. Hines, J. R. Sauer, F. W. Fallon, J. E. Fallon, and P. J. Heglund. 2000. A double observer approach for estimating detection probability and abundance from point counts. Auk 117(2): 393–408.
Norman, G. W., M. M. Conner, J. C. Pack, and G. C. White. 2004. Effects of fall hunting on survival of male wild turkeys in Virginia and West Virginia. J. Wildl. Manage. 68: 393–404.
Otis, D. L., K. P. Burnham, G. C. White, and D. R. Anderson. 1978. Statistical inference from capture data on closed animal populations. Wildl. Monogr. 62: 1–135.
Payne, R. W., Murray, D. A., Harding, S. A., Baird, D. B. & Soutar, D. M. 2006. GenStat for Windows, 9th Edition. Introduction. VSN International, Hemel Hempstead.
Peterjohn, B. G., J. R. Sauer, and W. A. Link. 1996. The 1994 and 1995 summary of the North American Breeding Bird Survey. Bird Popul. 3: 48–66.
Pollock, K. H., J. D. Nichols, C. Brownie, and J. E. Hines. 1990. Statistical inference for capture–recapture experiments. Wildl. Monogr. 107: 1–97.
R Development Core Team. 2006. R: a language and environment for statistical computing. R Foundation for Statistical Computing, Vienna, Austria. ISBN 3–900051–07–0, URL http://www.R-project.org
Ripley, B. D. 1996. Pattern Recognition and Neural Networks. Cambridge University, Cambridge.
Ripley, B. D. 2002. Statistical methods need software: a view of statistical computing. Opening Lecture, RSS Statistical Computing Section.
Robbins, C. S., D. Bystrack, and P. H. Geissler. 1986. The breeding bird survey: the first 15 years, 1965–1979. Resource Publication no. 157, U.S. Department of the Interior, Fish and Wildlife Service, Washington, DC.
Rosenstock, S. S., D. R. Anderson, K. M. Giesen, T. Leukering, and M. F. Carter. 2002. Landbird counting techniques: current practices and an alternative. Auk 119(1): 46–53.
Royle, J. A., and J. D. Nichols. 2003. Estimating abundance from repeated presence-absence data or point counts. Ecology 84: 777–790.
SAS Institute Inc. 2000. SAS language reference: dictionary, version 8. SAS Institute, Inc., North Carolina.
Sauer, J. R., B. G. Peterjohn, and W. A. Link. 1994. Observer differences in the North American Breeding Bird Survey. Auk 111: 50–62.
Schwarz, C. J. and G. A. F. Seber. 1999. Estimating animal abundance: review III. Stat. Sci. 14: 427–456.
Sinclair, A. R. E. 1991. Science and the practice of wildlife management. J. Wildl. Manage. 55: 767–773.
Skalski, J. R., and D. S. Robson. 1992. Techniques for Wildlife Investigations: Design and Analysis of Capture Data. Academic, San Diego, CA.
SPSS Inc. 1999. SPSS Base 10.0 for Windows User's Guide. SPSS Inc., Illinois.
StataCorp. 2005. Stata Statistical Software: Release 9. Texas.
StatSoft. 2003. STATISTICA data analysis software system, version 6. Oklahoma.
Seber, G. A. F. 1982. The Estimation of Animal Abundance and Related Parameters, 2nd Edition. Griffin, London.
Steidl, R. J. 2006. Model selection, hypothesis testing, and risks of condemning analytical tools. J. Wildl. Manage. 70: 1497–1498.
Steidl R. J., J. P. Hayes, and E. Schauber. 1997. Statistical power in wildlife research. J. Wildl. Manage. 61: 270–279.

Stephens, P. A., S. W. Buskirk, and C. M. del Rio. 2007a. Inference in ecology and evolution. Trends Ecol. Evol. 22: 192–197.
Stephens, P. A., S. W. Buskirk, G. D. Hayward, and C. M. Del Rio. 2007b. A call for statistical pluralism answered. J. Appl. Ecol. 44: 461–463.
Stewart-Oaten, A., W. W. Murdoch, and K. R. Parker. 1986. Environmental impact assessment: "Pseudoreplication" in time? Ecology 67: 929–940.
Sutherland, W. J. 2006. Planning a research programme, in W. J. Sutherland, Ed. Ecological Census Techniques, 2nd Edition, pp. 1–10. Cambridge University, Cambridge.
SYSTAT. 2002. SYSTAT for Windows, version 10.2. SYSTAT software Inc., California.
Thompson, S. K. 2002. Sampling, 2nd Edition. John Wiley and Sons, New York.
Thompson, W. L. 2002. Towards reliable bird surveys: accounting for individuals present but not detected. Auk 119(1): 18–25.
Thompson, S. K., and G. A. F. Seber. 1996. Adaptive Sampling. John Wiley and Sons, New York.
Thompson, W. L., G. C. White, and C. Gowan. 1998. Monitoring vertebrate populations. Academic, New York.
Tufte, E. R. 1983. The visual display of quantitative information. Graphics, Chesire, CT.
Tufte, E. R. 2001. The visual display of quantitative information, 2nd Edition. Graphics, Chesire, CT.
Venables, W. N., and B. D. Ripley. 2002. Modern applied statistics with S, 4th Edition. Springer-Verlag, New York.
Verner, J., and K. A. Milne. 1990. Analyst and observer variability in density estimates from spot mapping. Condor 92: 313–325.
Walters, C. J. 1986. Adaptive Management of Renewable Resources. Macmillan, New York.
Walters, C. J., and C. S. Holling. 1990. Large-scale management experiments and learning by doing. Ecology 71: 2060–2068.
White, G. C. 1996. NOREMARK: population estimation from mark-resighting surveys. Wildl. Soc. Bull. 24: 50–52.
White, G. C., and K. P. Burnham. 1999. Program MARK: survival estimation from populations of marked animals. Bird Study 46(Suppl.): 120–139.
Williams, B. K. 1996. Adaptive optimization and the harvest of biological populations. Math. Biosci. 136: 1–20.
Williams, B. K., J. D. Nichols, and M. J. Conroy. 2002. Analysis and Management of Animal Populations. Academic, San Diego, CA.

Chapter 3
Experimental Designs

3.1 Introduction

This chapter covers the fundamentals of experimental design as applied to wildlife studies. Milliken and Johnson (1984) defined *experimental design* as the combination of a design structure, treatment structure, and the method of randomization. We discuss most of the common design and treatment structures currently used in wildlife science from the relatively simple to the more complex. While we touch on sampling (randomization) plans because they are an integral part of experimental design, we delay detailed discussion of sampling until Chap. 4. Data analysis also is integral to study design but we leave this discussion to Chap. 5.

3.2 Principles

The relatively large geographic areas of interest, the amount of natural variability (noise) in the environment, the difficulty of identifying the target population, the difficulty of randomization, and the paucity of good controls make wildlife studies challenging. Wildlife studies typically focus on harvestable species and relatively scarce species of concern (e.g., threatened and endangered species) and factors that influence their abundance (e.g., death, reproduction, and use). In wildlife studies, the treatment is usually a management activity, land use change, or other perturbation contamination event potentially affecting a wildlife population. Additionally, this event could influence populations over an area much larger than the geographic area of the treatment. In most instances, quantification of the magnitude and duration of the treatment effects necessarily requires an observational study, because there usually is not a random selection of treatment and control areas. Early specification of the target population is essential in the design of a study. If investigators can define the target population, then decisions about the basic study design and sampling are much easier and the results of the study can be appropriately applied to the population of interest.

Hurlbert (1984) divided experiments into two classes: mensurative and manipulative. *Mensurative studies* involve making measurements of uncontrolled events at

M.L. Morrison et al., *Wildlife Study Design*.

one or more points in space or time with space and time being the only experimental variable or treatment. Mensurative studies are more commonly termed observational studies, a convention we adopt. *Observational studies* can include a wide range of designs including the BACI, line-transect surveys for estimating abundance, and sample surveys of resource use. The important point here is that all these studies are constrained by a specific protocol designed to answer specific questions or address hypotheses posed prior to data collection and analysis. *Manipulative studies* include much more control of the experimental conditions; there are always two or more treatments with different experimental units receiving different treatments and random application of treatments.

Eberhardt and Thomas (1991), as modified by Manly (1992) provided a useful and more detailed classification of study methods (Fig. 3.1). The major classes in their scheme are studies where the observer has control of events (manipulative experiments) and the study of uncontrolled events. Replicated and unreplicated manipulative experiments follow the classical experimental approach described in most statistics texts. Many of the designs we discuss are appropriate for these experiments. Their other category of manipulative experiment, sampling for modeling, deals with the estimation of parameters of a model hypothesized to represent the investigated process (see Chap. 4).

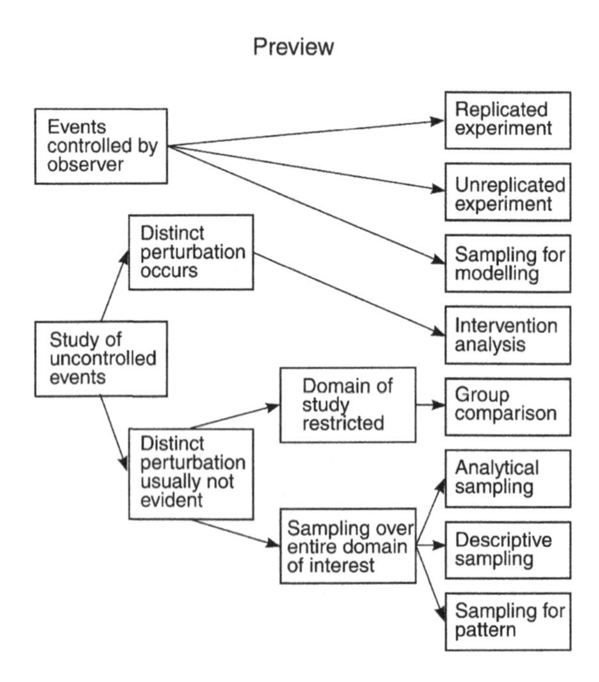

Fig. 3.1 Classification scheme of the types of research studies as proposed by Eberhardt and Thomas (1991) and modified by Manly (1992). Reproduced from Eberhardt et al. (1991) with kind permission from Springer Science + Business Media

The study of uncontrolled events can be broadly classified as observational studies. Observational studies also are referred to as "sample surveys" (Kempthorne 1966), "planned surveys" (Cox 1958), and "unplanned experiments/observational studies" (National Research Council 1985). We suggest Manly (1992) and McKinlay (1975) for additional discussion of the design and analysis of observational studies.

In dealing with observational studies, Eberhardt and Thomas (1991) distinguished between situations where some perturbation occurs and where this is not the case. The study of a perturbation is common in wildlife sciences, such as the study of some environmental contamination (e.g., the Exxon Valdez oil spill). Eberhardt and Thomas called these studies intervention analysis because they typically use time-series (Box and Tiao 1975) methods to study the effect of some distinct event. These *environmental impact studies* typically are large field studies as opposed to manipulative experiments, although manipulative experiments and smaller observational studies aid understanding of the mechanism of impact. In observational studies, data are collected by visual detection of an event in time and space. Many of the basic designs mentioned in this chapter (e.g., BACI) are covered in more detail in Chap. 6.

Eberhardt and Thomas (1991) identified four types of observational studies where no obvious perturbation exists. These studies correspond to investigations designed to develop better understanding of the biology of a system or population. Manly (1992) suggested, and we agree, that the "observational category" of Eberhardt and Thomas is really a special type of observational study where possible observations are limited to selected groups within the entire population of interest. The comparison of groups is another common form of wildlife study, often characterized by the study of representative study areas or groups of animals. The final three classes of study include the possibility of sampling the entire population or area of interest. The point of describing this scheme is that there is a variety of study types, and the design of each will determine the inferences that one can make with the resulting data (Manly 1992).

3.3 Philosophies

Scientific research is conducted under two broad and differing philosophies for making statistical inferences: *design/data-based* and *model-based*. These differing philosophies are often confused but both rely on current data to some degree and aim to provide *statistical inferences*. There is a continuum from strict design/data-based analysis (e.g., finite sampling theory [Cochran 1977] and randomization testing [Manly 1991]) to pure model-based analysis (e.g., global climate models and habitat evaluation procedures [HSI/HEP] using only historical data [USDI 1987]). A combination of these two types of analyses is often employed in wildlife studies, resulting in inferences based on a number of interrelated arguments. For more detailed discussion on design-based and model-based approaches see Chap. 4.

3.3.1 Design/Data-based Analysis

In the analysis of design/data-based studies, basic statistical inferences concerning the study areas or study populations are justified by the design of the study and data collected (Cochran 1977; Scheaffer et al. 1990). Computer-intensive statistical methods (e.g., randomization, permutation testing, etc.) are available that require no additional assumptions beyond the basic design protocol (e.g., Manly 1991). Design/data-based statistical conclusions stand on their own merits for the agreed-upon:

- Response variables
- Procedures to measure the variables
- Design protocol

Reanalysis of the data later does not mean the original statistical inferences were incorrect; instead, the original analysis stands if consensus still exists on the above study conditions.

3.3.2 Model-based Analysis

As the name implies, model-based analyses predict the outcome of experiments using models. In the extreme case of model-based analysis where no new data are available, all inferences are justified by assumption, are deductive, and are subject to counterarguments. The model-based approach usually involves the combination of new data with parameters from the literature or data from similar studies using a theoretical mathematical or statistical model. An example of this approach is the demographic modeling of wildlife populations combined with use of radio-telemetry data to estimate the influence of some perturbation on critical parameters in the model. This approach is illustrated by the telemetry studies of the golden eagle (*Aquila chrysaetos*) (Hunt 1995) in Altamont Pass, California, as described by Shenk et al. (1996).

3.3.3 Mixtures of Design/Data-based and Model-based Analyses

Inferences from wildlife studies often require mixtures of the strict design/data-based and pure model-based analyses. Examples of analyses using mixtures of study designs include:

1. Design/data-based studies conducted on a few target wildlife species
2. Manipulative tests using surrogate species to estimate the effect of exposure to some perturbation on the species of concern (Cade 1994)

3. Deductive professional judgment and model-based analyses used to quantify effects on certain components of the population or habitat in the affected area

Strict adherence to design/data-based analysis in wildlife studies may be impossible, but we recommend that the design/data-based analysis be adhered to as closely as possible. The value of indisputable design/data-based statistical inferences on at least a few response variables cannot be overemphasized in establishing confidence in the overall assessment of treatment effects. However, in some circumstances, model-based methods provide a suitable alternative to design/data-based methods. Additional discussion of the advantages, limitations, and appropriate applications of model-based methods exist in Chap. 4 and in Gilbert (1987), Johnson et al. (1989), and Gilbert and Simpson (1992).

3.4 Replication, Randomization, Control, and Blocking

Fisher (1966) defined the traditional design paradigm for the manipulative experiment in terms of the replication, randomization, control, and blocking, introduced in Chap. 2. Two additional methods are useful for increasing the precision of studies in the absence of increased replication:

1. Group randomly allocated treatments within homogeneous groups of experimental units (blocking)
2. *Use analysis of covariance* (ANCOVA) when analyzing the response to a treatment to consider the added influence of variables having a measurable influence on the dependent variable

3.4.1 Replication

Replication makes statistical inference possible by allowing the estimation of variance inherent in natural systems. Replication also reduces the likelihood that chance events will heavily influence the outcome of studies. In studies of wildlife populations, the experimental unit may be an animal, a group of animals, or all the animals within a specific geographic area. Using the wrong experimental unit can lead to errors in the identification of proper sample sizes and estimates of sample variance.

A good rule to follow when estimating the appropriate sample size in an experiment is that the analysis has only one value from each experimental unit. If five sample plots are randomly located in a study area, then statistical inferences to the area should be based on five values – regardless of the number of animals or plants that may be present and measured or counted in each plot. It becomes obvious that replication is difficult and costly in wildlife studies, particularly when the treatment is something as unique as an environmental perturbation, such as an oil spill, new wind plant, or dam.

3.4.2 Randomization

Like replication, an unbiased set of *independent data* is essential for estimating the error variance and for most statistical tests of treatment effects. Although truly unbiased data are unlikely, particularly in wildlife studies, a randomized sampling method can help reduce bias and dependence of data and their effects on the accuracy of estimates of parameters. A systematic sample with a random start is one type of randomization (Krebs 1989).

Collecting data from *representative locations* or *typical settings* is not random sampling. If landowners preclude collecting samples from private land within a study area, then sampling is not random for the entire area. In studies conducted on representative study areas, statistical inference is limited to the protocol by which the areas are selected. If private lands cannot be sampled and public lands are sampled by some unbiased protocol, statistical inference is limited to public lands. The selection of a proper sampling plan (see Chap. 4) is a critical step in the design of a project and may be the most significant decision affecting the utility of the data when the project is completed. If the objective of the study is statistical inference to the entire area, yet the sampling is restricted to a subjectively selected portion of the area, then there is no way to meet the objective with the study design. The inference to the entire area is reduced from a statistical basis to expert opinion.

3.4.3 Control and Error Reduction

Replication can increase the *precision* of an experiment (see Chap. 2), although this increased precision can be expensive. As discussed by Cochran and Cox (1957) and Cox (1958), the precision of an experiment can also be increased through:

1. Use of experimental controls
2. Refinement of experimental techniques, including greater sampling precision within experimental units
3. Improvement of experimental designs, including stratification and measurements of nontreatment factors (covariates) potentially influencing the experiment

Good experimental design should strive to improve confidence in cause and effect conclusions from experiments through the *control (standardization) of related variables* (Krebs 1989).

ANCOVA uses information measured on related variables as an alternative to standardizing variables (Green 1979). For example, understanding differences in predator use between areas improves when considered in conjunction with factors influencing use, such as the relative abundance of prey in each area. These factors are often referred to as *concomitant variables* or *covariates*. ANCOVA combines *analysis of variance* (ANOVA) and *regression* to assist interpretation of data when no specific experimental controls have been used (Steel and Torrie 1980). This analysis method allows adjustment of variables measured for treatment effects for differences in other independent variables also influencing the treatment response variable. ANCOVA assists in controlling error and increasing precision of experiments.

Precision can also be improved using *stratification*, or assigning treatments (or sampling effort) to homogeneous strata, or blocks, of experimental units. Stratification can occur in space (e.g., units of homogeneous vegetation) and in time (e.g., sampling by season). Strata should be small enough to maximize homogeneity, keeping in mind that smaller blocks may increase sample size requirements. For example, when stratifying an area by vegetation type, each stratum should be small enough to ensure a relatively consistent vegetation pattern within strata. Nevertheless, stratification requires some minimum sample size necessary to make estimates of treatment effects within strata. It becomes clear that stratification for a variable (e.g., vegetation type) in finer and finer detail will increase the minimum sample size requirement for the area of interest. If additional related variables are controlled for (e.g., treatment effects by season), then sample size requirements can increase rapidly. Stratification also assumes the strata will remain relatively consistent throughout the life of the study, an assumption often difficult to meet in long-term field studies.

3.5 Practical Considerations

Once the decision is made to conduct a wildlife study, several practical issues must be considered:

1. *Area of interest* (area to which statistical and deductive inferences will be made). Options include the study site(s), the region containing the study sites, the local area used by the species of concern, or the population potentially affected (in this case, population refers to the group of animals interbreeding and sharing common demographics).
2. *Time of interest*. The period of interest may be, for example, diurnal, nocturnal, seasonal, or annual.
3. *Species of interest*. The species of interest may be based on behavior, existing theories regarding species and their response to the particular perturbation, abundance, or legal/social mandate.
4. *Potentially confounding variables*. These may include landscape issues (e.g., large-scale habitat variables), biological issues (e.g., variable prey species abundance), land use issues (e.g., rapidly changing crops and pest control), weather, study area access, etc.

5. *Time available to conduct studies.* The time available to conduct studies given the level of scientific or public interest, the timing of the impact in the case of an accidental perturbation, or project development schedule in the case of a planned perturbation will often determine how studies are conducted and how much data can be collected.
6. *Budget.* Budget is always a consideration for potentially expensive studies. Budget should not determine what questions to ask but will influence how they are answered. Budget will largely determine the sample size, and thus the degree of confidence one will be able to place in the results of the studies.
7. *Magnitude of anticipated effect.* The magnitude of the perturbation or the importance of the effect to the biology of the species will often determine the level of concern and the required level of precision.

The remainder of this chapter is devoted to a discussion of some of the more common experimental designs used in biological studies. We begin with the simplest designs and progress toward the more complex while providing examples of practical applications of these designs to field studies. These applications usually take liberties with Fisher's requirements for designs of true experiments and thus we refer to them as quasiexperiments. Since the same design and statistical analysis can be used with either observational or experimental data, we draw no distinction between the two types of study. Throughout the remainder of this chapter, we refer to treatments in a general sense in that treatments may be manipulations by the experimenter or variables of interest in an observational study.

3.6 Single-factor Designs

Experiments are often classified based on the number of types of treatments that are applied to experimental units. A one-factor experiment uses one type of treatment or one classification factor in the experimental units in the study, such as all the animals in a specific area or all trees of the same species in a management unit. The treatment may be different levels of a particular substance or perturbation.

3.6.1 Paired and Unpaired

The simplest form of a biological study is the comparison of the means of two populations. An unpaired study design estimates the effect of a treatment by examining the difference in the population mean for a selected parameter in a treated and control population. In a paired study design, the study typically evaluates changes in study units paired for similarity. This may take the form of studying a population before and after a treatment is applied, or by studying two very similar study units. For example, one might study the effects of a treatment by randomly assigning treatment and control designation to each member of several sets of twins or to the right and left side of study animals, or study the effectiveness of two measurement methods by randomly applying each method to subdivided body parts or plant materials.

Comparison of population means is common in impact assessment. For example, as a part of a study of winter habitat use of mule deer (*Odocoileus hemionus*) in an area affected by gas exploration, development, and production, Sawyer et al. (2006) conducted quadrat counts of deer using the winter range from 2001 to 2005 and estimated a 49% decline in deer density after development. As Underwood (1997) points out, this is the classic "before–after" paired comparison where density is estimated before the treatment (gas development) and then compared to density estimates after development. Even though this rather dramatic decline in deer density is of concern, and represents a valid test of the null hypothesis that density will not change after development has occurred, the attribution of the change to development is not supported because of other influences potentially acting on the population. These other potential factors are usually referred to as *confounding* influences (Underwood 1997). In this case, other plausible explanations for the decline in density might be a regional decline in deer density due to weather or a response to competition with livestock for forage. Another approach to designing a study to evaluate the impacts of gas development on this group of deer is to measure density in both a treatment and a control area, where the comparison is the density in two independent groups of deer in the same region with similar characteristics except for the presence (treatment) or absence (control) of gas development.

While there is still opportunity for confounding, and cause and effect is still strictly professional judgment since this is a mensurative study, the presence or absence of a similar decline in the both the treatment and control groups of animals adds strength to the assessment of presence or absence of impact. This example illustrates a common problem in wildlife studies; that is, there is no statistical problem with the study, and there is confidence in not accepting the null hypothesis of no change in density after development. The dilemma is that there is no straightforward way of attributing the change to the treatment of interest (i.e., gas development). Fortunately, for Sawyer et al. (2006), contemporary estimates of habitat use made before and after gas development illustrated a rather clear reduction of available habitat resulting from gas development, which provides support for the conclusion that reduced density may be at least partially explained by development.

Another example of the value of paired comparisons is taken from the Coastal Habitat Injury Assessment (CHIA) following the massive oil spill when the Exxon Valdez struck Bligh Reef in Prince William Sound, Alaska in 1989 – the Exxon Valdez oil spill (EVOS). Many studies evaluated the injury to marine resources following the spill of over 41 million liters of Alaska crude oil. Pairing of oiled and unoiled sites within the area of impact of the EVOS was a centerpiece in the study of shoreline impacts by the Oil Spill Trustees' Coastal Habitat Injury Assessment (Highsmith et al., 1993; McDonald et al., 1995; Harner et al. 1995). In this case, beaches classified in a variety of oiled categories (none, light, moderate, and heavy) were paired based on beach substrate type (exposed bedrock, sheltered bedrock, boulder/cobble, and pebble/gravel). Measures of biological characteristics were taken at each site (e.g., barnacles per square meter, macroinvertebrates per square meter, intertidal fish, and algae per square meter) and comparisons were made between pairs of sites. The results were summarized as p-values (probabilities of observing differences as large as seen on the hypothesis that oiling had no effect) and p-values were combined using a meta-analysis approach (Manly 2001).

3.6.2 Completely Randomized Design

The simplest form of an experiment is the random application of two treatments to a group of experimental units known as the *completely randomized design*. This design is possible when experimental units are very similar (homogeneous) so blocking or other forms of partitioning of variance are of little benefit or sample sizes are large enough to be sure there is good representation of the target population in each treatment group. Allocation of treatments is by a random process such that each experimental unit has the same probability of receiving any treatment. Although it is preferable to have equal replication of each treatment across experimental units, it is not necessary.

The completely randomized design is a very flexible design. Analysis is simple and straightforward, allowing comparisons of means of different groups with the simple t-test or two or more treatments through ANOVA (Underwood 1997). Nonparametric equivalents of these tests are also readily available. The design maximizes the degrees of freedom (df) for estimating experimental error, increasing precision when df is <20. The loss of information due to missing data is small compared with other, more complicated designs. In addition, one can expand the design with more than two treatments without major alterations to the form of the experiment. The basic model for this design is:

Observed outcome = overall mean + treatment effect + experimental variation.

The completely randomized design is often inefficient, however, since experimental error contains all the variation among experimental units (i.e., measurement error and natural variation). The design may be acceptable for laboratory studies where experimental units are carefully controlled. In field situations, without considerable knowledge of the experimental units or a pretreatment test for differences among experimental units, there is a substantial leap of faith required to assume that experimental units are homogeneous. In the absence of homogeneous experimental units, an effect may be assumed when in reality the apparent treatment effects could actually be the result of pretreatment differences. The best way to deal with this naturally occurring heterogeneity is by true randomization of treatments (Manly 1992) and by maximization of sample size within the context of project goals and practical limitations (e.g., budget). However, as Hurlbert (1984) pointed out, we seldom encounter homogeneous experimental units in ecological studies and spatial segregation of experimental units can lead to erroneous results resulting from naturally occurring gradients (e.g., elevation and exposure effects on plant growth). This is especially problematic with small sample sizes common in field studies. A systematic selection of experimental units (see Chap. 4) may reduce the effects of spatial segregation of units for a given sample size while maintaining the mathematical properties of randomness. Regardless, the *natural gradients* existing in nature make application of the completely randomized design inappropriate for most field studies.

For a hypothetical example of the completely randomized design, assume the following situation. A farmer in Wyoming is complaining about the amount of alfalfa consumed by deer in his fields. Since the wildlife agency must pay for verified claims of damage by big game, there is a need to estimate the effect of deer use on production of alfalfa in the field. The biologist decides to estimate the damage by comparing production in plots used by deer vs. control plots not used by deer and divides the farmer's largest uniform field into a grid of plots of equal size. A sample of plots is then chosen by some random sampling procedure (see Chap. 4). Deer-proof fence protects half of the randomly selected plots, while the other half is unprotected controls. The effects of deer use is the difference between estimated alfalfa production in the control and protected plots, as measured either by comparing the two sample means by a simple t-test or the overall variation between the grazed and ungrazed plots by ANOVA (Mead et al. 1993).

An astute biologist who wanted to pay only for alfalfa consumed by deer could add an additional treatment to the experiment. That is, a portion of the plots could be fenced to allow deer use but exclude rabbits and other small herbivores that are not covered by Wyoming's damage law, without altering the design of the experiment. The analysis and interpretation of this expanded experiment also remains relatively simple (Mead et al. 1993).

In a real world example, Stoner et al. (2006) evaluated the effect of cougar (*Puma concolor*) exploitation levels in Utah. This study used a two-way factorial ANOVA in a completely randomized design with unequal variances to test for age differences among treatment groups (site and sex combinations) for demographic structure, population recovery, and metapopulation dynamics.

3.6.3 Randomized Complete Block Design

While the simplicity of the completely randomized design is appealing, the lack of any restriction in allocation of treatments even when differences in groups of experimental units are known seems illogical. In ecological experiments and even most controlled experiments in a laboratory, it is usually desirable to take advantage of blocking or stratification (see Chap. 4 for discussion) as a form of error control. In the deer example discussed earlier, suppose the biologist realizes there is a gradient of deer use with distance from cover. This variation could potentially bias estimates of deer damage, favoring the farmer if by chance a majority of the plots is near cover or favoring the wildlife agency if a majority of the plots is toward the center of the field. Dividing the field into strata or blocks and estimating deer use in each may improve the study. For example, the biologist might divide the field into two strata, one including all potential plots within 50 m of the field edge and one including the remaining plots. This stratification of the field into two blocks restricts randomization by applying treatments to groups of experimental units that are more similar and results in better estimates of the effect of deer use, resulting in an equitable damage settlement.

In the experiment where blocking is used and each treatment is randomly assigned within each block, the resulting design is called a *randomized complete block design* (Table 3.1). Blocking can be based on a large number of factors potentially affecting experimental variation. In animal studies, examples of blocks include things such as expected abundance, territoriality, individual animal weights, vegetation, and topographical features. Plant studies block on soil fertility, slope gradient, exposure to sunlight, individual plant parts, or past management. In ecological studies, it is common to block on habitat and across time. This form of grouping is referred to as local control (Mead et al. 1993). The typical analysis of randomized block designs is by ANOVA following the linear additive model

Observed outcome = overall mean + block effect + treatment effect + residual variation + block × treatment interaction

with the block × treatment interaction serving as the error estimate for hypothesis tests.

With proper blocking, no single treatment gains or loses advantage when compared with another because of the characteristics of the units receiving the treatment. If the units within blocks are homogeneous compared to units within other blocks, the blocking reduces the effects of random variation among blocks on the errors involved in comparing treatments. Notwithstanding, poorly designed blocking creates more problems than it solves (see Chap. 4 for a discussion of problems associated with stratification).

Volesky et al. (2005) provide an example of the randomized complete block design to determine the use and herbage production (of cool-season graminoids) in response to spring livestock grazing date and stocking rate in the Nebraska Sandhills. The study used spring grazing date as the main plot, stocking rate as the split plot (see Sect. 3.8.2), with a nongrazed control and grazing rate and stocking rate were factor combinations of treatments. The analysis combined treatments across years with years as fixed effects and blocks as random effects.

Table 3.1 A randomized complete block experiment with four blocks and three treatments (A, B, and C) applied to three plots in each block

Block	Treatment		
1	A	B	C
2	A	C	B
3	B	A	C
4	C	B	A

Reproduced from Morrison et al. (2001), with kind permissions from Springer Science + Business Media

Bates et al. (2005) also used the randomized complete block design in a long-term study of the successional trends following western juniper cutting. This study established four blocks with each block divided into two plots and one plot within each block randomly assigned the cutting treatment (CUT) and the remaining plot left as woodland (WOODLAND). ANOVA was used to test for treatment effect on herbaceous standing crop (functional group and total herbaceous), cover (species and functional group), and density (species and functional group). Cover and density of shrubs and juniper were analyzed by species with response variables analyzed as a randomized complete blocks across time. The final model included blocks (four blocks, df = 3), years (1991–1997 and 2003, df = 7), treatments (CUT, WOODLAND, df = 1), and year by treatment interaction (df = 7; with the error term df = 45).

3.6.4 Incomplete Block Design

A characteristic of the randomized block design discussed earlier was that each treatment was included in each block. In some situations, blocks or budgets may not be large enough to allow all treatments to be applied in all blocks. The *incomplete block design* results when each block has less than a full complement of treatments. In a balanced incomplete block experiment (Table 3.2), all treatment effects and their differences are estimated with the same precision, as long as every pair of treatments occurs together the same number of times (Manly 1992). However, analysis of incomplete block designs is considerably more complicated than complete block designs. It is important to understand the analysis procedures before implementing an incomplete block design. Example design and analysis methods are discussed in Mead et al. (1993).

3.6.5 Latin Squares Design

The randomized block design is useful when one source of local variation exists. When additional sources of variation exist, then the randomized block design can

Table 3.2 A balanced incomplete block experiment with four blocks and four treatments (A, B, C, and D) applied to three plots in each block

Block	Treatment		
1	A	B	C
2	A	B	D
3	A	C	D
4	B	C	D

Note that each treatment pair (i.e., AB, AC, BC, and CD) occurs the same number of times. Reproduced from Morrison et al. (2001), with kind permissions from Springer Science + Business Media

Table 3.3 A Latin square experiment with two blocking factors (X and Y) each with four blocks and four treatments (A, B, C, D)

	Blocking factor (Y)			
Blocking factor (X)	1	2	3	4
1	A	B	C	D
2	B	C	D	A
3	C	D	A	B
4	D	A	B	C

Reproduced from Morrison et al. (2001), with kind permissions from Springer Science + Business Media

be extended to form a *Latin square* (Table 3.3). For example, in a study of the effectiveness of some treatment, variation may be expected among plots, seasons, species, etc. In a Latin square, symmetry is required so that each row and column in the square is a unique block. The basic model for the Latin square design is as follows:

Observed outcome = row effect + column effect + treatment effect + random unit variation.

The Latin square design allows separation of variation from multiple sources at the expense of df, potentially reducing the ability of the experiment to detect effect. The Latin square design is useful when multiple causes of variation are suspected but unknown. However, caution should be exercised when adopting this design. As an example of the cost of the design, a 3 × 3 Latin square must reduce the mean square error by approximately 40% of the randomized block design of the same experiment to detect a treatment effect of a given size.

While the Latin square design is not a common study design in wildlife studies it can be useful in some situations. For example, with the aid of George Baxter and Lyman McDonald, both professors at the University of Wyoming, the Wyoming Game and Fish Department used the Latin square design on a commercial fisheries project involving carp (*Ctenopharyngodon idella*) in Wyoming. The Department wanted to determine the cause and frequency of "large" year classes and estimated abundance of young fish by different methods at beach sites to help answer this question. The study used three sites, three sampling periods separated by some time to let the fish settle down, and three types of gear (minnow seining, wing traps, and minnow traps). The design was set up in a balanced 3 × 3 Latin square and analysis was by ANOVA. The Latin square takes the form of rows as sites, columns as times, and three gear types with a response variable of p where p = proportion of young of the year fish caught. The Latin square is completed, where each gear type occurs once in each site and time. In addition to estimating the abundance of young fish, the Department was interested in correcting seining data collected elsewhere for biases relative to the "best" sampling method or the pooled proportions if there were significant differences.

3.6.6 Summary

Obviously the different levels of a single treatment in these designs are assumed to be independent and the treatment response assumed to be unaffected by interactions among treatment levels or between the treatment and the blocking factor. This might not present a problem if interaction is 0, an unlikely situation in ecological experiments. Heterogeneity in experimental units and strata (e.g., variation in weather, vegetation, and soil fertility) is common in the real world and results in the confounding of experimental error and interaction of block with treatment effects (Underwood 1997). This potential lack of independence with a corresponding lack of true replication can make interpretation of experiments very difficult, increasing the effect size necessary for significance (increase in Type II error).

3.7 Multiple-factor Designs

3.7.1 Factorial Designs

The preceding designs reduced the confounding effects of variance by blocking under the assumption that the different treatments of a single factor were unique and acted independently. In ecological studies, this independence of treatment effects is seldom encountered. Furthermore, studies usually deal with more than one factor or class of experimental units. Examples of factors include different treatments, such as temperature, diet, habitat, and water quality, or classifications of experimental units, such as season, time of day, sex, age, etc. *Factorial experiments* are more complex experiments where all possible combinations of factors of interest are tested and these tests are possibly replicated a number of times (Manly 1992) and with the resulting data typically analyzed with ANOVA (Underwood 1997).

3.7.2 Two-factor Designs

In a single-factor experiment, there is only one class of treatment. For example, a biologist is interested in the effects of a particular nutrient on the physical condition of deer. The biologist has 24 captive adult deer available for the study. By dividing the deer into three groups of eight deer each and feeding each group a diet with different amounts of the nutrient, the biologist has a single-factor experiment. This study becomes a *two-factor experiment* if the adult deer are divided into six groups of four deer each and a second class of treatment such as two different amounts of forage is added to the experiment. The deer could also be grouped by sex, e.g., three groups of four females and three groups of four males. The three levels of nutrient in the diet

Table 3.4 A 2 × 3 factorial experiment where factor A has three levels (a_1, a_2, and a_3) and factor B has two levels (b_1 and b_2)

		Factor A		
	Level	a_1	a_2	a_3
Factor	b_1	a_1b_1	a_2b_1	a_3b_1
B	b_2	a_1b_2	a_2b_2	a_3b_2

Reproduced from Morrison et al. (2001), with kind permissions from Springer Science + Business Media

and the two amounts of total forage (treatment factors) in the first example and the grouping by sex (classification factor) combined with the three levels of nutrients in the second example both result in a 2 × 3 factorial experiment (Table 3.4).

3.7.3 Multiple-factor Designs

Multiple-factor designs occur when one or more classes of treatments are combined with one or more classifications of experimental units. Continuing the deer feeding experiment, a multiple-factor experiment might include both classes of treatment and the classification of deer by sex resulting in a 2 × 2 × 3 factorial experiment (Table 3.5).

Classification factors, such as sex and age, are not random variables but are fixed in the population of interest and cannot be manipulated by the experimenter. On the other hand, the experimenter can manipulate treatment factors, usually the main point of an experiment (Manly 1992). It is not appropriate to think in terms of a random sample of treatments, but it is important to avoid bias by randomizing the application of treatments to the experimental units available in the different classes of factors. In the example above, a probabilistic sample of female deer selected from all females available for study receive different levels of the treatment.

In the relatively simple experiments with unreplicated single-factor designs, the experimenter dealt with treatment effects as if they were independent. In the real world, one would expect that different factors often interact. The ANOVA of factorial experiments allows the biologist to consider the effect of one factor on another. In the deer example, it is reasonable to expect that lactating females might react differently to a given level of a nutrient, such as calcium, than would male deer. Thus, in the overall analysis of the effect of calcium in the diet, it would be instructive to separate the effects of calcium and sex on body condition (main effects) from the effects of the interaction of sex and calcium. The linear model for the factorial experiment allows the subdivision of treatment effects into main effects and interactions, allowing the investigation of potentially interdependent factors. The linear model can be characterized as follows:

Observed outcome = main effect variable A + main effect variable B+(A)(B) interaction + Random unit variation

Table 3.5 An example of a 2 × 2 × 3 factorial experiment where the three levels of a mircronutrient (factor A) are applied to experimental deer grouped by sex (factor B), half of which are fed a different amount of forage (factor C)

Factor B, sex	Factor C, forage level	Factor A, micronutrient		
B_1	c_1	$a_1b_1c_1$	$a_2b_1c_1$	$a_3b_1c_1$
	c_2	$a_1b_1c_2$	$a_2b_1c_2$	$a_3b_1c_2$
B_2	c_1	$a_1b_2c_1$	$a_2b_2c_1$	$a_3b_2c_1$
	c_2	$a_1b_2c_2$	$a_2b_2c_2$	$a_3b_2c_2$

Reproduced from Morrison et al. (2001), with kind permissions from Springer Science + Business Media

Mead et al. (1993) considered this characteristic one of the major statistical contributions from factorial designs.

When interactions appear negligible, factorial designs have a second major benefit referred to as "hidden replication" by Mead et al. (1993). Hidden replication allows the use of all experimental units involved in the experiment in comparisons of the main effects of different levels of a treatment when there is no significant interaction. Mead et al. (1993) illustrated this increase in efficiency with a series of examples showing the replication possible when examining three factors, A, B, and C, each with two levels of treatment:

1. In the case of three independent comparisons, ($a_0b_0c_0$) with ($a_1b_0c_0$), ($a_0b_1c_0$), and ($a_0b_0c_1$) with four replications for each was possible, involving 24 experimental units. The variance of the estimate of the difference between the two levels of A (or B or C) is $2\sigma^2/4$, where σ^2 is the variance per plot.
2. Some efficiency is gained by reducing the use of treatment ($a_0b_0c_0$) by combining the four treatments ($a_0b_0c_0$), ($a_1b_0c_0$), ($a_0b_1c_0$), and ($a_0b_0c_1$) into an experiment with six replications each. Thus, the variance of the estimate of the difference between any two levels is $2\sigma^2/6$, reducing the variance by two-thirds.
3. There are eight factorial treatments possible from combinations of the three factors with their two levels. When these treatments are combined with three replications, each comparison of two levels of a factor includes 12 replicates. All 24 experimental units are involved with each comparison of a factor's two levels. Thus, in the absence of interaction, the factorial experiment can be more economical, more precise, or both, than experiments looking at a single factor at a time.

There is more at stake than simply an increase in efficiency when deciding whether to select a factorial design over independent comparisons. The counterargument for case 1 above is that the analysis becomes conditional on the initial test of interaction, with the result that main effect tests of significance levels may be biased. Perhaps the only situation where example 1 might be desirable is in a study where sample sizes are extremely limited.

Multiple-factor designs can become quite complicated, and interactions are the norm. Although there may be no theoretical limit to the number of factors that can be included in an experiment, it is obvious that sample size requirements increase dramatically as experimental factors with interactions increase. This increases the cost of experiments and makes larger factorial experiments impractical. Also, the more complicated the experiment is, the more difficulty one has in interpreting the results.

Factorial designs are reasonably common in ecology studies. Mieres and Fitzgerald (2006) used both two-factor and three-factor models in studying the monitoring and management of the harvest of tegu lizards (*Tupinambis* spp.) in Paraguay. The study applied general linear models (two-factor and three-factor ANOVA) to test the null hypothesis of no significant differences in mean size of males and females of each species among years and among check stations. To analyze data from tanneries, they used separate two-factor ANOVAs, with interaction (year and sex as factors), for each species to test the hypothesis that body size varied by year and sex. To test for size variation in tegu skins sampled in the field, the study used three-factor ANOVAs, with interaction (year, sex, and check station as factors), to test the hypothesis that body size varied by year, sex, and check station.

In a study of bandwidth selection for fixed-kernel analysis of animal utilization distributions, Gitzen et al. (2006) used mixtures of bivariate normal distributions to model animal location patterns. The study varied the degree of clumping of simulated locations to create distribution types that would approximate a range of real utilization distributions. Simulations followed a $4 \times 3 \times 3$ factorial design, with factors of distribution type (general, partially clumped, all clumped, nest tree), number of component normals (2, 4, 16), and sample size (20, 50, 150)

3.7.4 Higher Order Designs

The desire to include a large number of factors in an experiment has led to the development of complex experimental designs. For an illustration of the many options for complex designs, the biologist should consult textbooks with details on the subject (e.g., Montgomery 1991; Milliken and Johnson 1984; Mead et al. 1993; Underwood 1997). The object of these more complex designs is to allow the study of as many factors as possible while conserving observations. One such design is a form of the *incomplete block design* known as confounding. Mead et al. (1993) described confounding as the allocation of the more important treatments in a randomized block design so that differences between blocks cancel out the same way they do for comparisons between treatments in a randomized block design. The remaining factors of secondary interest, including those assumed to have negligible interactions are included as treatments in each block, allowing the estimate of their main effects while sacrificing the ability to include their effects on interactions. Thus, block effects are confounded with the effects of interactions. The resulting allocation of treatments becomes an incomplete block with a corresponding reduction in the number of treatment comparisons the experimenter must deal with. Mead et al. (1993) provided two examples that help describe the rather complicated blocking procedure. These complicated designs should not be attempted without consulting a statistician and unless the experimenter is confident about the lack of significant interaction in the factors of secondary interest.

3.8 Hierarchical Designs

3.8.1 Nested Designs

A *nested experimental design* is one that uses replication of experimental units in at least two levels of a hierarchy (Underwood 1997). Nested designs are also known as hierarchical designs and are common in biological studies. Milliken and Johnson (1984) lumped some *nested designs*, *split-plot designs*, and *repeated measures designs* into a category of designs "having several sizes of experimental units." In the earlier discussion of incomplete block experiments, the effects of confounding were dismissed because the experimenter presumably knew that the confounding effects of the interactions of some treatments were negligible. Unfortunately, as we have pointed out, the confounding effects of other variables are all too common in wildlife studies, making the estimation of treatment effects very difficult. Nested studies are a way to use replication to increase one's confidence that differences seen when comparing treatments are real and not just random chance or the effects of some other factor. Nested designs result in data from replicated samples taken from replicated plots receiving each treatment of interest. The only difference in the ANOVA of a nested design from the one-factor ANOVA is that total variation is identified as variation among sample replicates, variation among units (plots) within each treatment, and variation among treatments.

Berenbaum and Zangerl (2006) used a nested study design to study parsnip webworms (*Depressaria pastinacella*) and host plants at a continental scale by evaluating trophic complexity in a geographic mosaic and their role in coevolution. The study used a mixed/nested model (procedure UNIANOVA, SPSS 1999) to compare outcomes of the interaction between wild parsnip (*Pastinaca sativa*) in its indigenous area, Europe, to its area of introduction, the Midwestern United States. The study tested the hypothesis that increasing trophic complexity, represented by alternate host plants or the presence of natural enemies, reduces the selective impact of parsnip webworms and hence diminishes linkage between host plant chemistry and webworms that would be expected in coevolutionary hotspots (areas where webworms were common). The wild parsnip produces a phototoxic compound (furanocoumarins) that crosslink DNA and interfere with transcription in the webworm. Of interest in this study was the concentration of furanocoumarin in parsnip seeds as a function of continent and interaction of temperature and the density of webworms. The study treats the chemical characteristic of parsnip as a random factor nested within both continent and webworm density, and continent and webworm densities as fixed effects.

3.8.2 Split-plot Designs

Split-plot designs are a form of nested factorial design commonly used in agricultural and biological experiments. The study area is divided into blocks following

Table 3.6 An illustration of a two-factor split-plot experiment where factor A is considered at four levels in three blocks of a randomized complete block design and a second factor, B, is considered at two levels within each block

Block 1				Block 2				Block 3			
a_4b_2	a_1b_2	a_2b_1	a_3b_2	a_2b_1	A_1b_2	a_4b_1	a_3b_1	a_1b_1	a_2b_2	a_4b_2	a_3b_1
a_4b_1	a_1b_1	a_2b_2	a_3b_1	a_2b_2	A_1b_1	a_4b_2	a_3b_2	a_1b_2	a_2b_1	a_4b_1	a_3b_2

Note that each unit of factor A is divided into two subunits and randomization occurs for both factor A and factor B. Reproduced from Steel and Torrie (1980), with kind permission from The McGraw-Hill Company
Source: Steel and Torrie (1980)

the principles for blocking discussed earlier. The blocks are subdivided into relatively large plots called main plots, which are then subdivided into smaller plots called split plots, resulting in an incomplete block treatment structure. In a two-factor design, one factor is randomly allocated to the main plots within each block. The second factor is then randomly allocated to each split plot within each main plot. The design allows some control of the randomization process within a legitimate randomization procedure.

Table 3.6 illustrates a simple two-factor split-plot experiment. In this example, four levels of factor A are allocated as if the experiment were a single-factor completely randomized design. The three levels of factor B are then randomly applied to each level of factor A. It is possible to expand the split-plot design to include multiple factors and to generalize the design by subdividing split plots, limited only by the minimal practical size of units for measurements (Manly 1992).

The ANOVA of the split-plot experiment also occurs at multiple levels. At the main plot level, the analysis is equivalent to a randomized block experiment. At the split-plot level, variation is divided into variation among split-plot treatments, interaction of split-plot treatments with main effects, and a second error term for split plots (Mead et al. 1993). A thorough discussion of the analysis of split-plot experiments is presented in Milliken and Johnson (1984). It should be recognized that in the split-plot analysis, the overall precision of the experiment is the same as the basic design.

The split-plot design is useful in experiments when the application of one or more factors requires a much larger experimental unit than for others. For example, in comparing the suitability of different species of grass for revegetation of clear-cuts, the grass plots can be much smaller, e.g., a few square meters, as compared with the clear-cuts that might need to be several acres to be practical. The design can also be used when variation is known to be greater with one treatment vs. another, with the potential for using less material and consequently saving money. The design can be useful in animal and plant studies where litters of animals or closely associated groups of individual plants can be used as main plots and the individual animals and plants used as split plots.

Manly (1992) listed two reasons to use the split-plot design. First, it may be convenient or necessary to apply some treatments to whole plots at the same time. Second, the design allows good comparisons between the levels of the factor that is applied at the subplot level at the expense of the comparisons between the main plots, since experimental error should be reduced within main plots. However, Mead et al. (1993) pointed out that there is actually a greater loss of precision at the main plot level than is gained at the level of split-plot comparisons. They also indicate that there is a loss of replication in many of the comparisons of combinations of main plot and split treatments resulting in a loss of precision. These authors recommend against the split-plot design except where practically necessary. Underwood (1997) also warned against this lack of replication and the potential lack of independence among treatments and replicates. This lack of independence results because, in most layouts of split-plot designs, main plots and split plots tend to be spatially very close.

Barrett and Stiling (2006) used a split-plot design in a study of Key deer (*Odocoileus virginianus clavium*) impacts on hardwood hammocks near urban areas in the Florida Keys. The study used a split-plot ANOVA model to test each response variable (total basal area of large trees and percentage of canopy cover) with deer density (low and high) and distance (urban and exurban) as factors with island (Big Pine, No Name, Cudjoe, Sugarloaf) nested within levels of deer density. The study found evidence that deer density interacted with distance indicating differences in responses between urban and exurban hammock stands.

3.8.3 Repeated Measures Designs

Experiments where several comparable measurements are taken on each experimental unit are referred to as *repeated measures designs*. Repeated measures experiments are usually associated with nested and split-plot designs. However, repeated measures experiments may occur with the simple completely randomized design or the more complex designs involving blocking, Latin squares, incomplete blocks, or split blocks (Mead et al. 1993). The experimental design structure for the allocation of experimental units, upon which multiple measurements are recorded, must be clearly defined and understood. The significance of the basic design is that it defines the form the analysis must take. In every case, the analysis of repeated measures must consider the lack of independence of the multiple measures taken on the same unit.

Repeated measures usually occur because the experimenter concludes that multiple measurements on an experimental unit increases the knowledge gained in the study. Repeated measures experiments involve a step or steps where there is no randomization of treatment levels. The most common form of repeated measures experiment is the *longitudinal experiment* where observations are taken in the same order on each experimental unit (Manly 1992). Ordering can be a function of any condition that has an order that cannot be changed, but is usually a function of time or location.

A repeated measures experiment where each experimental unit is subjected to several treatments with the order varied for groups of units is called a *changeover experiment* (Manly 1992). In addition to the lack of independence of repeated measurements, the changeover experiment should consider the carryover effects of treatments on subsequent treatment effects. When possible, the ordering of treatments should be assigned at random to experimental units (Milliken and Johnson 1984).

Longitudinal studies are so common that the term repeated measures often refers only to this type of study (Manly 1992). Longitudinal studies are common in wildlife telemetry studies, environmental impact studies, habitat use and selection studies, studies of blood chemistry, and many other forms of wildlife research. Typically, in these studies, the logistics leads to a repeated measures experiment. For example, after going to the trouble and expense of capturing and radio-tagging an elk, deer, or golden eagle, a biologist correctly takes the position that taking repeated measurements of the marked animal improves the study. Nevertheless, it must be recognized that repeated measures on the same animal may improve the understanding of that animal's behavior but do not represent true replication leading to a better understanding of all animals. The appropriateness of repeated measures experiments is determined by the goal of the study. The biologist must guard against treating repeated measures as true replications and thus leading to what Hurlbert (1984) described as pseudoreplication.

The analysis of data from longitudinal experiments usually follows the same form as analysis of split-plot experiments. The only apparent difference between the longitudinal experiment and the split-plot design is that time, typically a split-plot factor, is beyond the control of the experimenter. Mead et al. (1993) provided a description of the more common approaches to the analysis of these studies. Manly (1992) listed the following analysis methods:

1. ANOVA of a summary variable, such as the mean of repeated measures among units or the difference between the first and last observations (Mead et al. 1993)
2. ANOVA of a response function fitted to observations on experimental units to see how they vary with the factors that describe the groups (Mead et al. 1993)
3. Multivariate ANOVA taking into account the lack of independence of observations taken at different times (Winer 1971)
4. ANOVA in the same way as with a split-plot experiment (Manly 1992)

Mead et al. (1993) pointed out that the form of the model chosen for analysis must always be biologically reasonable, recognizing that models will always require simplification of reality. When the form of analysis follows the split-plot design, there is a general assumption that observations on the same unit at different times will tend to be more similar than observations on different units. There is also the assumption that differences in observations on the same unit are independent of the time when the observations are made. That is, the first and last observations should have the same degree of similarity as consecutive observations, a seldom-valid assumption. This assumption of uniform similarity, also known as compound symmetry may be justified by the random allocation of treatment levels to split plots but should be formally tested (Manly 1992). Milliken and Johnson (1984) provided a detailed discussion of alternative models for analysis of repeated measures experiments when the assumption of compound symmetry is appropriate. Several statistical software programs (e.g., SAS Institute, Inc.) automatically test for compound symmetry and provide alternate models when the assumption is not met.

Mead et al. (1993) suggested that the split-plot analysis oversimplifies the true situation in its assumption of uniform similarity between times and fails to use the ordering of time. They point out that multivariate ANOVA makes no assumptions about patterns of similarity of observations at different times. Multivariate ANOVA estimates the relationships between times wholly from the data, ignoring the order of times (Crowder and Hand 1990). However, Underwood (1997) maintained that the *multivariate analysis* deals with only one of the problems with the analysis, namely the nonindependence among times of sampling. The other problem of lack of replication leading to unverifiable assumptions of no interaction between times of sampling and replicated plots in each treatment is not addressed. Underwood (1997) took the relatively hard line that proper independent replicates should be the design of choice unless interactions that must be assumed to be 0 are realistically likely to be 0. Obviously, repeated measures are likely to continue in wildlife studies.

Our best advice is to consider the implications of the nonindependence of the data when interpreting the meaning of the studies. The experimenter should consult a good reference, such as Crowder and Hand (1990), when considering repeated measures experiments or, better yet, consult with a statistician experienced in dealing with the design and analysis of such studies.

The repeated measures study design is one of the most common designs in wildlife studies, particularly in the evaluation of the impacts of management or environmental perturbations. An example of a repeated measures study design is provided by Martin and Wisley (2006) in their assessment of grassland restoration success as influenced by seed additions and native ungulate activities. The study used a randomized complete block split-plot design with unequal replication, with grazing or enclosures applied to main plots and seed addition treatments applied to subplots. The statistical analysis used randomized split-plot ANOVAs, with planting as a random block term; all grazing effects were tested with the main plot error term. A repeated measures ANCOVA was used to compare grazed and ungrazed plots for existing vegetation and resource variables, with time 0 data (measurements taken before exclosures were constructed) as a covariate and used repeated-measures ANOVA of corresponding data to analyze grazing effects on seedling enhancement over time. The study analyzed the exotic seedling and seedling diversity variables with repeated-measures ANOVA for the first seed addition, and with regular ANOVA for the second seed addition.

3.9 Analysis of Covariance

ANCOVA uses the concepts of *ANOVA* and *regression* (Huitema 1980; Winer et al. 1991; Underwood 1997) to improve studies by separating treatment effects on the response variable from the effects of *confounding variables* (covariates). ANCOVA can also be used to adjust response variables and summary statistics (e.g., treatment means), to assist in the interpretation of data, and to estimate missing data (Steel and Torrie 1980). It is appropriate to use ANCOVA in conjunction with most of the previously discussed designs.

Earlier in this chapter, we introduced the concept of increasing the precision of studies by the use of ANCOVA when analyzing the response to a treatment by considering the added influence of variables having a measurable influence on the dependent variable. For example, in the study of fatalities associated with different wind turbines, Anderson et al. (1999) recommended measuring bird use and the rotor-swept area as covariates. It seems logical that the more birds use the area around turbines and the larger the area covered by the turbine rotor, the more likely that bird collisions might occur. Figure 3.2 provides an illustration of a hypothetical example of how analysis of bird fatalities associated with two turbine types can be improved by the use of covariates. In the example, the average number of fatalities per turbine is much higher in the area with turbine type A vs. turbine type B. However, when the fatalities are adjusted for differences in bird use, the ratio of fatalities per unit of bird use is the same for both turbine types, suggesting no true difference in risk to birds from the different turbines. Normally, in error control, multiple regression is used to assess the difference between the experimental and control groups resulting from the treatment after allowing for the effects of the covariate (Manly 1992).

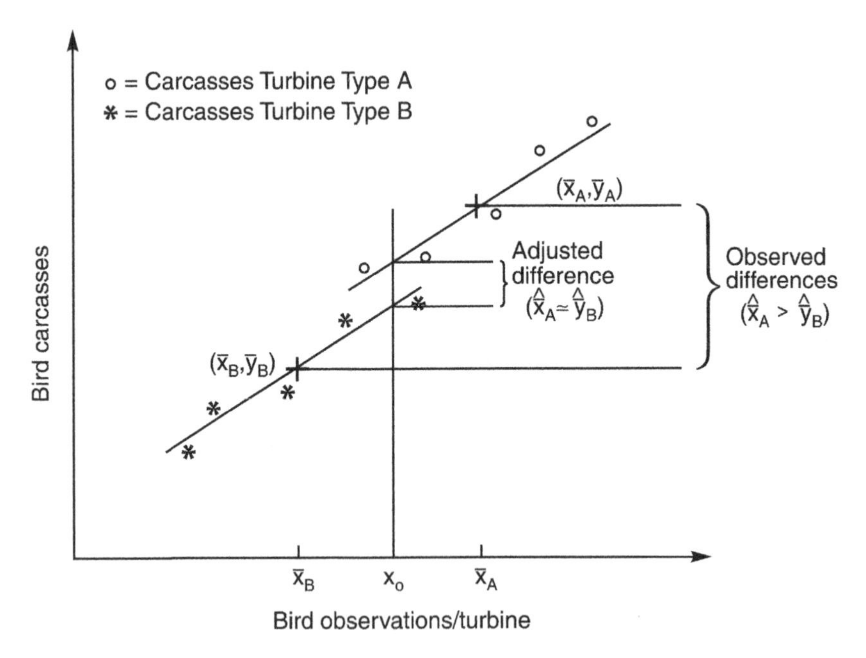

Fig. 3.2 Illustration of hypothetical example of bird fatalities associated with two turbine types (A and B) where the mean fatalities are adjusted for differences in bird use. The average number of fatalities per turbine is much higher associated with turbine type A vs. turbine type B, while the ratio of fatalities per unit of bird use is the same for both turbine types. Reproduced from Morrison et al. (2001) with kind permission from Springer Science + Business Media

ANCOVA adjusts estimates of response variables, such as treatment means. For example, when wildlife surveys record animals by habitat or behavior, these covariates adjust counts to estimate animal numbers more accurately. Strickland et al. (1994) used ANCOVA and logistic regression to adjust aerial counts of Dall's sheep in Alaska. To the authors' surprise, habitat had no effect on sightability but group size was quite important, resulting in significant upward adjustments of counts of individuals and small groups. Surveys of other large mammals (e.g., Gasaway et al. 1985; Samual et al. 1987) suggested that habitat and group size might influence the sightability of sheep. Normally, when using ANCOVA to control error and adjust parameter estimates, the experimenter measures covariates uninfluenced by treatments, such as environmental influences. When covariates are affected by treatments, then their interpretation can be misleading. For example, if one is interested in the effect on animal use in an area by the presence of wind turbines built in different habitats, the study is confounded somewhat because erecting turbines may change habitat characteristics. If this effect is relatively small or the data exist for its estimation, then ANCOVA is still preferable over ignoring the effects of the confounding variables. For example, if the tower pads and roads in the above example are the same size or are carefully measured in all habitats, their effect on bird use can be ignored or accounted for. Although measurements of covariates will have residual error, a violation of one of the necessary assumptions for ANCOVA, Glass et al. (1972) concluded that this is not a serious problem unless residual errors are large.

Manly (1992) also urged caution when using regression adjustment in ANCOVA. He points out that the linear model may be too simple, and a biased estimate of treatment effect may result or important confounding variables may not be measured. As an example, in the wildlife surveys example discussed earlier, we mentioned the propensity for surveys to include environmental covariates, such as habitat and animal behavior. However, it is our experience that variables associated with experimental methods, e.g., the type of aircraft, the experience of the observer, etc., may be far more important in determining the quality of the survey. As with repeated measures, the assumptions inherent in the basic design significantly influence ANCOVA, and good design principles (e.g., randomization and replication) are necessary even with a regression adjustment.

ANCOVA is useful in estimating missing values (Steel and Torrie 1980), and recently, in a model-based analysis of spatial data (e.g., kreiging) discussed in more detail in Chap. 4. The latter application uses the correlations between neighboring sampling units to estimate the variable of interest at points not sampled. Generally, these studies adopt a completely randomized design using a systematic grid of sampling points with a random starting point. Confidence intervals can be calculated for estimates of variables of interest indicating where increased precision is desirable. In environmental contamination studies, these initial samples may be used retrospectively for blocking or stratifying the area of interest so that additional samples can be taken where more precision is desired. We suggest more extensive reading if this form of study is of interest, starting with the summary discussion provided by Borgman et al. (1996).

Flemming et al. (2006) used ANCOVA in their study to test for the effects of embedded lead shot on body condition of common eiders (*Somateria mollissima*).

The assumptions of normality required a log-transformation (using the Andersen–Darling test) of the dependent variable total carcass lipids (TCL) and homogeneity of variances (using Bartlett's test).

Herring and Collazo (2006) used ANCOVA in the study of lesser scaup (*Aythya affinis*) winter foraging and nutrient reserve acquisition in east-central Florida. The study used ANCOVA to examine the effects of season on each of the response variables (CBM, protein, lipids, minerals) for each sex and year separately; in the models, winter period was the treatment and PC1 (first principal component) the covariate to adjust contrasts between season by size of birds.

3.10 Multivariate Analyses

To this point, we dealt with designs that are concerned with the effect of a treatment on one response variable (*univariate methods*). The point of *multivariate analysis* is to consider several related random variables simultaneously, each one being considered equally important at the start of the analysis (Manly 1986). There is a great deal of interest in the simultaneous analysis of multiple indicators (multivariate analysis) to explain complex relationships among many different kinds of response variables over space and time. This is particularly important in studying the impact of a perturbation on the species composition and community structure of plants and animals (Page et al. 1993; Stekoll et al. 1993). Multivariate techniques include multidimensional scaling and ordination analysis by methods such as principal component analysis and detrended canonical correspondence analysis (Gordon 1981; Dillon and Goldstein 1984; Green 1984; Seber 1984; Pielou 1984; Manly 1986; Ludwig and Reynolds 1988; James and McCulloch 1990; Page et al. 1993). If sampling units are selected with equal probability by simple random sampling or by systematic sampling (see Chap. 4) from treatment and control areas, and no quasiexperimental design is involved (e.g., no pairing), then the multivariate procedures are applicable.

It is unlikely that multivariate techniques will directly yield indicators of effect (i.e., combinations of the original indicators) that meet the criteria for determination of effect. Nevertheless, the techniques certainly can help explain and corroborate impact if analyzed properly within the study design. Data from many recommended study designs are not easily analyzed by those multivariate techniques, because, for example,

- In stratified random sampling, units from different strata are selected with unequal weights (unequal probability).
- In matched pair designs, the inherent precision created by the pairing is lost if that pair bond is broken.

A complete description of multivariate techniques is beyond the scope of this book and is adequately described in the sources referenced earlier. Multivariate analysis has intuitive appeal to wildlife biologists and ecologists because it deals simultaneously with variables, which is the way the real world works (see Morrison et al.

2006). However, complexity is not always best when trying to understand natural systems. We think it is worth repeating Manly's (1986) precautions:

1. Use common sense when deciding how to analyze data and remember that the primary objective of the analysis is to answer the questions of interest.
2. The complexity of multivariate analysis usually means that answers that are produced are seldom straightforward because the relationship between the observed variables may not be explained by the model selected.
3. As with any method of analysis, a few extreme observations (outliers) may dominate the analysis, especially with a small sample size.
4. Finally, missing values can cause more problems with multivariate data than with univariate data.

The following are examples of multivariate designs in wildlife studies. Miles et al. (2006) used multivariate models to study the multiscale roost site selection by evening bats on pine-dominated landscapes in southwest Georgia. The study developed 16 a priori multivariate models to describe day-roost selection by evening bats, with pooling data across gender and age classes. Model sets included all possible additive combinations of categories that described tree, plot, stand, and landscape scales. The study used logistic regression to create models and the second-order Akaike's Information Criteria (AIC_c) to identify the most parsimonious model and to predict variable importance. Kristina et al. (2006) evaluated habitat use by sympatric mule and white-tailed deer in Texas using *multivariate analysis of variance* (MANOVA) to test for differences and interactions in habitat composition of home ranges, core areas, among years, and between species for males, and among years, seasons, and species for females. Cox et al. (2006) evaluated Florida panther habitat use using a MANOVA to test the hypothesis that overall habitat selection did not differ from random with sex as a main effect and individual panthers as the experimental unit. The study used the same procedure to test for differences in habitat selection between Florida panthers and introduced Texas cougars. Lanszki et al. (2006) evaluated feeding habits and trophic niche overlap between sympatric golden jackal (*Canis aureus*) and red fox (*Vulpes vulpes*) in the Pannonian ecoregion (Hungary). They used a MANOVA to compare the canids in consumption of fresh biomass of prey based on the prey's body mass as the dependent variable, carnivore species as the fixed factor, and seasons and mass categories as covariates.

3.11 Other Designs

3.11.1 Sequential Designs

It is always desirable to use research dollars and time as efficiently as possible. In the study designs covered so far there is an a priori decision on the number of samples taken and there are two potential statistical inferences, accept or reject the null hypothesis. Sequential designs have been proposed as a way of optimizing research dollars. Sequential designs are unique in that the sample size is not fixed before the study begins and there are now three potential statistical inferences, accept, reject, or uncertainty (more data are needed). After each sampling event, the available data are analyzed to determine if conclusions can be reached without additional sampling. The obvious advantage to this approach is the potential savings in dollars and time necessary to conclude a study.

Sequential sampling can be very useful when data are essentially nonexistent on a study population and a priori sample size estimation is essentially a guess. As an example, suppose in a regulatory setting the standard for water quality below a waste treatment facility is survival time for a particular fish species (e.g., fathead minnow). The null hypothesis is that mean survival time is less than the regulatory standard and the alternate hypothesis is greater than equal to the regulatory standard. The primary decision criterion is the acceptable risks for Type I and II errors. Typically, in a regulatory setting the emphasis is placed on reducing the Type I errors (i.e., rejecting a true null hypothesis). Sequential sampling continues until a decision regarding whether the facility is meeting the regulatory standard is possible within the acceptable risk of error.

Biological studies commonly use computer-intensive methods (see Manly 1997). Randomization tests, for example, involve the repeated sampling of a randomization distribution (say 5,000 times) to determine if a sample statistic is significant at a certain level. Manly (1997) suggests that a sequential version of a randomization test offers the possibility of reducing the number of randomizations necessary, potentially saving time and reducing the required computing power. Nevertheless, Manly (1997) advocates the use of a fixed number of randomizations to estimate the significance level rather than determining if it exceeds some prespecified level.

The above discussion of the sequential study design presumes there is comprehensive knowledge of the biology of the population of interest. That is, we know which variables are most important, the range of variables that should be studied, the proper methods and metrics to use, and potential interactions. However, the sequential study can also be thought of at a more global scale. That is, an investigation could begin with a moderately sized experiment followed by reassessment after the first set of results is obtained. The obvious advantage to this approach is that the a priori decisions made regarding the biology of populations and the resulting initial study design are modified based on new information. Adaptive resource management (Walters 1986; see Chap. 2) is popularizing this method of scientific study. Box et al. (1978) advocate "the 25% rule," that is not more than one quarter of the experimental effort (budget) should be invested in a first design. The bottom-line is that when there is a great deal of uncertainty regarding any of the necessary components of the study one should not put all of the proverbial eggs (budget and time) into one basket (study).

3.11.2 Crossover Designs

The crossover design is a close relative of the Latin square and in some instances the analysis is identical (Montgomery 1991). Simply put, crossover designs involve the random assignment of two or more treatments to a study population during the first study period and then the treatments are switched during subsequent study periods so that all study units receive all treatments in sequence. Contrast this with the above designs where treatments are assigned in parallel groups where some subjects get the first treatment and different subjects get the second treatment. The crossover design is typically implemented with a single treatment and control, and represents a special situation where there is not a separate comparison group. In effect, each study unit serves as its own control. In addition, since the same study unit receives both treatments, there is no possibility of covariate imbalance. That is, by assigning all treatments to each of the units crossover designs eliminate effects of variation between experimental units (Williams et al. 2002).

The crossover design can be quite effective when spatially separated controls are unavailable but temporal segregation of treatments is a possibility. However, a key requirement is that the treatments must not have a lasting effect on the study units such that the response in the second allocation of treatments is influenced by the first. This potential for a carry-over effect limits to some extent the type of treatments and study units that can be used in crossover experiments. Typically study units are given some time for recovery (i.e., overcome any potential effects of the first treatment application) before the second treatment phase begins. Williams et al. (2002) describes an analysis procedure that includes a treatment effect, time effect, carry-over effect, and two random terms, one for replication and one that accounts for the sequencing of treatments.

Wolfe et al. (2004) provide a straightforward example of the application of the crossover design in the study of the immobilization of mule deer with the drug Thiafentanil (A-3080). This study utilized a balanced crossover design where each deer was randomly assigned one of two Thiafentanil dose treatments. One treatment was the existing study protocol dose (0.1 mg kg^{-1}), and the other treatment was 2× the protocol dose (0.2 mg kg^{-1}). Treatment assignments were switched for the second half of the experiment so that each animal eventually received both treatments. The first half of the crossover experiment occurred on day 0 of the study and the second half occurred 14 days later to allow the mule deer to recover from the application of the first treatment dose. As another example, a study currently being implemented at the Altamont Pass Wind Resource Area in central California, where a high (>40 per year) number of golden eagles are being killed by wind turbines. The study uses a crossover design to determine if a seasonal shutdown of turbines can be effective in reducing eagle fatalities. A set of turbines are operated during the first half of the winter season while another set is shut down and eagle fatalities are quantified; the on–off turbines are reversed for the second half of the season; and, the same protocol is followed for a second year. The objectives are to see if the overall fatalities in the area decline because of a winter shutdown, to see if winter fatalities decline due to partial shutdown, and to see if variation in fatalities occurs within seasons of operation. Thus, the treatment has been "crossed-over" to the other elements. Power remains low in such experiments, and the experimenter draws conclusions using a weight of evidence approach (where "weight of evidence" simply means you see a pattern in the response).

3.11.3 Quasiexperiments

To this point, we have concentrated on designs that closely follow the principles Fisher (1966) developed for agricultural experiments where the observer can control the events. These principles are the basis for most introductory statistics courses and textbooks. In such courses, there is the implication that the researcher will have a great deal of latitude in the control of experiments. The implication is that experimental controls are often possible and blocking for the partitioning of sources of variance can commonly be used, and the designs of experiments often become quite complicated. The principles provide an excellent foundation for the study of uncontrolled events that include most wildlife studies. However, when wildlife students begin life in the real world, they quickly learn that it is far messier than their statistics professors led them to believe.

Wildlife studies are usually observational with few opportunities for the conduct of replicated manipulative experiments. Studies usually focus on the impact of a perturbation on a population or ecosystem, and fall into the category classified by Eberhardt and Thomas (1991) as studies of uncontrolled events (see Fig. 3.1). The perturbation may be a management method or decision with some control possible or an environmental pollutant with no real potential for control. Even when some control is possible, the ability to make statistical inference to a population is limited. The normal circumstance is for the biologist to create relatively simple models of the real world, exercise all the experimental controls possible, and then, based on the model-based experiments, make subjective conjecture (Eberhardt and Thomas 1991) to the real world.

Regardless of the approach, most of the fundamental statistical principles still apply, but the real world adds some major difficulties, increasing rather than diminishing the need for careful planning. Designing observational studies require the same care as the design of manipulative experiments (Eberhardt and Thomas 1991). Biologists should seek situations in which variables thought to be influential can be manipulated and results carefully monitored (Underwood 1997). When combined with observational studies of intact ecosystems, the results of these experiments increase our understanding of how the systems work. The usefulness of the information resulting from research is paramount in the design of studies and, if ecologists are to be taken seriously by decision-makers, they must provide information useful for deciding on a course of action, as opposed to addressing purely academic questions (Johnson 1995).

The need for quasiexperiments is illustrated by using the existing controversy over the impact of wind power development on birds (Anderson et al. 1999). There is a national desire by consumers for more environmentally friendly sources of energy from so-called "Green Power." Some industry analysts suggest that as much as 20% of the energy needs in the United States could be met by electricity produced by wind plants. As with most technology development, power from wind apparently comes with a cost to the environment. Early studies of the first large wind resource areas in the Altamont Pass and Solano County areas of California by the California Energy Commission (Orloff and Flannery 1992) found unexpectedly high levels of bird fatalities. The resulting questions about the significance of these fatalities to the impacted populations were predictable and led to independent research on wind/bird interactions at these two sites and other wind plants throughout the country (Strickland et al. 1998a,b; Anderson et al. 1996; Howell 1995; Hunt 1995; Orloff and Flannery 1992; Erickson et al. 2002). While these studies look at project-specific impacts, the larger question is what these studies can tell us about potential impacts to birds as this technology expands. The study of the impact of wind power on birds is a classic example of the problems associated with study of uncontrolled events.

First, the distribution of wind plants is nonrandom with respect to bird populations and windy sites. Four conditions are necessary for a wind project to be feasible. There must be a wind resource capable of producing power at rates attractive to potential customers. There must be access to the wind. There must be a market for the power, usually in the form of a contract. Finally, there must be distribution lines associated with a power grid in close proximity. Thence, randomization of the treatment is not possible. Wind plants are large and expensive, and sites with favorable wind are widely dispersed. As a result, replication and contemporary controls are difficult to achieve. Nevertheless, public concern will not allow the industry, its regulators, or the scientific community to ignore the problem simply because Fisher's principles of experimental design are difficult to implement.

A second and more academic example of a quasiexperiment is illustrated by Bystrom et al. (1998) in their whole-lake study of interspecific competition among young predators and their prey. Before their study, most research on the issue occurred on a much smaller scale in enclosures or ponds. Bystrom et al. sought to evaluate the effect of competition from a prey fish (roach, *Rutilus rutilus*) on the recruitment of a predatory fish (perch, *Perca fluviatilis*). The study introduced roach to two of four small, adjacent unproductive lakes inhabited by natural populations of perch. After the introduction, the investigators collected data on diet, growth, and survival of the newborn cohorts of perch during a 13-month period. Several complications were encountered, including the incomplete removal of a second and larger predator (pike, *Esox lucius*) in two of the four lakes and an unfortunate die-off of adult perch in the roach-treatment lakes. A second unreplicated enclosure experiment was conducted in one of the lakes to evaluate intraspecific vs. interspecific competition.

Bystrom et al. (1998) attempted to follow good experimental design principles in their study. The problems they encountered illustrate how difficult experiments in nature really are. They were able to replicate both treatment and control environments and blocked treatment lakes. However, the experiment was conducted with a bare minimum of two experimental units for each treatment. They attempted to control for the effects of the pike remaining after the control efforts by blocking. They also attempted to control for intraspecific competition, but with a separate unreplicated study. It could be argued that a better study would have included replications of the enclosure study in some form of nested design or a design that considered the density of perch as a covariate in their blocked experiment. In spite of a gallant effort, they are left with a study utilizing four subjectively selected lakes from what is likely a very large population of oligotrophic lakes in Sweden and somewhat arbitrary densities of prey and other natural predators. In addition, the two "control" lakes were not true experimental controls and some of the differences seen between the control and treatment conditions no doubt resulted from preexisting differences. It is doubtful that a sample size of two is sufficient replication to dismiss the possibility that differences attributed to the treatment could have occurred by chance. Any extrapolation of the results of this study to other lakes and other populations of perch is strictly a professional judgment; it is subject to the protocols and unique environmental conditions of the original study and is not an exercise of statistical inference.

3.11.3.1 Characteristics of Quasiexperimental Designs

In observational studies of treatment effects, conclusions concerning cause-and-effect relationships are limited. Practically speaking, identical control areas seldom exist and similar *reference* areas must be used instead. Moreover, there is seldom random assignment of treatment, and replication is usually impossible. Oil spills only occur along shipping lanes and power plant sites tend to be unique topographically, geographically, and biologically, and no one would duplicate an oil spill for the sake of science. In the case of an industrial development, where most of the potential construction sites are known, the decision regarding where to locate a new facility never includes a random element in the process. The expense of a new facility or the potential damage caused by a contaminant spill makes replication impractical. Thus, one does not have a true experiment.

Wildlife investigators usually design studies to learn something about some treatment that leads to the prediction of outcomes at unstudied contemporary or future sites with the same or similar treatment (see Sect. 1.2.3.2). For example, from data generated from a probabilistic sample of study plots throughout all oiled areas resulting from an oil spill, the biologist can make statistical (*inductive*) inference to the entire oiled area. The practice of extending the conclusions of wildlife studies beyond the specific study areas to unstudied areas is acceptable, as long as study assumptions are specified and it is clear that the extrapolation is based on expert opinion (*deductive inference*). For example, one can make deductive predictions of the impact of future oil spills in similar areas based on the data from a study of an oil spill. When the extrapolation is presented as an extension of statistical conclusions, it is an improper form of data analysis. In the wind power example, deductive inferences that extend beyond the specific study areas to draw general conclusions about cause-and-effect aspects of operating a wind plant may be possible if enough independent studies of different wind plants identify similar effects. However, *statistical inferences* beyond the study areas are not possible; nor should this be the primary objective of *quasiexperiments*, given the unique aspects of any particular development or ecological inquiry.

3.11.3.2 Examples of Quasiexperimental Designs

The following discussion deals primarily with the study of a distinct treatment or perturbation. These designs fall into the category of *intervention analysis* in Eberhardt and Thomas's (1991) classification scheme. Because these designs typically result in data collected repeatedly over time they are also called an *interrupted time series* (Manly 1992). We do not specifically discuss designs for studies when no distinct treatment or perturbation exists, as these depend on sampling and may be characterized by the way samples are allocated over the area of interest. Sampling plans are covered in detail in Chap. 4.

There are several alternative methods of observational study when estimating the impact of environmental perturbations or the effects of a treatment. The following is a brief description of the preferred designs, approximately in order of reliability for sustaining confidence in the scientific conclusions. A more complete description of these designs can be found in Chap. 6 under the discussion of impact studies and in Manly (1992) and Anderson et al. (1999).

3.11.3.3 Before–After/Control-Impact Design

The *before–after/control-impact* (*BACI*) design is a common design reported in the ecological literature (e.g., Stewart-Oaten 1986), and has been called the "optimal impact study design" by Green (1979). The term *BACI* is so common that we retain the letter *C* in the name, even though we use the term *reference area* rather than *control area*, as true control areas rarely exist. In the BACI design, experimental units are randomly allocated to both treatment and reference areas and populations before the treatment is applied.

The BACI design is desirable for studies of impact or treatment effects because it addresses two major quasiexperimental design problems:

1. Response variables, such as the abundance of organisms, vary naturally through time, so any change observed in a study area between the pretreatment and posttreatment periods could conceivably be unrelated to the treatment (e.g., the construction and operation of a wind plant). Large natural changes are expected during an extended study period.
2. There are always differences in the random variables between any two areas. Observing a difference between treatment and reference areas following the treatment does not necessarily mean that the treatment was the cause of the difference. The difference may have been present prior to treatment. Conversely, one would miss a treatment effect if the levels of the response variable on the reference and treatment areas were the same after the treatment, even though they were different before the treatment.

By collecting data at both reference and treatment areas using exactly the same protocol during both pretreatment and posttreatment periods one can ask the question:

Did the average difference in abundance between the reference area(s) and the treatment area change after the treatment?

The BACI design is not always practical or possible. Adequate reference areas are difficult to locate, the perturbation does not always allow enough time for study before the impact, and multiple times and study areas increase the cost of study. Additionally, alterations in land use or disturbance occurring before and after treatment complicate the analysis of study results. We advise caution when employing this method in areas where potential reference areas are likely to undergo significant changes that potentially influence the response variable of interest. If advanced knowledge of a study area exists, the area of interest is somewhat varied, and the response variable of interest is wide ranging, then the BACI design is preferred for observational studies for treatment effect.

3.11.3.4 Matched Pairs in the BACI Design

Matched pairs of study sites from treatment and reference areas often are subjectively selected to reduce the natural variation in impact indicators (Skalski and Robson 1992; Stewart-Oaten et al. 1986). Statistical analysis of this form of quasiexperiment is dependent on the sampling procedures used for site selection and the amount of information collected on concomitant site-specific variables. For example, sites may be randomly selected from an assessment area and each subjectively matched with a site from a reference area.

When matched pairs are used in the BACI design to study a nonrandom treatment (perturbation), the extent of statistical inferences is limited to the assessment area, and the reference pairs simply act as an indicator of baseline conditions. Inferences also are limited to the protocol by which the matched pairs are selected. If the protocol for selection of matched pairs is unbiased, then statistical inferences comparing the assessment and reference areas are valid and repeatable. For example, McDonald et al. (1995) used this design to evaluate the impacts of the *Exxon Valdez* oil spill on the intertidal communities in Prince William Sound, Alaska. Since the assessment study units were a random sample of oiled units, statistical inferences were possible for all oiled units. However, since the reference units were subjectively selected to match the oiled units, no statistical inferences were possible or attempted to nonoiled units. The selection of matched pairs for extended study contains the risk that sites may change before the study is completed, making the matching inappropriate (see discussion of stratification in Chap. 4). The presumption is that, with the exception of the treatment, the pairs remain very similar – a risky proposition in long-term studies.

3.11.3.5 Impact-Reference Design

The *impact-reference design* quantifies treatment effects through comparison of response variables measured on a treatment area with measurements from one or more reference areas. Studies of the effect of environmental perturbations frequently lack "before" baseline data from the assessment area and/or a reference area requiring an alternative to the BACI, such as the impact-reference design. Assessment and reference areas are censused or randomly subsampled by an appropriate sampling design. Design and analysis of treatment effects in the absence of preimpact data follow Skalski and Robson's (1992) (see Chap. 6) recommendations for accident assessment studies.

Differences between assessment and reference areas measured only after the treatment might be unrelated to the treatment, because site-specific factors differ. For this reason, differences in natural factors between assessment and reference areas should be avoided as much as possible. Although the design avoids the added cost of collecting preimpact data, reliable quantification of treatment effects must include as much temporal and spatial replication as possible. Additional study components, such as the measurement of other environmental covariates that might influence response variables, may help limit or explain variation and the confounding effects of these differences. ANCOVA may be of value to adjust the analysis of a random variable to allow for the effect of another variable.

3.11.3.6 Response-Gradient Design

The *response-gradient design* is useful for quantifying treatment effects in relatively small study areas with homogeneous environments. If the distribution of experimental units is relatively restricted (e.g., small home ranges of passerines) and a response is expected to vary relative to the distance or time from the application of the treatment (gradient of response), this design is an excellent choice for observational studies. When this design is appropriate, treatment effects can usually be estimated with more confidence and associated costs should be less than for those designs requiring baseline data and/or reference areas.

Analysis of the response-gradient design considers the relationship between the response variable and the gradient of treatment levels. For example, in the analysis of an environmental impact, the analysis considers the relationship between the impact indicator and distance from the hypothesized impact source. In effect, the study area includes the treatment area with a reference area on its perimeter. This design does not require that the perimeter of the treatment area be free of effect, only that the level of effect be different. If a gradient of biological response(s) is identified, the magnitude of differences can be presumed to represent at least a minimum estimate of the amount of effect. This response-gradient design would be analogous to a laboratory toxicity test conducted along a gradient of toxicant concentrations. An example might be an increasing rate of fledgling success in active raptor nests or a decrease in passerine mortality as a function of distance to a wind plant.

As in any field study, treatment effects will likely be confounded by the effect of naturally varying factors on response variables. Thus, it is important to have supporting measurements of covariates to help interpret the observed gradient of response. In the example of decreased mortality in passerines associated with a wind plant, an obvious covariate to consider would be level of use of the species of interest.

If one discovers a gradient of response is absent but a portion of the study area meets the requirements of a reference area, data analysis compares the response variables measured in the treatment and control portions of the study area. The impact-gradient design can be used in conjunction with BACI, impact reference, and before–after designs.

3.11.3.7 Before–After Design

The *before–after design* is a relatively weak design, which is appropriate when measurements on the study area before the treatment are compared with measurements on the same area following the treatment. Wildlife managers use long-term monitoring programs to track resources within an area and periodically analyze the resulting data as a before–after designed study. However, observed changes might be unrelated to the treatment, because confounding factors also change with time (see the earlier discussion of the BACI design). Reliable quantification of treatment effects usually include additional study components to limit variation and the confounding effects of natural factors that may change with time.

Because of the difficulty in relating posttreatment differences to treatment effects in the absence of data from reference areas, indirect indicators of treatment effect can be particularly useful in detecting impacts using the before–after design. The correlation of exposure to toxic substances and a physiological response in wildlife has been documented well enough for some substances to allow the use of the physiological response as a *biomarker* for evidence of effect. Examples of biomarkers used in impact studies include the use of blood plasma dehydratase in the study of lead exposure, acetylcholinesterase levels in blood plasma in the study of organophosphates, and the effect of many organic compounds on the microsomal mixed-function oxidase system in liver (Peterle 1991).

Costs associated with conducting the before–after design should be less than those for designs requiring reference areas. Statistical analysis procedures include the time-series method of intervention analysis (Box and Tiao 1975). An abrupt change in the response variable at the time of the treatment may indicate that the response is due to the treatment (e.g., an oil spill) and confidence in this interpretation increases if the response variables return to baseline conditions through time after removal of the treatment. Interpretation of this type of response without reference areas or multiple treatments is difficult and more subjective than the other designs discussed. This type of design is most appropriate for study of short-term perturbations rather than for long-term and ongoing perturbations, such as an industrial development or the study of some persistent contaminant.

3.11.4 Improving Reliability of Study Designs

When studies using reference areas are possible, the use of more than one reference area increases the reliability of conclusions concerning quantification of a treatment response in all designs (Underwood 1994). Multiple reference areas help deal with the frequently heard criticism that the reference area is not appropriate for the treatment area. Consistent relationships among several reference areas and the treatment area will generate far more scientific confidence than if a single reference area is used. In fact, scientific confidence is likely increased more than would be expected given the increase in number of reference areas. This confidence comes from the *replication in space* of the baseline condition. Multiple reference areas also reduce the impact on the study if one reference area is lost, e.g., due to a change in land use affecting response variables.

Collection of data on study areas for several time periods before and/or after the treatment also will enhance reliability of results. This *replication in time* allows the detection of convergence and divergence in the response variables among reference and treatment areas. The data can be tested for interaction among study sites, time, and the primary indicator of effect (e.g., mortality), assuming the data meet the assumptions necessary for ANOVA of repeated measures. The specific test used depends on the response variable of interest (e.g., count data, percentage data, continuous data, categorical data) and the subsampling plan used (e.g., point counts, transect counts, vegetation collection methods, GIS [Geographic Information System] data available, radio-tracking data, capture–recapture data). Often, classic ANOVA procedures will be inappropriate and nonparametric, Bayesian, or other computer-intensive methods will be required.

3.11.5 Model-based Analysis and Use of Site-Specific Covariates

The conditions of the study may not allow a pure design/data-based analysis, particularly in impact studies. For example, animal abundance in an area might be estimated on matched pairs of impacted and reference study sites. However, carefully the matching is conducted, uncontrolled factors always remain that may introduce too much variation in the system to allow one to statistically detect important differences between the assessment and reference areas. In a field study, there likely will be naturally varying factors whose effects on the impact indicators are confounded with the effects of the incident. Data for easily obtainable random variables that are correlated with the impact indicators (covariates) will help interpret the gradient of response observed in the field study. These variables ordinarily will not satisfy the criteria for determining impact, but are often useful in model-based analyses for the prediction of impact (Page et al. 1993; Smith 1979). For example, in the study of bird use on the Wyoming wind plant site, Western Ecosystems Technology, Inc. (1995) developed indices to prey abundance (e.g., prairie dogs [*Cynomys*], ground squirrels [*Spermophilus*], and rabbits [*Lagomorpha*]). These ancillary variables are used in model-based analyses to refine comparisons of avian predator use in assessment and reference areas. Land use also is an obvious covariate that could provide important information when evaluating differences in animal use among assessment and reference areas and time.

Indicators of degree of exposure to the treatment also should be measured on sampling units. As in the response-gradient design, a clear effect–response relationship between response variables and level of treatment will provide corroborating evidence of effect. These indicators are also useful with other concomitant variables in model-based analyses to help explain the "noise" in data from natural systems. For example, in evaluating the effect of an oil spill, the location of the site with regard to prevailing winds and currents or substrate of the oiled site are useful indicators of the degree of oil exposure.

3.12 Meta-analyses

A common practice when embarking on a new investigation is to review the literature on the subject and subjectively assess knowledge about the research question of interest. Typically, in the wildlife research field, one finds numerous independent quasiexperiments. For example, if one is interested in the impact of antler restrictions on deer populations, hunting effects on prairie grouse, or herbicide effects on sagebrush, it might be possible to find studies conducted in several states, or even several studies within states. The resulting review of the literature usually produces a subjective evaluation of what all these independent studies mean, and in a sense is a form of *meta-analysis*. Alternatively, the investigator could compare these independent studies statistically in a quantitative meta-analysis.

A number of procedures exist for statistical meta-analysis. Manly (2001) describes two methods for comparing studies by combining the p-values from several independent studies (Fisher 1970; Folks 1984) using a chi-square analysis for tests of significance. Fisher's approach is simple and provides a test of whether the null hypothesis is false for any of the studies. However, other methods are more appropriate when addressing the more interesting question usually asked by wildlife scientists; that is, is the null hypothesis generally supported when considering all the studies. One common concern when conducting meta-analysis is the potential variation in studies related to the methods and metrics used, independent of the treatment effects (i.e., are we comparing apples and oranges).

An alternative form of meta-analysis used in medical research involves a statistical analysis of data pooled from independent studies on the response to a particular management action. This approach is appealing, but is most appropriate when study methods and metrics are similar among the studies included in the analysis. In both forms of meta-analysis, the rules for deciding to include or exclude studies are of paramount importance.

Conducting meta-analysis on observational studies, the common form of wildlife study, can be useful, but also controversial because of the inherent variability among studies.

Egger et al. (1998) suggest that while formal meta-analysis of observational studies can be misleading if insufficient attention is not given to heterogeneity, it is a desirable alternative to writing highly subjective narrative reviews. They make the logical recommendation that meta-analysis of observational studies should follow many of the principles of systematic reviews: a study protocol should be written in advance, complete literature searches carried out, and studies selected and data extracted in a reproducible and objective fashion. Following this systematic approach exposes both differences and similarities of the studies, allows the explicit formulation and testing of hypotheses, and allows the identification of the need for future studies. Particular with observational studies, meta-analysis should carefully consider the differences among studies and stratify the analysis to account for these differences and for known biases.

Erickson et al. (2002) provide a nice example of a meta-analysis using pooled data from a relatively large group of independent observational studies of the impacts of wind power facilities on birds and bats. The meta-analysis evaluated data on mortality, avian use, and raptor nesting for the purpose of predicting direct impacts of wind facilities on avian resources, including the amount of study necessary for those predictions. The authors considered approximately 30 available studies in their analysis of avian fatalities. In the end, they restricted the fatality and use components of the meta-analysis to the 14 studies that were conducted consistent with recommendations by Anderson et al. (1999). They also restricted their analysis to raptors and waterfowl/waterbird groups because the methods for estimating use appeared most appropriate for the larger birds.

Based on correlation analyses, the authors found that overall impact prediction for all raptors combined would typically be similar after collection of one season of raptor use data compared to a full year of data collection. The authors cautioned that this was primarily the case in agricultural landscapes where use estimates were relatively low, did not vary much among seasons, and mortality data at new wind projects indicated absent to very low raptor mortality. Furthermore, the authors recommended more than one season of data if a site appears to have relatively high raptor use and in landscapes not yet adequately studied.

Miller et al. (2003) reviewed results of 56 papers and subjectively concluded that current data (on roosting and foraging ecology of temperate insectivorous bats) were unreliable due to small sample sizes, short-term nature of studies, pseudoreplication, inferences beyond scale of data, study design, and limitations of bat detectors and statistical analyses. To illustrate the value of a quantitative meta-analysis, Kalcounis-Ruppell et al. (2005) used a series of meta-analyses on the same set of 56 studies to assess whether data in this literature suggested general patterns in roost tree selection and stand characteristics. The authors also repeated their analyses with more recent data, and used a third and fourth series of meta-analyses to separate the studies done on bat species that roost in cavities from those that roost in foliage. The quantitative meta-analysis by Kalcounis-Ruppell et al. (2005) provided a much more thorough and useful analysis of the available literature compared to the more subjective analysis completed by Miller et al. (2003).

3.13 Power and Sample Size Analyses

Traditionally in the analysis of an experiment, a null hypothesis (H_0) is the *straw man* that must be rejected to infer statistically that a response variable has changed or that a cause-and-effect relationship exists. The typical H_0 is that there is no difference in the value of a response variable between control areas and assessment areas or that there is a zero correlation between two response variables along their gradients. In the regulatory setting and in impact studies, this approach usually places the burden of scientific proof of impact on regulators.

The classical use of a H_0 protects only against the probability of a Type I error (also called α, concluding that impact exists when it really does not, i.e., a false positive). By convention the significance level is set at $\alpha = 0.05$ before the conclusion of effect is considered to be valid, although there is nothing magic about 0.05. The probability of a Type II error (also called β, concluding no effect when in fact effect does exist, i.e., a false negative) is almost always unknown, commonly ignored and is often much larger than 0.05. At a given α-level, the risk of a Type II error can be decreased by increasing sample size, reducing sampling error, or, in some situations, through use of better experimental design and/or more powerful types of analysis. In general, the power of a statistical test of some hypothesis is the probability that it rejects the H_0 when it is false $1 - \beta$. An experiment is said to be very powerful if the probability of a Type II error is very small.

As Underwood (1997) points out, it makes intuitive sense to design a study to make equal the probability of making either a Type I or II error. However, he introduces the precautionary principle that the willingness to accept a type of error will depend on the nature of the study. For example, in testing drugs or in environmental monitoring it may be more acceptable to commit a Type I error much more often than a type Type II error. Thus, one would want to design a more powerful study to decrease the probability of concluding no effect when one actually exists.

In summary, four interrelated factors determine statistical power: power increases as sample size, α-level, and effect size increase; power decreases as variance increases. Understanding statistical power requires an understanding of Type I and Type II error, and the relationship of these errors to null and alternative hypotheses. It is important to understand the concept of power when designing a research project, primarily because such understanding grounds decisions about how to design the project, including methods for data collection, the sampling plan, and sample size. To calculate power the researcher must have established a hypothesis to test, understand the expected variability in the data to be collected, decide on an acceptable α-level, and most importantly, a biologically relevant response level.

3.13.1 Effect Size

Effect size is the difference between the null and alternative hypotheses. That is, if a particular management action is expected to cause a change in abundance of an organism by 10%, then the effect size is 10%. Effect size is important in designing experiments for obvious reasons. At a given α-level and sample size, the power of an experiment increases with effect size and, conversely, the sample size necessary to detect an effect typically increases with a decreasing effect size.

Given that detectable effect size decreases with increasing sample size, there comes a condition in most studies that a finding of a statistically significant difference has no biological meaning (for example, a difference in canopy cover of 5% over a sampling range of 30–80%; see Sect. 1.5.3). As such, setting a biologically meaningful effect size is the most difficult and challenging aspect of power analysis and this "magnitude of biological effect" is a hypothetical value based on the researcher's biological knowledge. This point is important in designing a meaningful research project. Nevertheless, the choice of effect size is important and is an absolute necessity before it is possible to determine the power of an experiment or to design an experiment to have a predetermined power (Underwood 1997).

3.13.2 Simple Effects

When the question of interest can be reduced to a single parameter (e.g., differences between two population means or the difference between a single population and a fixed value), establishing effect size is in its simplest form. There are three basic types of simple effects:

- *Absolute effect size* is set when the values are in the same units; for example, looking for a 10 mm difference in wing length between males and females of some species.
- *Relative effect size* is used when temporal or spatial control measures are used and effects are expressed as the difference between in response variable due. As expected, relative effect sizes are expressed as percentages (e.g., the percent increase in a population due to a treatment relative to the control).
- *Standardized effect sizes* are measures of absolute effect size scaled by variance and therefore combine these two components of hypothesis testing (i.e., effect size and variance). Standardized effect sizes are unit-less and are thus comparable across studies. They are, however, difficult to interpret biologically and it is thus usually preferable to use absolute or relative measures of effect size and consider the variance component separately.

3.13.3 Complex Effects

Setting an effect size when dealing with multiple factors or multiple levels of a single factor is a complex procedure involving and examination of the absolute effect size based on the variance of the population means:

$$\sigma^2 = 1/k\sum(\mu_i - \mu_{\text{mean}})^2.$$

Steidl and Thomas (2001) outlined four approaches for establishing effect size in complex situations:

- *Approach 1.* Specify all cell means. In an experiment with three treatments and a control, you might state that you are interested in examining power given a control value of 10 g and treatment yields of 15, 20, and 25 g. Although these statements are easy to interpret, they are also difficult to assign.
- *Approach 2.* Delineate a measure of effect size based on the population variances through experimenting with different values of the means. That is, you experiment with different values of the response variable and reach a conclusion based on what a meaningful variance would be.
- *Approach 3.* Simplify the problem to one of comparing only two parameters. For example, in a one-factor ANOVA you would define a measure of absolute effect size ($\mu_{max} - \mu_{min}$), which places upper and lower bounds on power, each of which can be calculated.
- *Approach 4.* Assess power at prespecified levels of standardized effect size for a range of tests. In the absence of other guidance, it is possible to calculate power at three levels as implied by the adjectives small, medium, and large. This approach is seldom applied in ecological research and is mentioned here briefly only for completeness.

In sum, power and sample size analyses are important aspects of study design, but only so that we can obtain a reliable picture of the underlying distribution of the biological parameters of interest. The statistical analyses that follow provide additional guidance for making conclusions. By setting effect size or just your expectation regarding results (e.g., in an observational study) a priori, the biology drives the process rather than the statistics. That is, the proper procedure is to use statistics to first help guide study design, and later to compliment interpretations. The all too common practice of collecting data, applying a statistical analysis, and then interpreting the outcome misses the needed biological guidance necessary for an adequate study. What you are doing, essentially, is agreeing a priori to accept whatever guidance the statistical analyses provide and then trying to force a biological explanation into that framework. Even in situations where you are doing survey work to develop a list of species occupying a particular location, stating a priori what you expect to find, and the relative order by abundance, provides a biological framework for later interpretation (and tends to reduce the fishing expedition mentality).

Sensitivity analysis can be used to help establish an appropriate effect size. For example, you can use the best available demographic information – even if it is from surrogate species – to determine what magnitude of change in, say, reproductive success will force λ (population rate of increase) above or below 1.0. This value then sets the effect size for prospective power analysis or for use in guiding an observational study (i.e., what difference in nest success for a species would be of interest when studying reproduction along an elevation gradient?). For a primarily observational study, there will usually be information – sometimes qualitative – on the likely distribution and relative abundance of the element of interest (e.g., previous studies, field guides and natural history reports, expert opinion).

3.14 Prospective Power Analysis

A primary defense against weak tests of hypotheses is to perform a prospective power analysis at the start of the research, hopefully following a pilot study (Garton et al. 2005). The first step in the prospective power analysis is to decide on the null hypothesis, alternate hypothesis, and significance level before beginning the investigation (Zar 1998). Power analysis can be used to help make a decision regarding the necessary sample size, or at least inform the investigator of the chances of detecting the anticipated effect size with the resources available. Zar (1998) is a useful reference for methods for estimating the required sample size for most common sampling and experimental designs.

Prospective power analysis is used to:

- Determine the number of replicates or samples necessary to achieve a specified power given the specified effect size, alpha, and variance (scenario 1)
- The power of a test likely to result when the maximum number of replicates or samples that you think can be obtained are gathered (scenario 2)
- The minimum effect size that can be detected given a target power, alpha, variance, and sample size (scenario 3)

Below, we discuss each of these topics as applied to ecological field research:

1. *Scenario 1*. In this scenario you are able to specify the effect size, set α (an easy task relative to setting effect size), and estimate the population variance. We have previously discussed how to establish effect size. Estimating the population variance can be accomplished either through previous work on the element of interest (pilot test or existing literature), or by using estimates from a similar element (e.g., congeneric species). Remember that power analysis is used to provide a starting point for research and is not intended to set a final sample size. Thus, using a range (min, max) of estimates for population variance provides you with a method to estimate what your sample size should be, given the effect size and alpha you have selected.
   If you determine you cannot achieve the desired number of samples using α and effect size you initially selected, then your primary option is to change α and effect size; variance can seldom be modified. Many papers are available that discuss selection of α; we do not review them here. Effect size can be modified, but remember that you must be able to justify the effect size that you set in the publication that follows the research.
2. *Scenario 2*. Here you are asking what you can achieve given the available sampling situation. This is often the situation encountered in wildlife research where

a funding entity (e.g., agency) has developed a request for a study (i.e., Request for Proposal or RFP) that includes a specific sampling location(s), sampling conditions, and a limit to the amount of funding available. By accepting such funding, you are in essence accepting what the resulting power and effect size. You often have the ability, however, to adjust the sampling protocol to ensure that you can address at least part of the study objectives with appropriate rigor. In this scenario you conduct power analysis in an iterative manner using different effect sizes and α-levels to determine what you can achieve with the sample size limits in place (Fig. 3.3).

3. *Scenario 3*. Here you are determining what effect size you can achieve given a target power, α-level, variance, and sample size. As discussed earlier, α can be changed within some reasonable bounds (i.e., a case can usually be made for ≤ 0.15) and variance is set. Here you also are attempting to determine what role sample size has in determining effect size.

In summary, the advantage of prospective power analysis is the insight you gain regarding the design of your study. Moreover, even if you must conduct a study given inflexible design constraints, power analysis provides you with knowledge of the likely rigor of your results.

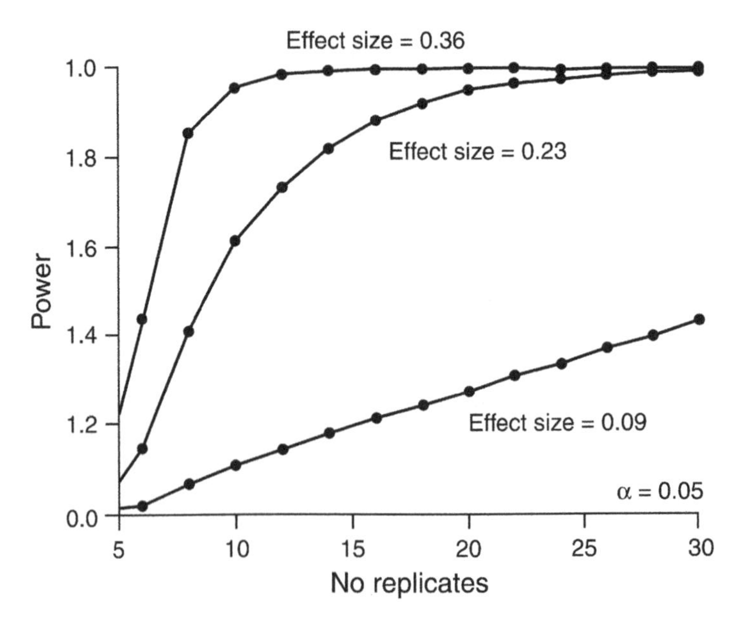

Fig. 3.3 The influence of number of replicates on statistical power to detect small (0.09), medium (0.23), and large (0.36) effect sizes (differences in the probability of predation) between six large and six small trout using a Wilcoxon signed-ranks test. Power was estimated using a Monte Carlo simulation. Reproduced from Steidel et al. (2001) with kind permission from Springer Science + Business Media

3.15 Retrospective Power Analysis

As the names implies, retrospective power analysis is conducted after the study is completed, the data have been collected and analyzed, and the outcome is known. Statisticians typically dismiss retrospective power analysis as being uninformative and perhaps inappropriate and its application is controversial (Gerard et al. 1998). However, in some situations retrospective power analysis can be useful. For example, if a hypothesis was tested and not rejected you might want to know the probability that a Type II error was committed (i.e., did the test have low power?). As summarized by Steidl and Thomas (2001), retrospective power analysis is useful in distinguishing between two reasons for failing to reject the null hypothesis:

- The true effect size was not biologically significant.
- The true effect size was biologically significant but you failed to reject the null hypothesis (i.e., you committed a Type II error).

To make this distinction, you calculate the power to detect a minimally biologically significant effect size given the sample size, α, and variance used in the study. If the resulting power at this effect size is large, then the magnitude of the minimum biologically significant effect would likely lead to statistically significant results. Given that the test was actually not significant, you can infer that the true effect size is likely not this large. If, however, power was small at this effect size, you can infer that the true effect size could be large or small and that your results are inconclusive.

Despite the controversy, retrospective power analysis can be a useful tool in management and conservation. Nevertheless, retrospective power analysis should never be used when power is calculated using the observed effect size. In such cases, the resulting value for power is simply a reexpression of the *p*-value, where low *p*-values lead to high power and vice versa.

3.16 Power Analysis and Wildlife Studies

In practice, observational studies generally have low statistical power. In the case of environmental impact monitoring, the H_0 will usually be that there is no impact to the variable of interest. Accepting a "no impact" result when an experiment has low statistical power may give regulators and the public a false sense of security. The α-level of the experiment is usually set by convention and the magnitude of the effect in an observational study is certainly not controllable. In the case of a regulatory study, the regulation may establish the α-level. Thus, sample size and estimates of variance usually determine the power of observational studies. Many of the methods discussed in this chapter are directed toward reducing variance in observational studies. In properly designed observational studies, the ultimate determinant of statistical power is sample size.

The lack of sufficient sample size necessary to have reasonable power to detect differences between treatment and reference (control) populations is a common problem in observational studies. For example, reasonably precise estimates of direct animal death from a given environmental perturbation may be made through carcass searches. However, tests of other parameters indicating indirect effects for any given impact (e.g., avoidance of a particular portion of their range by a species) may have relatively little power to detect an effect on the species of concern. Most field studies will result in data that must be analyzed with an emphasis on detection of biological significance when statistical significance is marginal. For a more complete study of statistical power, see Cohen (1973), Peterman (1989), Fairweather (1991), Dallal (1992), and Gerard et al. (1998).

The trend of differences between reference and treatment areas for several important variables may detect effects, even when tests of statistical significance on individual variables have marginal confidence. This deductive, model-based approach is illustrated by the following discussion. The evaluation of effects from wind energy development includes effects on individual birds (e.g., reduction or increase in use of the area occupied by the turbines) and population effects such as mortality (e.g., death due to collision with a turbine). Several outcomes are possible from impact studies. For example, a decline in bird use on a new wind plant without a similar decline on the reference area(s) may be interpreted as evidence of an effect of wind energy development on individual birds. The presence of a greater number of carcasses of the same species near turbines than in the reference plots increases the weight of evidence that an effect can be attributed to the wind plant. However, a decline in use of both the reference area(s) and the development area (i.e., area with wind turbines) in the absence of large numbers of carcasses suggests a response unrelated to the wind plant. Data on covariates (e.g., prey) for the assessment and reference area(s) could be used to further clarify this interpretation.

The level at which effects are considered biologically significant is subjective and will depend on the species/resource involved and the research question of interest. Additionally, we note that a biologically significant effect, although not statistically significant, can have population level implications (see Sect. 1.5.3). In the case of bird fatalities, even a small number of carcasses of a rare species associated with the perturbation may be considered significant, particularly during the breeding season. A substantial number of carcasses associated with a decline in use relative to the reference area, particularly late in the breeding season during the dispersal of young, may be interpreted as a possible population effect. The suggestion of a population effect may lead to additional, more intensive studies.

3.17 Sequential Sample Size Analysis

Sequential sample size analysis is primarily a graphical method of evaluating sample size as data are collected, and attempting to justify the sample size collected after the study is completed. While a study is ongoing, you can easily plot the values of any variable of interest as the sample size increases. For example, one might calculate means and variance as every ten vegetation (or habitat use) plots are gathered

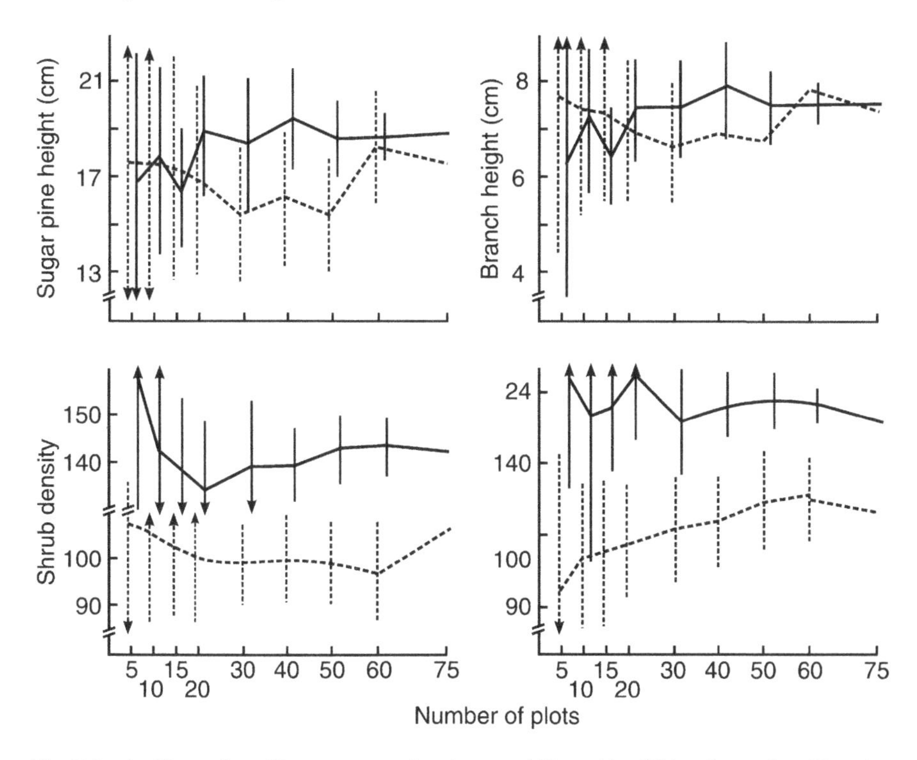

Fig. 3.4 An illustration of how means and variance stabilize with additional sampling. Note that in the all four examples the means (*horizontal solid* and *dashed lines*) and variance (*vertical solid* and *dashed lines*) stabilize with 20–30 plots. Knowledge of the behavior of means and variance influences the amount of sampling in field studies

for a species of interest. You can justify ceasing sampling when the means and variance stabilize (i.e., asymptote; see Fig. 3.4). In a similar fashion, you can take increasingly large random subsamples from a completed data set, calculate the mean and variance, and determine if the values reached an asymptote.

3.18 Bioequivalence Testing

Much has been written criticizing null hypothesis significance testing including applications to wildlife study (see Sect. 1.4.1; Johnson 1999; Steidl and Thomas 2001). McDonald and Erickson (1994), and Erickson and McDonald (1995) describe an alternative approach often referred to as bioequivalence testing. Bioequivalence testing reverses the burden of proof so that a treatment is considered biologically significant until evidence suggests otherwise; thus the role of the null and alternative hypotheses are switched. As summarized by Steidl and Thomas (2001), a minimum effect size that is considered biologically significant is defined.

Then, the alternative hypothesis is stated such that the true effect size is greater than or equal to the minimum effect size that was initially selected. Lastly, the alternative hypothesis is that the true effect size is less than the initial minimum effect size. Thus, Type I error occurs when the researcher concludes incorrectly that no biologically significant difference exists when one does. Recall that this is the type of error addressed by power analysis within the standard hypothesis-testing framework. Bioequivalence testing controls this error rate a priori by setting the α-level of the test. Type II error, however, does remain within this framework when the researcher concludes incorrectly that an important difference exists when one does not.

For a real world example of the significance of value of this alternative approach, consider testing for compliance with a regulatory standard for water quality. In the case of the classic hypothesis testing, poor laboratory procedure resulting in wide confidence intervals could easily lead to a failure to reject the null hypothesis that a water quality standard had been exceeded. Conversely, bioequivalence testing protects against this potentiality and is consistent with the precautionary principle. While this approach appears to have merit, it is not currently in widespread use in wildlife science.

3.19 Effect Size and Confidence Intervals

As discussed earlier, null hypothesis significance testing is problematic because any two samples will usually, show a statistical difference if examined finely enough, such as through increasing sample size (see Sect. 1.4.1). Conversely, no statistical significance will be evident if the sample size was too small or the variance in the data is too great even when differences are biologically important (see Sect. 1.5.3). These scenarios can be distinguished by reporting an estimate of the effect size and its associated confidence interval, thus providing far more biological information than available through a p-value.

Confidence intervals (CI) may be used to test a null hypothesis. When estimated with the data for an observed effect size, a CI represents the likely range of numbers that cannot be excluded as possible values of the true effect size if the study were repeated infinitely into the future with probability $1 - \alpha$. If the $100(1 - \alpha)\%$ CI for the observed effect does not include the value established by the null hypothesis, you can conclude with $100(1 - \alpha)\%$ confidence that a hypothesis test would be statistically significant at level α.. Additionally, CIs provide more information than a hypothesis test because they establish approximate bounds on the likely value of the true effect size. Figure 3.5 (from Steidl and Thomas 2001) presents the possible various hypothetical observed effects and their associated $100(1 - \alpha)\%$ CI. Note that when the vertical CI line crosses the solid horizontal line (zero effect), no statistically significant effect has occurred.

Case A – the CI for the estimated effect excludes 0 effect and includes only biologically significant effects; the study is both statistically and biologically significant.

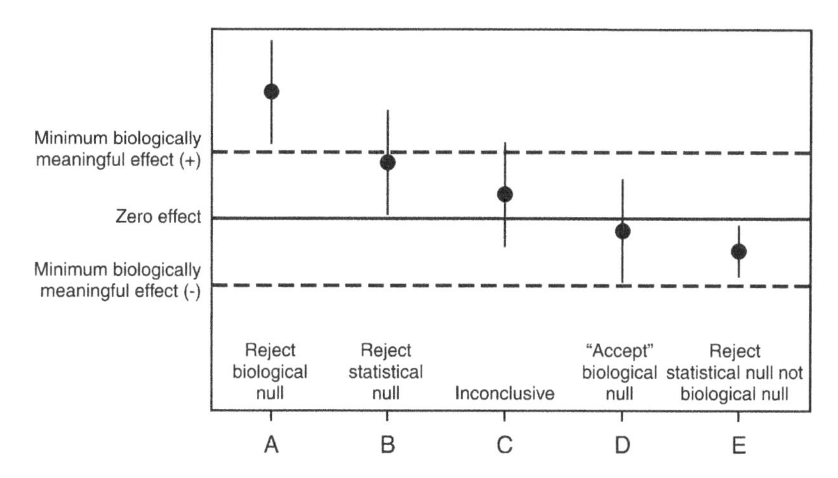

Fig. 3.5 Hypothetical observed effects (*circles*) and their associated $100(1-\alpha)\%$ confidence intervals. The *solid line* represents zero effect, and *dashed lines* represent minimum biologically important effects. In case A, the confidence interval for the estimated effect excludes zero effect and includes only biologically important effects, so the study is both statistically and biologically important. In case B, the confidence interval excludes zero effect, so the study is statistically significant; however, the confidence interval also includes values below those thought to be biologically important, so the study is inconclusive biologically. In case C, the confidence interval includes zero effect and biologically important effects, so the study is both statistically and biologically inconclusive. In case D, the confidence interval includes zero effect but excludes all effects considered biologically important, so the "practical" null hypothesis of no biologically important effect can be accepted with $100(1-\alpha)\%$ confidence. In case E, the confidence interval excludes zero effect but does not include effects considered biologically important, so the study is statistically but not biologically important. Reproduced from Steidel et al. (2001) with kind permission from Springer Science + Business Media

Case B – the CI excludes 0 so it is statistically significant, but includes values that are below that thought to be biologically significant; the study is thus inconclusive biologically.

Case C – the CI includes 0 effect and biologically significant effects, so it is inconclusive statistically.

Case D – the CI includes 0 effect but excludes all effects considered biologically significant; thus the null hypothesis of no biologically significant effect cannot be rejected.

Case E – the CI excludes 0 effect but does not include effects considered biologically significant; the study is statistically but not biologically significant. This situation often occurs when you have very large sample sizes – note now the CI has narrowed.

In CI estimation, the focus is on establishing plausible bounds on the true effect size and determining whether biologically significant effect sizes are contained within those bounds. In retrospective power analysis, however, the focus is on the probability of obtaining a statistically significant result if the effect sizes were truly biologically significant. Steidl and Thomas (2001) concluded that the CI approach was preferable because interpretation of results is relatively straightforward, more informative, and viewed from a biological rather than a probabilistic context.

3.20 Making Adjustments When Things Go Wrong

As in much of life, things can and often do go wrong in the best-designed studies. The following are a few case studies that illustrate adjustments that salvage a study when problems occur.

Case 1 – As previously discussed, Sawyer (2006) conducted a study to determine the impact of gas development on habitat use and demographics of mule deer in southwestern Wyoming. Although the study of habitat use clearly demonstrated a decline in use of otherwise suitable habitat, the lack of a suitable control hampered identification of the relationship of this impact to population demographics. Sawyer (2006) established a reference area early in the study based on historical data supplemented by aerial surveys during a pilot study period. While the impact area boundary remained suitable over the course of the 4-year study, the boundary around the control area turned out to be inadequate. That is, each year the deer distribution was different, resulting in the need for continually expanding the area being surveyed as a control. Thus, even though the numbers of deer remained relatively unchanged in the reference area, the fact that the boundaries continued to change made a comparison of abundance and other demographic characteristics between the control and impact area problematic. Demographic data for the deer within the impact area did show declines in reproductive rate and survival, although the reductions were not statistically different from 0. Additionally, emigration rates did not satisfactorily explain the decline in deer numbers in the impact area. Finally, simulations using the range of reproduction and survival measured in the impact area suggested that those declines, while not statistically significant could, when combined with emigration rates explain the decline in deer numbers. While there is still opportunity for confounding and cause and effect is still strictly professional judgment, the weight of evidence suggests that the loss in effective habitat caused by the gas development may have resulted in a decline in deer abundance and supports a closer look at the impact of gas development on mule deer in this area.

Case 2 – McDonald (2004) surveyed statisticians and biologists, and reported successes and failures in attempts to study rare populations. One of the survey respondents, Lowell Diller (Senior Biologist, Green Diamond Resource Company, Korbel, California, USA) suggested that "A rare population is one where it is difficult to find individuals, utilizing known sampling techniques, either because of small numbers, secretive and/or nocturnal behavior, or because of clumped distribution over large ranges, i.e., a lot of zeros occur in the data. Therefore, a rare population is often conditional on the sampling techniques available." Lowell provided an illustration of his point by describing surveys conducted for snakes during the mid-1970s on the Snake River Birds of Prey Area in southern Idaho. Surveys were being conducted for night snakes (*Hypsiglena torquata*), which were thought to be one of the rarest snakes in Idaho with only four known records for the state. His initial surveys, using standard collecting techniques for the time (turning rocks and such along some transect, or driving roads at night), confirmed that night snakes were very rare. In the second year of his study, however, he experimented with drift fences and funnel traps and suddenly began capturing numerous night snakes. They turned out to be the most common snakes in certain habitats and were the third most commonly captured snake within the entire study area. This case study illustrates two points, unsuccessful surveys may be the result of "common wisdom" being incorrect, and/or standard techniques may be ineffective for some organisms and/or situations.

Case 3 – The Coastal Habitat Injury Assessment started immediately after the EVOS in 1989 with the selection of heavily oiled sites for determining the rate of recovery. To allow an estimate of injury, the entire oiled area was divided into 16 strata based on the substrate type (exposed bedrock, sheltered bedrock, boulder/cobble, and pebble/gravel) and degree of oiling (none, light, moderate, and heavy). Four sites were then selected from each of the 16 strata for sampling to estimate the abundance of more than a thousand species of animals and plants. The stratification and site selection were all based on the information in a geographical information system (GIS). Unfortunately, some sites were excluded from sampling because of their proximity to active eagle nests, and more importantly, many of the oiling levels were misclassified and some of the unoiled sites were under the influence of freshwater dramatically reducing densities of marine species. So many sites were misclassified by the GIS system that the initial study design was abandoned in 1990. Alternatively, researchers matched each of the moderately and heavily oiled sites sampled in 1989 with a comparable unoiled control site based on physical characteristics, resulting in a paired comparison design. The Trustees of Natural Resources Damage Assessment, the state of Alaska and the US Government, estimated injury by determining the magnitude of difference between the paired oiled and unoiled sites (Highsmith et al. 1993; McDonald et al. 1995; Harner et al. 1995). Manley (2001) provides a detailed description of the rather unusual analysis of the resulting data.

McDonald (2004) concluded that the most important characteristics of successful studies are (1) they trusted in random sampling, systematic sampling with a random start, or some other probabilistic sampling procedure to spread the initial sampling effort over the entire study area and (2) they used appropriate field procedures to increase detection and estimate the probability of detection of individuals on sampled units. It seems clear that including good study design principles in the initial study as described in this chapter increases the chances of salvaging a study when things go wrong.

3.21 Retrospective Studies

As the name implies, a retrospective study is an observational study that looks backward in time. Retrospective studies can be an analysis of existing data or a study of events that have already occurred. For example, we find data on bird fatalities from several independent surveys of communications towers and we figure out why they died. Similarly, we design a study to determine the cause of fatalities in an area that has been exposed to an oil spill. A retrospective study can address specific statistical hypotheses relatively rapidly, because data are readily available or already in hand; all we need to do is analyze the data and look for apparent treatment effects and correlations. In the first case, the birds are already dead; we just have to tabulate all the results and look at the information available for each communications tower. Numerous mensurative experiments used to test hypotheses are retrospective in nature (See Sinclair 1991; Nichols 1991); and, medical research on human diseases is usually a retrospective study. Retrospective studies are opposed to prospective studies, designed studies based on a priori hypotheses about events that have not yet occurred.

Retrospective studies are common in ecology and are the only option in most post hoc impact assessments. Williams et al. (2002) offer two important caveats to the interpretation of retrospective studies. First, inferences from retrospective studies are weak, primarily because response variables may be influenced by unrecognized and unmeasured covariates. Second, patterns found through mining the data collected during a retrospective study are often used to formulate a hypothesis that is then tested with the same data. This second caveat brings to mind two comments Lyman McDonald heard Wayne Fuller make at a lecture at Iowa State University. The paraphrased comments are that "the good old data are not so good" and "more will be expected from the data than originally designed." In general, data mining should be avoided or used as to develop hypotheses that are tested with newly obtained empirical data. Moreover, all the above study design principles apply to retrospective studies.

3.22 Summary

Wildlife studies may include manipulative experiments, quasiexperiments, or mensurative or observational studies. With manipulative experiments there is much more control of the experimental conditions; there are always two or more different experimental units receiving different treatments; and there is a random application of treatments. Observational studies involve making measurements of uncontrolled events at one or more points in space or time with space and time being the only experimental variable or treatment. Quasiexperiments are observational studies where some control and randomization may be possible. The important point here is that all these studies are constrained by a specific protocol designed to answer specific questions or address hypotheses posed prior to data collection and analysis.

Once a decision is made to conduct research there are a number of practical considers including the area of interest, time of interest, species of interest, potentially confounding variables, time available to conduct studies, budget, and the magnitude of the anticipated effect.

Single-factor designs are the simplest and include both paired and unpaired experiments of two treatments or a treatment and control. Adding blocking, including randomized block, incomplete block, and Latin squares designs further complicates the completely randomized design. Multiple designs include factorial experiments, two-factor experiments and multifactor experiments. Higher order designs result from the desire to include a large number of factors in an experiment. The object of these more complex designs is to allow the study of as many factors as possible while conserving observations. Hierarchical designs as the name implies increases complexity by having nested experimental units, for example split-plot and repeated measures designs. The price of increased complexity is a reduction in effective sample size for individual factors in the experiment.

ANCOVA uses the concepts of ANOVA and regression to improve studies by separating treatment effects on the response variable from the effects of covariates. ANCOVA can also be used to adjust response variables and summary statistics (e.g., treatment means), to assist in the interpretation of data, and to estimate missing data.

Multivariate analysis considers several related random variables simultaneously, each one considered equally important at the start of the analysis. This is particularly important in studying the impact of a perturbation on the species composition and community structure of plants and animals. Multivariate techniques include multidimensional scaling and ordination analysis by methods such as principal component analysis and detrended canonical correspondence analysis.

Other designs frequently used to increase efficiency, particularly in the face of scarce financial resources, or when manipulative experiments are impractical include sequential designs, crossover designs, and quasiexperiments. Quasiexperiments are designed studies conducted when control and randomization opportunities are possible, but limited. The lack of randomization limits statistical inference to the study protocol and inference beyond the study protocol is usually expert opinion. The BACI study design is usually the optimum approach to quasiexperiments. Meta-analysis of a relatively large number of independent studies improves the confidence in making extrapolations from quasiexperiments.

An experiment is statistically very powerful if the probability of concluding no effect when in fact effect does exist is very small. Four interrelated factors determine statistical power: power increases as sample size, α-level, and effect size increase; power decreases as variance increases. Understanding statistical power requires an understanding of Type I and Type II error, and the relationship of these errors to null and alternative hypotheses. It is important to understand the concept of power when designing a research project, primarily because such understanding grounds decisions about how to design the project, including methods for data collection, the sampling plan, and sample size. To calculate power the researcher must have established a hypothesis to test, understand the expected variability in the data to be collected, decide on an acceptable α-level, and most importantly, a biologically relevant response level. Retrospective power analysis occurs after the study is completed, the data have been collected and analyzed, and with a known outcome. Statisticians typically dismiss retrospective power analysis as being uninformative and perhaps inappropriate and its application is controversial, although it can be useful in some situations.

Bioequivalence testing, an alternative to the classic null hypothesis significance testing reverses the burden of proof and considers the treatment biologically significant until evidence suggests otherwise; thus switching the role of the null and alternative hypotheses. The use of estimation and confidence intervals to examine treatment differences is also an effective alternative to null hypothesis testing and often provides more information about the biological significance of a treatment.

Regardless of the care taken, the best-designed experiments can and many will go awry. The most important characteristics of successful studies include (1) they trusted in random sampling, systematic sampling with a random start, or some other probabilistic sampling procedure to spread the initial sampling effort over the entire study area and (2) they used an appropriate field procedures to increase detection and estimate the probability of detection of individuals on sampled units. It seems clear that including good study design principles in the initial study as described in this chapter increases the chances of salvaging a study when things go wrong.

Study designs must be study-specific. The feasibility of different study designs will be strongly influenced by characteristics of the different designs and by the available opportunities for applying the treatment (i.e., available treatment structures). Other, more practical considerations include characteristics of study subjects, study sites, the time available for the study, the time period of interest, the existence of confounding variables, budget, and the level of interest in the outcome of the study by others. Regardless of the study environment, all protocols should follow good scientific methods. Even with the best of intentions, though, study results will seldom lead to clear-cut statistical inferences.

There is no single combination of design and treatment structures appropriate for all situations. Our advice is to seek assistance from a statistician and let common sense be your guide.

References

Anderson, R. L., J. Tom, N. Neumann, and J. A. Cleckler. 1996. Avian Monitoring and Risk Assessment at Tehachapi Pass Wind Resource Area, California. Staff Report to California Energy Commission, Sacramento, CA, November, 1996.

Anderson, R. L., M. L. Morrison, K. Sinclair, and M. D. Strickland. 1999. Studying Wind Energy/ Bird Interactions: A Guidance Document. Avian Subcommittee of the National Wind Coordinating Committee, Washington, DC.

Barrett, M. A., and P. Stiling. 2006. Key deer impacts on hardwood hammocks near urban areas. J. Wildl. Manage. 70(6): 1574–1579.

Bates, J. D., R. F. Miller, and T. Svejcar. 2005. Long-term successional trends following western juniper cutting. Rangeland Ecol. Manage. 58(5): 533–541.

Berenbaum, M. R., and A. R. Zangerl. 2006. Parsnip webworms and host plants at home and abroad: Trophic complexity in a geographic mosaic. Ecology 87(12): 3070–3081.

Borgman, L. E., J. W. Kern, R. Anderson-Sprecher, and G. T. Flatman. 1996. The sampling theory of Pierre Gy: Comparisons, implementation, and applications for environmental sampling, in

L. H. Lawrence, Ed. Principles of Environmental Sampling, 2nd Edition, pp. 203–221. ACS Professional Reference Book, American Chemical Society, Washington, DC.
Box, G. E. P., and B. C. Tiao. 1975. Intervention analysis with applications to economic and environmental problems. J. Am. Stat. Assoc. 70: 70–79.
Box, G. E. P., W. G. Hunter, and J. S. Hunter. 1978. Statistics for Experimenters, an Introduction to Design, Data Analysis, and Model Building. Wiley, New York.
Bystrom, P., L. Persson, and E. Wahlstrom. 1998. Competing predators and prey: Juvenile bottlenecks in whole-lake experiments. Ecology 79(6): 2153–2167.
Cade, T. J. 1994. Industry research: Kenetech wind power. In Proceedings of National Avian-Wind Power Planning Meeting, Denver, CO, 20–21 July 1994, pp. 36–39. Rpt. DE95-004090. Avian Subcommittee of the National Wind Coordinating Committee, % RESOLVE Inc., Washington, DC, and LGL Ltd, King City, Ontario.
Cochran, W. G. 1977. Sampling Techniques, 3rd Edition. Wiley, New York.
Cochran, W. G., and G. Cox. 1957. Experimental Designs, 2nd Edition. Wiley, New York.
Cohen, J. 1973. Statistical power analysis and research results. Am. Educ. Res. J. 10: 225–229.
Cox, D. R. 1958. Planning of Experiments (Wiley Classics Library Edition published 1992). Wiley, New York.
Cox, J. J., D. S. Maehr, and J. L. Larkin. 2006. Florida panther habitat use: New approach to an old problem. J. Wildl. Manage. 70(6): 1778–1785.
Crowder, M. J., and D. J. Hand. 1990. Analysis of Repeated Measures. Chapman and Hall, London.
Dallal, G. E. 1992. The 17/10 rule for sample-size determinations (letter to the editor). Am. Stat. 46: 70.
Dillon, W. R., and M. Goldstein. 1984. Multivariate analysis methods and applications. Wiley, New York.
Eberhardt, L. L., and J. M. Thomas. 1991. Designing environmental field studies. Ecol. Monogr. 61: 53–73.
Egger, M., M. Schneider, and D. Smith. 1998. Meta-analysis spurious precision? Meta-analysis of observations studies. Br. Med. J. 316: 140–144.
Erickson, W. P., and L. L. McDonald. 1995. Tests for bioequivalence of control media and test media in studies of toxicity. Environ. Toxicol. Chem. 14: 1247–1256.
Erickson, W., G. Johnson, D. Young, D. Strickland, R. Good, M. Bourassa, K. Bay, and K. Sernka. 2002. Synthesis and Comparison of Baseline Avian and Bat Use, Raptor Nesting and Mortality Information from Proposed and Existing Wind Developments. Prepared by Western EcoSystems Technology, Inc., Cheyenne, WY, for Bonneville Power Administration, Portland, OR. December 2002 [online]. Available: http://www.bpa.gov/Power/pgc/wind/Avian_and_Bat_Study_12-2002.pdf
Fairweather, P. G. 1991. Statistical power and design requirements for environmental monitoring. Aust. J. Mar. Freshwat. Res. 42: 555–567.
Fisher, R. A. 1966. The Design of Experiments, 8th Edition. Hafner, New York.
Fisher, R. A. 1970. Statistical Methods for Research Workers, 14th Edition. Oliver and Boyd, Edinburgh.
Flemming, R. M., K. Falk, and S. E. Jamieson. 2006. Effect of embedded lead shot on body condition of common eiders. J. Wildl. Manage. 70(6): 1644–1649.
Folks, J. L. 1984. Combination of independent tests, in P. R. Krishnaiah and P. K. Sen, Eds. Handbook of Statistics 4, Nonparametric Methods, pp. 113–121. North-Holland, Amsterdam.
Garton, E. O., J. T. Ratti, and J. H. Giudice. 2005. Research and experimental design, in C. E. Braun, Ed. Techniques for Wildlife Investigation and Management, 6th Edition, pp. 43–71. The Wildlife Society, Bethesda, Maryland, USA.
Gasaway, W. C., S. D. Dubois, and S. J. Harbo. 1985. Biases in aerial transect surveys for moose during May and June. J. Wildl. Manage. 49: 777–784.
Gerard, P. D., D. R. Smith, and G. Weerakkody. 1998. Limits of retrospective power analysis. J. Wildl. Manage. 62: 801–807.

Gilbert, R. O. 1987. Statistical Methods for Environmental Pollution Monitoring. Van Nostrand Reinhold, New York.

Gilbert, R. O., and J. C. Simpson. 1992. Statistical methods for evaluating the attainment of cleanup standards. Vol. 3, Reference-Based Standards for Soils and Solid Media. Prepared by Pacific Northwest Laboratory, Battelle Memorial Institute, Richland, WA, for U.S. Environmental Protection Agency under a Related Services Agreement with U.S. Department of Energy, Washington, DC. PNL-7409 Vol. 3, Rev. 1/UC-600.

Gitzen, R. A., J. J. Millspaugh, and B. J. Kernohan. 2006. Bandwidth selection for fixed-kernel analysis of animal utilization distributions. J. Wildl. Manage. 70(5): 1334–1344.

Glass, G. V., P. D. Peckham, and J. R. Sanders. 1972. Consequences of failure to meet assumptions underlying the fixed effects analyses of variance and covariance. Rev. Educ. Res. 42: 237–288.

Gordon, A. D. 1981. Classification. Chapman and Hall, London.

Green, R. H. 1979. Sampling Design and Statistical Methods for Environmental Biologists. Wiley, New York.

Green, R. H. 1984. Some guidelines for the design of biological monitoring programs in the marine environment, in H. H. White, Ed. Concepts in Marine Pollution Measurements, pp. 647–655. University of Maryland, College Park. MD.

Harner, E. J., E. S. Gilfillan, and J. E. O'Reilly. 1995. A comparison of the design and analysis strategies used in assessing the ecological consequences of the Exxon Valdez. Paper presented at the International Environmetrics Conference, Kuala Lumpur, December 1995.

Herring, G., and J. A. Collazo. 2006. Lesser scaup winter foraging and nutrient reserve acquisition in east-central Florida. J Wildl. Manage. 70(6): 1682–1689.

Highsmith, R. C., M. S. Stekoll, W. E. Barber, L. Deysher, L. McDonald, D. Strickland, and W. P. Erickson. 1993. Comprehensive assessment of coastal habitat, final status report. Vol. I, Coastal Habitat Study No. 1A. School of Fisheries and Ocean Sciences, University of Fairbanks, AK.

Howell, J. A. 1995. Avian Mortality at Rotor Swept Area Equivalents. Altamont Pass and Montezuma Hills, California. Prepared for Kenetech Windpower (formerly U.S. Windpower, Inc.), San Francisco, CA.

Huitema, B. E. 1980. The Analysis of Covariance and Alternatives. Wiley, New York.

Hunt, G. 1995. A Pilot Golden Eagle population study in the Altamont Pass Wind Resource Area, California. Prepared by Predatory Bird Research Group, University of California, Santa Cruz CA, for National Renewable Energy Laboratory, Golden, CO. Rpt. TP-441-7821.

Hurlbert, S. H. 1984. Pseudoreplication and the design of ecological field experiments. Ecol. Monogr. 54: 187–211.

James, F. C., and C. E. McCulloch. 1990. Multivariate analysis in ecology and systematics: Panacea or Pandora's box? Annu. Rev. Ecol. Syst. 21: 129–166.

Johnson, D. H. 1995. Statistical sirens: The allure of nonparametrics. Ecology 76: 1998–2000.

Johnson, D. H. 1999. The insignificance of statistical significance testing. J. Wildl. Manage. 63(3): 763–772.

Johnson, B., J. Rogers, A. Chu, P. Flyer, and R. Dorrier. 1989. Methods for Evaluating the Attainment of Cleanup Standards. Vol. 1, Soils and Solid Media. Prepared by WESTAT Research, Inc., Rockville, MD, for U.S. Environmental Protection Agency, Washington, DC. EPA 230/02-89-042.

Kalcounis-Ruppell, M. C., J. M. Psyllakis, and R. M. Brigham. 2005. Tree roost selection by bats: An empirical synthesis using meta-analysis. Wildl. Soc. Bull. 33(3): 1123–1132.

Kempthorne, O. 1966. The Design and Analysis of Experiments. Wiley, New York.

Krebs, C. J. 1989. Ecological Methodology. Harper and Row, New York.

Kristina, J. B., B. B. Warren, M. H. Humphrey, F. Harwell, N. E., Mcintyre, P. R. Krausman, and M. C. Wallace. 2006. Habitat use by sympatric mule and white-tailed deer in Texas. J. Wildl. Manage. 70(5): 1351–1359.

Lanszki, J. M., M. Heltai, and L. Szabo. 2006. Feeding habits and trophic niche overlap between sympatric golden jackal (Canis aureus) and red fox (Vulpes vulpes) in the Pannonian ecoregion (Hungary). Can. J. Zool. 84: 1647–1656.

Ludwig, J. A., and J. F. Reynolds. 1988. Statistical Ecology: A Primer on Methods and Computing. Wiley, New York.
Manly, B. F. J. 1986. Multivariate Statistical Methods: A Primer. Chapman and Hall, London.
Manly, B. F. J. 1991. Randomization and Monte Carlo Methods in Biology. Chapman and Hall, London.
Manly, B. F. J. 1992. The Design and Analysis of Research Studies. Cambridge University Press, Cambridge.
Manly, B. F. J. 1997. Randomization, Bootstrap and Monte Carlo Methods in Biology, 2nd Edition, 300 pp. Chapman and Hall, London (1st edition 1991, 2nd edition 1997).
Manly, B. F. J. 2001. Statistics for environmental science and management. Chapman and Hall/CRC, London.
Martin, L. M., and B. J. Wisley. 2006. Assessing grassland restoration success: Relative roles of seed additions and native ungulate activities. J. Appl. Ecol. 43: 1098–1109.
McDonald, L. L. 2004. Sampling rare populations, in W. L. Thompson, Ed. Sampling Rare or Elusive Species, pp. 11–42. Island Press, Washington, DC.
McDonald, L. L., and W. P. Erickson. 1994. Testing for bioequivalence in field studies: Has a disturbed site been adequately reclaimed?, in D. J. Fletcher and B. F. J. Manly, Eds. Statistics in Ecology and Environmental Monitoring, pp. 183–197. Otago Conference Series 2, Univ. Otago Pr., Dunedin, New Zealand.
McDonald, L. L., W. P. Erickson, and M. D. Strickland. 1995. Survey design, statistical analysis, and basis for statistical inferences in Coastal Habitat Injury Assessment: *Exxon Valdez* Oil Spill, in P. G. Wells, J. N. Butler, and J. S. Hughes, Eds. *Exxon Valdez* Oil Spill: Fate and Effects in Alaskan Waters. ASTM STP 1219. American Society for Testing and Materials, Philadelphia, PA.
McKinlay, S. M. 1975. The design and analysis of the observational study – A review. J. Am. Stat. Assoc. 70: 503–518.
Mead, R., R. N. Curnow, and A. M. Hasted. 1993. Statistical Methods in Agriculture and Experimental Biology, 2nd Edition. Chapman and Hall, London.
Mieres, M. M., and L. A. Fitzgerald. 2006. Monitoring and managing the harvest of tegu lizards in Paraguay. J. Wildl. Manage. 70(6): 1723–1734.
Miles, A. C., S. B. Castleberry, D. A. Miller, and L. M. Conner. 2006. Multi-scale roost site selection by evening bats on pine-dominated landscapes in southwest Georgia. J. Wildl. Manage. 70(5): 1191–1199.
Miller, D. A., E. B. Arnett, and M. J. Lacki. 2003. Habitat management for forest-roosting bats of North America: A critical review of habitat studies. Wildl. Soc. Bull. 31: 30–44.
Milliken, G. A., and D. E. Johnson. 1984. Analysis of Messy Data. Van Nostrand Reinhold, New York.
Montgomery, D. C. 1991. Design and Analysis of Experiments, 2nd Edition. Wiley, New York.
Morrison, M. L., G. G. Marcot, and R. W. Mannan. 2006. Wildlife–Habitat Relationships: Concepts and Applications, 2nd Edition. University of Wisconsin Press, Madison, WI.
National Research Council. 1985. Oil in the Sea: Inputs, Fates, and Effects. National Academy, Washington, DC.
Nichols, J. D. 1991. Extensive monitoring programs viewed as long-term population studies: The case of North American waterfowl. Ibis 133(Suppl. 1): 89–98.
Orloff, S., and A. Flannery. 1992. Wind Turbine Effects on Avian Activity, Habitat Use, and Mortality in Altamont Pass and Solano County Wind Resource Areas. Prepared by Biosystems Analysis, Inc., Tiburon, CA, for California Energy Commission, Sacramento, CA.
Page, D. S., E. S. Gilfillan, P. D. Boehm, and E. J. Harner. 1993. Shoreline ecology program for Prince William Sound, Alaska, following the *Exxon Valdez* oil spill: Part 1 – Study design and methods [Draft]. Third Symposium on Environmental Toxicology and Risk: Aquatic, Plant, and Terrestrial. American Society for Testing and Materials, Philadelphia, PA.
Peterle, T. J. 1991. Wildlife Toxicology. Van Nostrand Reinhold, New York.
Peterman, R. M. 1989. Application of statistical power analysis on the Oregon coho salmon problem. Can. J. Fish. Aquat. Sci. 46: 1183–1187.

Pielou, E. C. 1984. The Interpretation of Ecological Data: A Primer on Classification and Ordination. Wiley, New York.

Samual, M. D., E. O. Garton, M. W. Schlegel, and R. G. Carson. 1987. Visibility bias during aerial surveys of elk in northcentral Idaho. J. Wildl. Manage. 51: 622–630.

Sawyer, H., R. M. Nielson, F. Lindzey, and L. L. McDonald. 2006. Winter habitat selection of mule deer before and during development of a natural gas field. J. Wildl. Manage. 70: 396–403.

Scheaffer, R. L., W. Mendenhall, and L. Ott. 1990. Elementary Survey Sampling. PWS-Kent, Boston.

Seber, G. A. F. 1984. Multivariate Observations. Wiley, New York.

Shenk, T. M., A. B. Franklin, and K. R. Wilson. 1996. A model to estimate the annual rate of golden eagle population change at the Altamont Pass Wind Resource Area. In Proceedings of National Avian-Wind Power Planning Meeting II, Palm Springs, California, 20–22 September 1995, pp. 47–54. Proceedings prepared for the Avian Subcommittee of the National Wind Coordinating Committee Washington, DC, by LGL Ltd, King City, Ontario.

Sinclair, A. R. E. 1991. Science and the practice of wildlife management. J. Wildl. Manage. 55: 767–773.

Skalski, J. R., and D. S. Robson. 1992. Techniques for Wildlife Investigations: Design and Analysis of Capture Data. Academic, San Diego, CA.

Smith, W. 1979. An oil spill sampling strategy, in R. M. Cormack, G. P. Patil, and D. S. Robson, Eds. Sampling Biological Populations, pp. 355–363. International Co-operative Publishing House, Fairland, MD.

Steel, R. G. D., and J. H. Torrie. 1980. Principles and Procedures of Statistics: A Biometrical Approach, 2nd Edition. McGraw-Hill, New York.

Steidl, R. J. and L. Thomas. 2001. Power analysis and experimental design, in Scheiner, S. M. and J. Gurevitch, Eds. Design and Analysis of Ecological Experiments, 2nd Edition, pp 14–36. Oxford University Press, New York.

Stekoll, M. S., L. Deysher, R. C. Highsmith, S. M. Saupe, Z. Guo, W. P. Erickson, L. McDonald, and D. Strickland. 1993. Coastal Habitat Injury Assessment: Intertidal communities and the *Exxon Valdez* oil spill. Presented at the *Exxon Valdez* Oil Spill Symposium, February 2–5, 1993, Anchorage, AK.

Stewart-Oaten, A. 1986. The Before–After/Control-Impact-Pairs Design-for Environmental Impact. Prepared for Marine Review Committee, Inc., Encinitas, CA.

Stewart-Oaten, A., W. W. Murdoch, and K. R. Parker. 1986. Environmental impact assessment: "Pseudoreplication" in time? Ecology 67: 929–940.

Stoner, D. C., M. L. Wolfe, and D. M. Choate. 2006. Cougar exploitation levels in Utah: Implications for demographic structure, population recovery, and metapopulation dynamics. J. Wildl. Manage. 70(6): 1588–1600.

Strickland, M. D., L. McDonald, J. W. Kern, T. Spraker, and A. Loranger. 1994. Analysis of 1992 Dall's sheep and mountain goat survey data, Kenai National Wildlife Refuge. Bienn. Symp. Northern Wild Sheep and Mountain Goat Council.

Strickland, M. D., G. D. Johnson, W. P. Erickson, S. A. Sarappo, and R. M. Halet. 1998a. Avian use, flight behavior and mortality on the Buffalo Ridge, Minnesota, Wind Resource Area. In Proceedings of National Avian-Wind Power Planning Meeting III. Avian Subcommittee of the National Wind Coordinating Committee, % RESOLVE, Inc., Washington, DC.

Strickland, M. D., D. P. Young, Jr., G. D. Johnson, W. P. Erickson, and C. E. Derby. 1998b. Wildlife monitoring studies for the SeaWest Windpower Plant, Carbon County, Wyoming. In Proceedings of National Avian-Wind Power Planning Meeting III. Avian Subcommittee of the National Wind Coordinating Committee, % RESOLVE, Inc., Washington, DC.

Underwood, A. J. 1994. On beyond BACI: Sampling designs that might reliably detect environmental disturbances. Ecol. Appl. 4: 3–15.

Underwood, A. J. 1997. Experiments in Ecology. Cambridge University Press, Cambridge.

United States Department of the Interior [USDI]. 1987. Type B Technical Information Document: Guidance on Use of Habitat Evaluation Procedures and Suitability Index Models for CERCLA

Application. PB88-100151. U.S. Department of the Interior, CERCLA 301 Project, Washington, DC.

Volesky, J. D., W. H. Schacht, P. E. Reece, and T. J. Vaughn. 2005. Spring growth and use of cool-season graminoids in the Nebraska Sandhills. Rangeland Ecol. Manage. 58(4): 385–392.

Walters, C. 1986. Adaptive Management of Renewable Resources. Macmillan, New York.

Western Ecosystems Technology, Inc. 1995. Draft General Design, Wyoming Windpower Monitoring Proposal. Appendix B in Draft Kenetech/PacifiCorp Windpower Project Environmental Impact Statement. FES-95-29. Prepared by U.S. Department of the Interior, Bureau of Land Management, Great Divide Resource Area, Rawlins, WY, and Mariah Associates, Inc., Laramie, WY.

Williams, B. K., J. D. Nichols, and M. J. Conroy. 2002. Analysis and Management of Animal Populations, Modeling, Estimation, and Decision Making. Academic, New York.

Winer, B. J. 1971. Statistical Principles in Experimental Design, 2nd Edition. McGraw-Hill, New York.

Winer, B. J., D. R. Brown, and K. M. Michels. 1991. Statistical Principles in Experimental Design, 3rd Edition. McGraw-Hill, New York.

Wolfe, L. L., W. R. Lance, and M. W. Miller. 2004. Immobilization of mule deer with Thiafentanil (A-3080) or Thiafentanil plus Xylazine. J. Wildl. Dis. 40(2): 282–287.

Zar, J. H. 1998. Biostatistical analysis, 2nd Edition. Prentice-Hall, Englewood Cliffs, NJ.

Chapter 4
Sample Survey Strategies

4.1 Introduction

The goal of wildlife ecology research is to learn about wildlife populations and their use of habitats. The objective of this chapter is to provide a description of the fundamentals of sampling for wildlife and other ecological studies. We discuss a majority of sampling issues from the perspective of design-based *observational studies* where empirical data are collected according to a specific study design. We end the chapter with a discussion of several common model-based sampling approaches that combine collection of new data with parameters from the literature or data from similar studies by way of a theoretical mathematical/statistical model. This chapter draws upon and summarizes topics from several books on applied statistical sampling and wildlife monitoring and we would encourage interested readers to see Thompson and Seber (1996), Thompson (2002b), Thompson et al. (1998), Cochran (1977), and Williams et al. (2002).

Typically, the availability of resources is limited in wildlife studies, so researchers are unable to carry out a *census* of a population of plants or animals. Even in the case of fixed organisms (e.g., plants), the amount of data may make it impossible to collect and process all relevant information within the available time. Other methods of data collection may be destructive, making measurements on all individuals in the population infeasible. Thus, in most cases wildlife ecologists must study a subset of the population and use information collected from that subset to make statements about the population as a whole. This subset under study is called a *sample* and is the focus of this section. We again note that there is a significant difference between a statistical population and a biological population (Chap. 1).

All wildlife studies should involve random selection of units for study through sample surveys. This will result in data that can be used to estimate the biological parameters of interest. Studies that require a sample must focus on several different factors. What is the appropriate method to obtain a sample of the population of interest? Once the method is determined, what measurements will be taken on the characteristics of the population? Collecting the sample entails questions of sampling design, plot delineation, sample size estimation, enumeration (counting) methods, and determination of what measurements to record (Thompson 2002b).

M.L. Morrison et al., *Wildlife Study Design.*

Measurement of population characteristics allows the calculation of summary values called *parameters* that aid in describing the population or its habitat. The most common values estimated in studies of animal or plant populations are population size, density, survival, and recruitment. Each of these values is characterized by a set of parameters of interest or estimators (means, variances, and standard errors). These estimators (e.g., mean abundance per sampling plot) then allow the scientists to draw inferences about the population under study (Williams et al. 2002). For example, in the study of a deer mouse (*Peromyscus* spp.) population, the parameters of interest might be total number of mice (population size), population survival (or mortality rate) age structure and sex ratio, and mean range size. Habitat parameters of interest might include the abundance of preferred forage each season, niche parameters such as the utilization of key food items, and the standing crop of those food items.

Design-based studies are those that have a predetermined sampling and treatment structure, usually probability based. Most studies in wildlife ecology are design-based observational studies as it is frequently difficult to assign treatments and controls randomly to wildlife populations. However, design based do differ from observational studies in that under design-based studies scientists can deliberately select a sample, avoiding unrepresentativeness (Anderson 2001; Thompson 2002). In design-based studies, basic statistical inferences concerning the study areas are justified by the design of the study and data collected (Cochran 1977). However, sampling is usually distinguished from the related field of true experimental design (Kuehl 2000) where the researchers deliberately applies a specific treatment to a randomly selected portion of the population to see what impact the treatment has on the population (Thompson 2002b). Additionally, we will discuss some of the more common model-based observational sampling approaches; these approaches use assumptions to account for patterns within the populations of interest.

4.1.1 Basic Sampling Estimators

Sampling in wildlife studies is used to obtain parameter estimates for individuals within the population of interest. The goal of any ecological study is to provide estimates that are accurate as discussed in Sect. 2.5.4. If the design is appropriate and implemented correctly, wildlife ecologists can obtain estimates that satisfy these requirements with few assumptions about the underlying population.

In order to determine estimates for the population characteristics of interest, we must use an *estimator*. The most common estimators are those for means, variances, and other associated measures of central tendency and precision. The primary measure of central tendency collected in ecological studies is the sample mean ($\bar{x}$). Consider a simple random sample taken from all potential plots of a statistical population to measure some characteristic x (no. of individuals per plot) and our interest is in estimating the mean number of individuals per plot. The sample mean ($\bar{x}$) will be an unbiased estimator for the population mean (μ) or the average population size for each randomly selected sample. While the population mean is the average measurement for each of N samples (after Cochran 1977; Thompson 2002b) defined as

$$\mu = \frac{1}{N}(X_1 + X_2 + \cdots + X_N) = \frac{1}{N}\sum_{i=1}^{N} X$$

the sample mean $\bar{x}_i$ is the average count from those surveyed plots selected under the simple random sampling design (n) estimated by

$$\overline{X} = \frac{1}{N}(X_1 + X_2 + \cdots + X_N) = \frac{1}{n}\sum_{i=1}^{n} X_i.$$

In this situation, we assumed we had a *finite* population of known size N. Thus, within a simple random sampling framework, the sample variance (s^2) is an unbiased estimator for the finite population variance σ^2. Thus,

$$s^2 = \frac{1}{n-1}\sum_{i=1}^{n}\left(X_i - \overline{X}\right)^2$$

This approach holds true for estimation of subpopulation means (mean of a statistical population based on stratification). A subpopulation mean is one where we wish to estimate the mean of a subsample of interest. For example, consider the situation where we want to estimate the abundance of mice (*Mus or Peromyscus*) across an agricultural landscape. After laying out our sampling grid, however, we determine that abundance of the two species should be estimated for both fescue (Family Poaceae) and mixed warm-season grass fields. Thus, we are interested in both the mean number of mice per sample plot and the mean number of mice per sample plot within a habitat type, e.g., a subpopulation. For habitat type h, our sample mean subpopulation estimates would be

$$\overline{X}_{\mathrm{h}} = \frac{1}{N}\sum_{i=1}^{n_{\mathrm{h}}} X_{\mathrm{h}}$$

with sample variance

$$s_{\mathrm{h}}^2 = \frac{1}{n_{\mathrm{h}}-1}\sum_{i=1}^{n_{\mathrm{h}}}\left(X_{\mathrm{h}i} - \overline{X}_{\mathrm{h}}\right)^2$$

As many ecological researchers wish to estimate the total population size based on sample data, under a situation with no subpopulation estimates, our estimator for total population size (T) would be

$$\hat{T} = \frac{N}{n}\sum_{i=1}^{n} X_i.$$

Additional information on estimation of population total and means for more complex designs can be found in Cochran (1977) and Thompson (2002b).

4.1.2 Plot Construction

We use sampling designs to ensure that the data collected are as accurate as possible for a given cost. Thus, plot construction necessitates that researchers evaluate the impacts of different plot sizes and shapes have on estimator precision. Although the importance of determining optimal sizes and shape for sampling plots [for consistency within the text, we are using "plots" rather than "quadrats" as defined by Krebs (1999)] is obvious. With the exception of work by Krebs (1989, 1999) and general discussion by Thompson et al. (1998), there has been little research on plot construction in wildlife science. Wildlife tend to be nonrandomly distributed across the landscape and are influenced by inter- and intraspecific interactions (Fretwell and Lucas 1970; Block and Brennan 1993). When developing a sampling design to study a population, the researcher must decide what size of plots should be used and what shape of plots would be most appropriate based on the study question and the species life history (Thompson et al. 1998; Krebs 1999). Most frequently, plot size and shape selection is based on statistical criteria (e.g., minimum standard error), although in studies of ecological scale, the shape and size will be dependent upon the process under study (Krebs 1999). Additionally, it is important to realize that estimates of precision (variance) are dependent upon the distribution of the target organism(s) in the plots to be sampled (Wiegert 1962).

Krebs (1999) listed three approaches to determine which plot shape and size would be optimal for a given study:

1. Statistically, or the plot size which has the highest precision for a specific area or cost
2. Ecologically, or the plot sizes which are most efficient to answering the question of interest
3. Logistically, or the plot size which is the easiest to construct and use

Plot shape is directly related to both the precision of the counts taken within the plot and potential coverage of multiple habitat types (Krebs 1999). Four primary factors influence plot shape selection: (1) detectability of individuals, (2) distribution of individuals, (3) edge effects, and (4) data collection methods. Shape relates to count precision because of the *edge effect*, which causes the researcher to decide whether an individual is within the sample plot or not, even when total plot size is equal (Fig. 4.1). Given plots of equal area, long and narrow rectangular plots will have greater edge effect than square or circular plots. Thompson (1992) concluded that rectangular plots were more efficient than other plots for detecting individuals. Note that, in general, long and narrow rectangular plots will have a greater chance of intersecting species with a clumped distribution. Previous research in vegetation science has shown that rectangular plots are more efficient (higher precision) than square plots (Kalamkar 1932; Hasel 1938; Pechanec and Stewart 1940; Bormann 1953). Size is more related to efficiency in sampling (Wiegert 1962), in that we are trying to estimate population parameters as precisely as possible at the lowest cost (Schoenly et al. 2003). Generally, larger plots have a lower ratio of edge to interior,

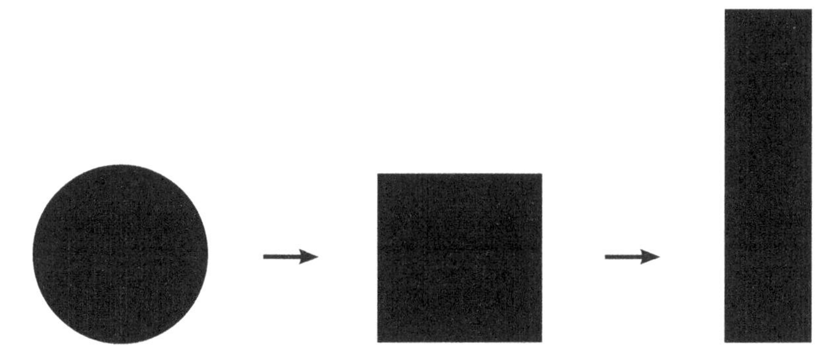

Fig. 4.1 An example of three different types of plot shapes, each with the same area, but with different perimeter to edge ratios. Reproduced from Thompson et al. (1998) with kind permission from Elsevier

limiting potential edge effects. Large plots, however, are typically more difficult to survey based on cost and logistics. Thus, under a fixed budget, there is a general trade off between plot size and number of plots to sample. A method developed by Hendricks (1956) found that as sample area increased, variance decline, but this method is less flexible as this approach had several assumptions such as proportionality of sampling cost per unit area.

4.2 Basic Sample Structure, Design, and Selection

Wildlife studies are limited by fundamental principles of inferential statistics when using sample survey data to make predictions about the population of interest. Within the population or study area boundaries, statistical inference is limited by the protocol by which study sites and/or study specimens are selected. Thus, sampling is an example of inductive logic wherein the conclusions are determined based on a limited number of events (Foreman 1991; see Sect. 1.2.3.2 and Table 1.1). A sample is a subset of the population of interest, where the population encompasses every individual located in a particular place at a particular time. Sampling entails selecting sample units (unique collection of elements; Scheaffer et al. 1990) from a sampling frame from a population and then collecting measurements on the sampling unit (Foreman 1991). Note that sampling units and elements can represent the same quantity (Thompson et al. 1998). Essentially, our purpose in sampling is to make inferences to our target population or those individuals within the population study boundaries at a specific time.

One of the primary functions of statistic and sampling is to make inductive inference and measure the degree of uncertainty around such inferences (Mood et al. 1974). Scientific progress in ecological studies is often credited to experiments that randomize and replicate treatments (Johnson 2002). However, ecologists are frequently unable to randomize treatments and must use natural experiments or descriptive studies consisting of observations of an organism's response to a perturbation.

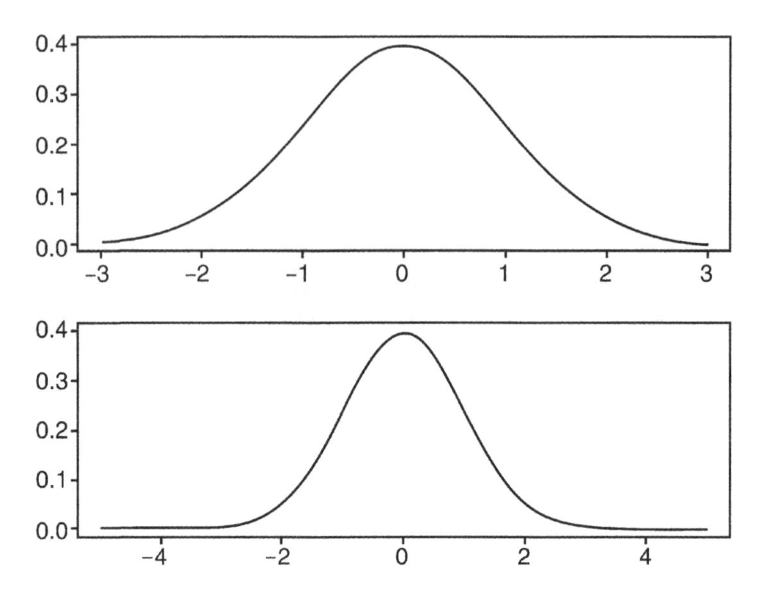

Fig. 4.2 Two normal distributions with different the same mean and different variances

Methods for sample selection typically fall into two general categories: nonrandom sampling and random sampling. In random sampling, also called probability sampling, the selection of units for inclusion in the sample has a known probability of occurring. If the sample is selected randomly, then based on sample survey theory (Cochran 1977) the sample estimates will be normally distributed. With normally distributed estimates, knowledge of the sample mean and variance specifies the shape of the normal distribution (Fig. 4.2). There is considerable literature justifying the need for probabilistic sampling designs from a standpoint of statistical inference (Cochran 1977; Thompson and Seber 1996; Thompson 2002b), but little evidence exists that nonprobabilistic samples can be inferentially justified (Cochran 1977; Anderson 2001; Thompson 2002a). In wildlife ecology, nonprobabilistic sampling designs are likely to be divided into several (overlapping) categories which we generalize as convenience/haphazard sampling (hereafter convenience) or judgment sampling/search sampling (hereafter judgment) while probabilistic sampling is the other category used in wildlife ecology. For the rest of the chapter, we will discuss these different sampling designs and their application to wildlife ecology research.

4.2.1 Nonprobability Sampling

Convenience sampling has historically been the most common approach to sampling wildlife populations. A convenience sample is one where the samples chosen are based on an arbitrary selection procedure, often based on accessibility, and justified because of constraints on time, budgets, or study logistics. Gilbert (1987, p. 19) noted in discussion of *haphazard sampling*, that:

> Haphazard sampling embodies the philosophy of "any sampling location will do." This attitude encourages taking samples at convenient locations (say near the road) or times, which can lead to biased estimates of means and other population characteristics. Haphazard sampling is appropriate if the target population is completely homogeneous. This assumption is highly suspect in most wildlife studies.

Examples of convenience sampling approaches are abundant in wildlife ecology: abundance and sex ratio estimates from spotlight surveys from roads for white-tailed deer (Collier et al. 2007), point counts along roads for birds (Peterjohn et al. 1996), surveys for mammal tracks near roads, habitat sampling in only locations where individuals were detected, to name a few. In these situations, the location of the individual(s) of interest determines the location and number of samples collected, but with no scheme to infer to the larger population (Thompson and Seber 1996). Certain kinds of surveys, such as report card harvest surveys, may have an element of convenience sampling in them if the sample is self-selected by individuals volunteering to complete the survey. One of the limitations of convenience sampling is that it cannot provide data for valid statistical inferences, because results are not repeatable. Information obtained by this type of sampling may be appropriate for preliminary inventory of an area but should not be used for formal discussion of parameter estimates.

Judgment or search sampling is another common approach used in wildlife studies. This form of sampling is based on the presumption that the wildlife scientist can select studies representative of the study area or population based on expert knowledge of the system, often requiring historical knowledge or data indicating where the resources of interest exist. Gilbert (1987) argued that judgment sampling results in subjective selection of population units by the researcher resulting in the following outcome:

> If the [researcher] is sufficiently knowledgeable, judgment can result in accurate estimates of population parameters such as means and totals even if all population units cannot be visually assessed. But, it is difficult to measure the accuracy of the estimated parameters. Thus, subjective sampling can be accurate, but the degree of accuracy is difficult to quantify.

Judgment sampling may be appropriate for preliminary inventory of an area, but is not useful for statistical inferences because results are not repeatable. Judgment sampling may have a role to play in understanding the mechanisms in force in a biological system. For example, several study areas may be selected to investigate the magnitude and duration of an environmental impact or the effect of some management action under a specific set of conditions. Judgment sampling can also be used to develop data for models of natural systems (see capture–recapture model discussion later in this chapter). However, statistical inferences from sites selected for study are strictly limited to the study sites selected and any inference beyond those sites is deductive, depending on the professional judgment of the individual making the selection and the rules by which the sites are selected.

Note that all of the above sampling approaches are based on nonprobabilistic designs and rely either on observations of the organism or expert opinion to select locations for sample data collection. Consequently, while many convenience sampling procedures are often justified based on their economics (e.g., easier to sample roads than contact landowners for access), this is often not the case as these samples do not allow for wide ranging inferences, thus limiting their applicability. Probabilistic samples allows the researcher to design a study and be confident that the results are sufficiently accurate and economical (Cochran 1977). Nonprobabilistic sampling, while common, do not lend themselves to valid statistical inference or estimation of variability and often more cost is incurred attempting to validate convenience samples than would be spent developing and applying probabilistic designs.

4.2.2 Probability Sampling

Random sampling is the process by which samples are selected from a set of n distinct sampling units, where each sample has a known likelihood of selection predetermined by the sampling methods chosen (Cochran 1977; Foreman 1991). Samples selected probabilistically provide a basis for inference (estimation of means and variances) from the data collected during the sampling process; samples from nonprobability designs do not have this characteristic.

4.2.3 Single-Level and Multilevel Probability Sampling

The simplest form of random sampling is sampling at a single level or scale. That is, the study area is divided into a set of potential units from which a sample is taken. For example, a study area could be divided into a grid of sample plots all of the same size from which a simple random sample is drawn (Fig. 4.3). The organisms of interest in each cell in the selected sample are then counted. In its simplest sense, single level sampling for a simple random sample, assume that we have n = 100 distinct samples, $S_1, S_2, \ldots, S_n$, where each sample S_i has a known probability of selection (π_i) or the probability that the ith sample is taken (Cochran 1977). Assuming that each sample unit (plot) is of equal size, then the probability that a single plot is chose to be sampled is 1/100 or $\pi_i = 0.01$. In the application of single-level probability sampling we assume that each unit in the population has the same chance of being selected. Although this assumption may be modified by other probabilistic sampling schemes (e.g., *stratified sampling* or *unequal probability* sampling), the decisions regarding sample selection satisfy this assumption. Sampling at more than one level, however, often is beneficial in wildlife studies. Multilevel sampling can be simple, such as selecting subsamples of the original probability sample for additional measurements as described in ranked set sampling (Sect. 4.3.5). Multilevel sampling can be more complicated, such as double sampling to estimate animal abundance (Sect. 4.3.6). In the correct circumstances, multilevel sampling can increase the quality of field data, often at a lower cost.

1	2	3	4	5	6	7	8	9	10
11	12	13	14	15	16	17	18	19	20
21	22	23	24	25	6	7	8	9	10
31	32	33	34	35	36	37	38	39	40
41	42	43	44	45	46	47	48	49	50
51	52	53	54	55	56	57	58	59	60
61	62	63	64	65	66	67	68	69	70
71	72	73	74	75	76	77	78	79	80
81	82	83	84	85	86	87	88	89	90
91	92	93	94	95	96	97	98	99	100

Fig. 4.3 A simple sampling frame of 100 sample plots that can be used for selecting a simple random sample

4.3 Sampling Designs

Although a simple random sample is the most basic method for sample selection, there are others that are relevant to wildlife ecology studies, including stratified random sampling, systematic sampling, sequential random sampling, cluster sampling, adaptive sampling, and so on. These sampling plans (and others) can be combined or extended to provide a large number of options for study designs, which can include concepts like unequal probability sampling. Many sampling designs are complicated, thus statistical guidance is suggested to select the appropriate design and analysis approaches. Below we discuss several sampling scales and then appropriate designs for each scale.

4.3.1 Simple Random Sampling

Simple random sampling is the selection of n units from a population of N units in a manner such that each of the n units has the same chance (probability) of being selected (Cochran 1977; Foreman 1991). Simple random sampling requires that the location of each sample site (unit) be selected independently of all other sites (units). Typically in ecology studies, a given unit appears at most once in the sample when sampling without replacement (Thompson 2002b). Samples can be replaced after measurements are taken so that sampling is with replacement but sampling without replacement results in a more precise estimate (Caughley 1977; Thompson 2002b).

A simple random sample may be obtained by following the basic steps in the following list (Cochran 1977; Thompson 2002b):

1. The population of sampling units is assumed to be *finite*.
2. Units (n) selected in the sample can be located and the measurement of the attribute of interest (e.g., count of animals) on the unit is possible. Also, the error in measuring the attribute of interest should be small compared with the differences in the attribute (counts) from unit to unit.
3. The study region, also known as the sampling frame, must be completely covered by distinct and nonoverlapping sampling units.
4. Sampling units need not be of equal size nor selected with equal probability, but differences in size and selection probability increase the complexity of those parameter estimation formulas.
5. Sample units are normally sampled without replacement.

Random sampling plans have straightforward mathematical properties (Sect. 4.1.1), but random locations are often more clumped and patchy than expected. In studies with small sample sizes, which are common in wildlife studies, entire regions of a sampling frame may be under- or overrepresented. Thus, random sampling is not always the best procedure. Random sampling should be used only if the area of interest is homogeneous with respect to the *elements* and *covariates* of interest. Because this is seldom the case, researchers should try to avoid relying solely on simple random sampling.

4.3.2 Stratified Random Sampling

In *stratified sampling*, the sampling frame is separated into different regions (*strata*) comprising the population to be surveyed and a sample of units within stratum are selected for study, usually by a random or systematic process. Ideally, strata should be homogeneous with respect to the variable of interest itself (e.g., animal density), but in practice, stratification is usually based on covariates that scientists hope are highly correlated with the variable of interest (e.g., habitat types influences animal density). Stratification may be used to increase the likelihood that the sampling effort will be spread over important subdivisions or strata of the study area, population, or study period (Fig. 4.4). Similarly, units might also be stratified for subsampling. For example, when estimating the density of forest interior birds, the wildlife biologist might stratify the study area into regions of high, medium, and low canopy cover and sample each independently, perhaps in proportion to area size.

Stratification is common in wildlife studies, as it often is used to estimate parameters within strata and for contrasting parameters among strata. This type of analysis is referred to using "strata as domains of study ... in which the primary purpose is to make comparisons between different strata" (Cochran 1977, p. 140). Under

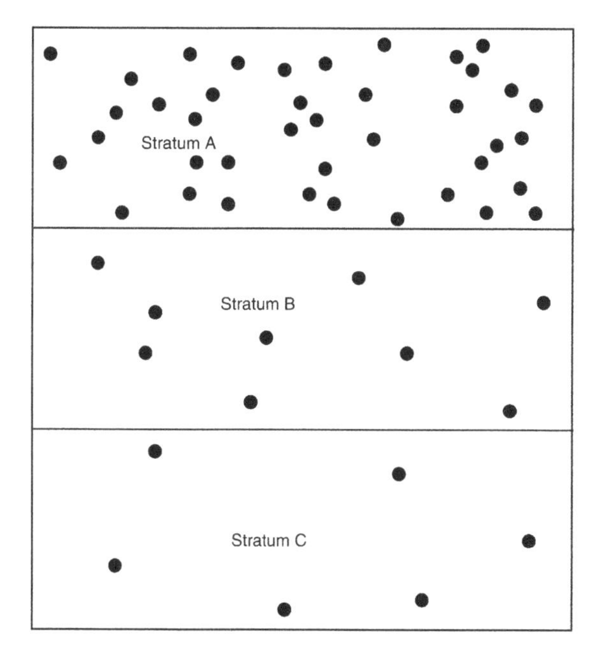

Fig. 4.4 Stratification based on the density of a population. Reproduced from Krebs (1999) with kind permission from Pearson Education

stratified designs, the formulas for analysis and for allocation of sampling effort (Cochran 1977, pp. 140–141) are quite different from formulas appearing in introductory texts such as Scheaffer et al. (1990), where the standard objective is to minimize the variance of summary statistics for all strata combined.

The primary objective of stratification is improved precision based on optimal allocation of sampling effort into more homogeneous strata. In practice, it may be possible to create homogeneous strata with respect to one or a few primary indicators, but there are often many indicators measured, and it is not likely that the units within strata will be homogeneous for all of them. For example, one could stratify a study area based on vegetative characteristics and find that the stratification works well for indicators of effect associated with trees. But, because of management (e.g., grazing), the grass understory might be completely different and make the stratification unsatisfactory for indicators of effect measured in the understory. Differences in variance among strata for the primary indicators may not occur or may not be substantially better than random sampling. Factors used to stratify an area should be based on the spatial location of regions where the population is expected to be relatively homogeneous, the size of sampling units, and the ease of identifying strata boundaries. Strata should be of obvious biological significance for the variables of interest.

A fundamental problem is that strata normally are of unequal sizes; therefore, units from different strata have different weights in any overall analysis. The formulas for computing an overall mean and its standard error based on stratified sampling are relatively complex (Cochran 1977). Formulas for the analysis of subpopulations (subunits of a study area) that belong to more than one stratum (Cochran 1977, pp. 142–144; Thompson 2002b) are even more complex for basic statistics such as means and totals. Samples can be allocated to strata in proportion to strata size or through some optimal allocation process (Thompson 2002b). When using the stratification with proportional allocation, the samples are self-weighting in that estimates of the overall mean and proportion are the same as for estimates of these parameters from simple random sample. Although proportional allocation is straightforward, it may not make the most efficient use of time and budget. If it is known that within strata variances differ, samples can be allocated to optimize sample size. Detailed methods for optimizing sample size are described in Cochran (1977) and Thompson (2002b).

Stratification has some inherent problems. In any stratification scheme, some potential study sites will be misclassified in the original classification (e.g., a dark area classified as a pond on the aerial photo was actually a parking lot). Stratification is often based on maps that are inaccurate, resulting in misclassification of sites that have no chance of selection. Misclassified portions of the study area can be adjusted once errors are found, but data analysis becomes much more complicated, primarily because of differences in the probability of selecting study units in the misclassified portions of the study area. Short-term studies usually lead to additional research questions requiring longer term research and a more complicated analysis of subpopulations (Cochran 1977, pp. 142–144) that cross strata boundaries. However, strata may change over the course of a study. Typical strata for wildlife studies include physiography/topography, vegetative community, land use, temporal frame, or management action of interest. Note, however, that the temporal aspect of a study is of particular significance when stratifying on a variable that will likely change with time (e.g., land use). Stratified sampling works best when applied to short-term studies, thus reducing the likelihood that strata boundaries will change. In long-term studies, initial stratification procedures at the beginning of the study are likely to be the most beneficial to the investigators.

4.3.3 Systematic and Cluster Sampling

In *systematic sampling*, the sampling frame is partitioned into primary units where each primary unit consists of a set of secondary units (Thompson 2002b). Sampling then entails selecting units spaced in some systematic fashion throughout the population based on a random start (Foreman 1991). A systematic sample from an ordered list would consist of sampling every *k*th item in the list. A spatial sample typically utilizes a systematic grid of points. Systematic sampling distributes the locations of samples (units) uniformly through the list or over the area (site). Mathematical properties of systematic samples are not as straightforward as for random sampling, but the statistical precision generally is better (Scheaffer et al. 1990).

Systematic sampling has been criticized for two basic reasons. First, the arrangement of points may follow some unknown cyclic pattern in the response variable. Theoretically, this problem is addressed a great deal, but is seldom a problem in practice. If there are known cyclic patterns in the area of interest, the patterns should be used to advantage to design a better systematic sampling plan. For example, in a study of the cumulative effects of proposed wind energy development on passerines and shore birds in the Buffalo Ridge area of southwestern Minnesota, Strickland et al. (1996) implemented a grid of sampling points resulting in observations at varying distances from the intersection of roads laid out on section lines.

Second, in classical finite sampling theory (Cochran 1977), variation is assessed in terms of how much the result might change if a different random starting point could be selected for the uniform pattern. For a single uniform grid of sampling points (or a single set of parallel lines) this is impossible, and thus variation cannot be estimated in the classical sense. Various model-based approximations have been proposed for the elusive measure of variation in systematic sampling (Wolter 1984). Sampling variance can be estimated by replicating the systematic sample. For example, in a study requiring a 10% sample it would be possible to take multiple smaller samples (say a 1% sample repeated ten times), each with a random starting point. Inference to the population mean and total can be made in the usual manner for simple random sampling.

Systematic sampling works very well in the following situations:

1. Analyses of observational data conducted as if random sampling had been conducted (effectively ignoring the potential correlation between neighboring locations in the uniform pattern of a systematic sample)
2. Encounter sampling with unequal probability (Overton et al. 1991; Otis et al. 1993)
3. The model-based analysis commonly known as spatial statistics, wherein models are proposed to estimate treatment effects using the correlation between neighboring units in the systematic grid (kriging)

The design and analysis in case 1 above is often used in evaluation of indicators of a treatment response (e.g., change in density) in relatively small, homogeneous study areas or small study areas where a gradient is expected in measured values of the indicator across the area. Ignoring the potential correlation and continuing the analysis as if it is justified by random sampling can be defended (Gilbert and Simpson 1992), especially in situations where a conservative statistical analysis is desired (e.g., impact assessment). Estimates of variance treating the systematic sample as a random sample will tend to overestimate the true variance of the sample (Hurlbert 1984; Scheaffer et al. 1990; Thompson 2002). Thus, systematic sampling in relatively small impact assessment study areas following Gilbert and Simpson's (1992) formulas for analysis makes a great deal of sense. This applies whether systematic sampling is applied to compare two areas (assessment and reference), the same area before and following the incident, or between strata of a stratified sample.

In wildlife studies, populations tend to be aggregated or clustered, thus sample units closer to each other will be more likely to be similar. For this reason, systematic sampling tends to overestimate the variance of parameter estimates. A uniform grid of points or parallel lines may not encounter rare units. To increase the likelihood of capturing some of these rare units, scientists may stratify the sample such that all units of each distinct type are joined together into strata and simple random samples are drawn from each stratum. Nevertheless, stratification works best if the study is short term, no units are misclassified and no units change strata during the study. In longer term studies, such as the US Environmental Protection Agency's (EPA's) long-term Environmental Monitoring and Assessment Program (EMAP), as described by Overton et al. (1991), systematic sampling has been proposed to counter these problems.

Cluster sampling is closely related to systematic sampling. A cluster sample is a probabilistic sample in which each sampling unit is a collection, or cluster, of elements such as groups of animals or plants (Scheaffer et al. 1990; Thompson 2002b). One of the most common uses of cluster sampling is the two-stage cluster sample. First, the researcher selects a probabilistic sample of plots, each of the primary plots having eight secondary plots. Then, within those primary plots, we either select another probability sample of plots from the eight secondary plots, or consider the cluster of eight secondary plots of our sample and conduct our enumeration method within each of those plots (Fig. 4.5). The selection of progressively smaller subsets of elements within the original set of sample clusters leads to a multistage cluster sample. Cluster sampling methods can become considerably complex, depending on sampling design, study question, and phenology of the species under study (Christman 2000). For example, consider an ecologist interested in estimating Greater Prairie-chicken (*Tympanuchus cupido*) lek numbers in the

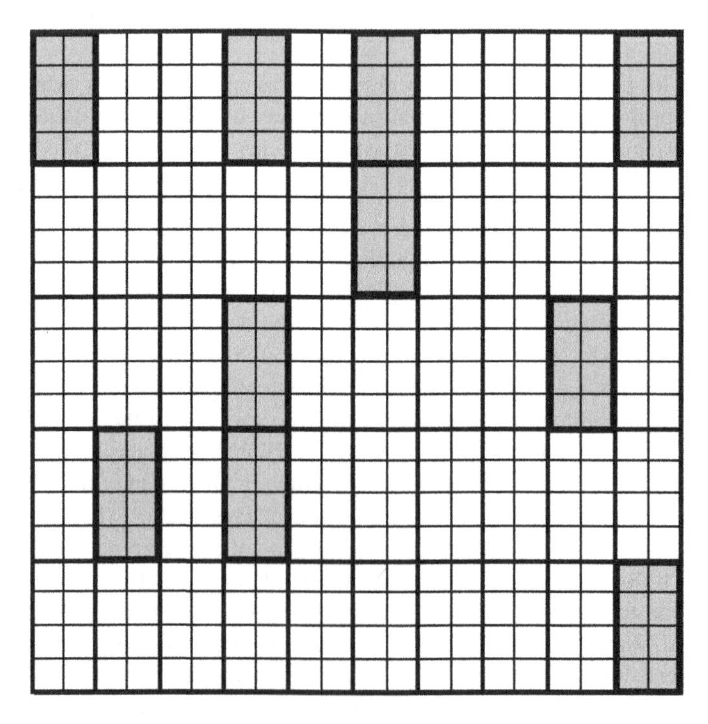

Fig. 4.5 (**a**) Cluster sample of ten primary units with each primary unit consisting of eight secondary units; (**b**) systematic sample with two starting points. Reproduced from Thompson (2002) with kind permission from Wiley

plains during the breeding season. Lek sites are typically close spatially, relative to the size of grasslands matrix these birds inhabit, thus we would expect that if a lek is located within a primary sample plot, there are other leks in the vicinity. For this reason, the researcher would randomly sample primary plots across a landscape of Greater Prairie-chicken habitat, then, within those large plots, conduct enumeration of lek numbers within the secondary plots.

Thompson (2002b, pp. 129–130) lists several features that systematic and cluster sampling that make these designs worth evaluating for ecological studies:

- In systematic sampling, it is not uncommon to have a sample size of 1, that is, a single primary unit (see Fig. 4.5).
- In cluster sampling, the size of the cluster may serve as auxiliary information that may be used either in selecting clusters with unequal probabilities or in forming ratio estimators.
- The size and shape of clusters may affect efficiency.

4.3.4 Adaptive Sampling

Numerous sampling designs integrate stratified, systematic, and cluster sampling – commonly under a framework called *adaptive sampling* – where, following an initial probabilistic sample of units, additional units are added to the sample in the neighborhood of original units that satisfy a specified condition (Thompson and Seber 1996). Thus, methods for adaptive sampling differ from most other sampling designs as the sample selection procedure is not determined before sampling, but is fluid and changes as successive samples are taken. Given the wide range of adaptive techniques available, we refer the interested readers to Thompson and Seber (1996), Christman (2000), and Thompson (2002).

Wildlife biologists are often bothered by probability sampling plans because sampling is limited to a set of previously selected units to the exclusion of units adjacent to, but not in, the sample. Adaptive sampling offers biologists a way to augment the probability sample with samples from adjacent units without losing the benefits of the original probabilistic design. Because animal populations usually are aggregated, adaptive methods take advantage of this tendency and uses information on these aggregations to direct future sampling. Adaptive sampling may yield more precise estimates of population abundance or density for given sample size or cost and may increase the yield of interesting observations resulting in better estimates of population parameters of interest (Thompson and Seber 1996).

Under a general adaptive sampling framework, a sample of units is first selected by any probabilistic sampling design. Rules for selection of additional samples are established based on some characteristic of the variable of interest (e.g., presence/absence, age, sex, and height). The values of the variables of interest are then noted on the original probabilistic sample of units and rules for selection of additional samples are applied (Thompson and Seber 1996). In a sense, adaptive sampling is a method for systematically directing biologists' tendency toward search sampling.

4.3.4.1 Definitions

We provide a brief and general description of the theory of adaptive sampling; a comprehensive discussion of the mathematics and theory is beyond the scope of this book. See Thompson and Seber (1996) for a complete discussion of this subject. We adopted the notations used by Thompson (2002b) for this discussion. To understand adaptive sampling it is useful to label sampling units and aggregations of units. The following definitions assume a simple random sample of units from a study area.

A neighborhood is a cluster of units grouped together based on some common characteristic. Typical definitions of a neighborhood include spatially contiguous units or a systematic pattern of surrounding units. For example, a neighborhood of 1-m^2 units in a grid might include each unit and the eight adjacent units (i.e., units at the four sides and corners). However, neighborhoods of units may be defined in many other ways including social or institutional relationships among units. For every unit, if unit i (u_i) is in the neighborhood of unit k (u_k) then u_k is in the neighborhood of u_i. Likewise, if neighborhood u_{ij} belongs to the neighborhood of u_{kl} then u_{kl} belongs to the neighborhood of u_{ij}.

The condition of interest (C) is the characteristic of the variable of interest (y) that determines if a unit is added to the neighborhood of units in the sample. Thus, u_i satisfies the condition and is added to the neighborhood if $y_i \in C$ where C is a specified interval or set of y_i. For example, C might be a carcass search plot containing ≥ 1 carcass. When a selected unit satisfies the condition, then all units within its neighborhood are added to the sample.

All the units added to the sample as the result of the selection of u_i are considered a cluster. A cluster may combine several neighborhoods. All the units within a cluster that satisfy the condition are considered to be in the same network. A population can be uniquely partitioned into K networks. An originally selected unit that does not satisfy the condition forms a unique network by itself. Units in the neighborhood that do not satisfy the condition are defined to be edge units and are not included in networks.

4.3.4.2 Adaptive Cluster Sampling Example

Adaptive sampling refers to those designs where selection of sample plots is dependent upon variables of interest observed (or not observed) within the sample during the survey. Adaptive sampling provides a method for using the clustering tendencies of a population when locations and shapes of clusters can generally be predicted (i.e., they are not known in the physical landscape but can be predicted based on existing information). Therefore, adaptive designs allow the researchers to add nearby plots under the assumption that if the species of interest is located in a plot, then it is likely that there are more members of the species within the immediate vicinity. Probably the most frequently used adaptive approach in wildlife ecology is that of adaptive cluster sampling (Smith et al. 1995, 2004, Noon et al. 2006). For example, consider a survey of mule deer across a range that is divided into 400 study units (Fig. 4.6a). In an effort to estimate the number of dead deer

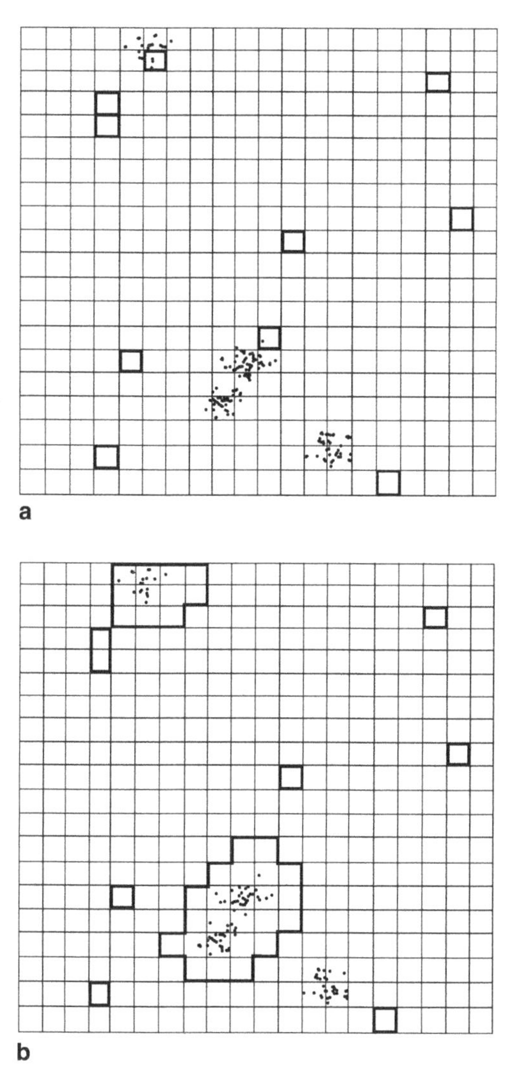

Fig. 4.6 A hypothetical example of adaptive sampling, illustrating a mule deer winter range that is divided into 400 study units (*small squares*) with simple random sample of ten units selected (*dark squares* in (**a**)) potentially containing deer carcasses (*black dots*). Each study unit and all adjacent units are considered a neighborhood of units. The condition of including adjacent units in the adaptive sample is the presence of one or more carcasses (*black dots*) in the unit. Additional searches result in the discovery of additional carcasses in a sample of 45 units in ten clusters (*dark squares* in (**b**)). (Thompson 1990. Reprinted with permission from the Journal of the American Statistical Association, Copyright 1990 by the American Statistical Association. All rights reserved)

following a severe winter, a survey for deer carcasses is conducted. An initial simple random sample of ten units is selected (see Fig. 4.6a). Each study unit and all adjacent units are considered a neighborhood of units. The condition of including adjacent units is the presence of one or more carcasses in the sampled unit. With the adaptive design, additional searches are conducted in those units in the same neighborhood of a unit containing a carcass in the first survey. Additional searches are conducted until no further carcasses are discovered, resulting in a sample of 45 units in ten clusters (see Fig. 4.6b).

The potential benefits of adaptive sampling are obvious in the mule deer example. The number of carcasses (point-objects in Fig. 4.6) is relatively small in the initial sample. The addition of four or five more randomly selected sample units probably would not have resulted in the detection of the number of carcasses contained in the ten clusters of units. Thus, the precision of the estimates obtained from the cluster sample of 45 units is greater than from a random sample of 45 units. This increase in precision could translate into cost savings by reducing required samples for a given level of precision. Cost savings also could result from reduced cost and time for data collection given the logistics of sampling clusters of sampled units vs. potentially a more widely spread random sample of units. This cost saving, however, is partially offset by increased record keeping and increased training costs. Although there are numerous adaptive sampling options, design efficiency depends upon several factors, including initial sample size, population distribution, plot shape, and selection conditions (Smith et al. 1995; Thompson 2002b). Thus we recommend that adaptive designs be pilot tested before implementation to ensure that estimate precision and sampling efficiency is increased over alternate designs.

4.3.4.3 Unbiased Estimators for Simple Random Samples

The potential efficiencies of precision and cost associated with adaptive sampling come with a price. Computational complexities are added because of sample size uncertainty and unequal probability associated with the sample unit selection. Units within the neighborhood of units meeting the condition enter the sample at a much higher probability than the probability of any one unit when sampled at random, resulting in potentially biased estimates of the variable of interest. For example, u_i is included if selected during the initial sample, if it is in the network of any unit selected, or if it is an edge unit to a selected network. In sampling with replacement, repeat observations in the data may occur either due to repeat selections in the initial sample or due to initial selection of more than one unit in a cluster.

The Horvitz–Thompson (H–T) estimator (Horvitz and Thompson 1952) provides an unbiased estimate of the parameter of interest when the probability α_i that unit i is included in the sample is known. The value for each unit in the sample is divided by the probability that the unit is included in the sample. Inclusion probabilities are seldom known in field studies, and modifying the Horvitz–Thompson estimator, where estimates of inclusion probabilities are obtained from the data, as described by Thompson and Seber (1996) forms an unbiased estimator (modified H–T).

Implementing the adaptive sampling procedure described above results in an initial sample of n_1 primary units selected by a systematic or random procedure (without replacement). If a secondary unit satisfies the condition, then all units in the neighborhood are added to the sample. If any of the new units satisfies the condition, then their neighbors also are added. In the modified H–T estimator, the final sample consists of all units in the initial primary units and all units in the neighborhood of any sample unit satisfying the condition. Edge units must be surveyed, but are used in the modified H–T estimator only if they belong to the initial primary units. Thus, an edge unit in the initial sample of primary units is weighted more than individual units in networks and edge units within a cluster are given a weight of 0. Formulas for the modified H–T estimator may be found in Thompson and Seber (1996).

4.3.4.4 Other Adaptive Designs

Thompson and Seber (1996) and Thompson (2002b) summarized a variety of other adaptive sampling designs. Strip adaptive cluster sampling includes sampling an initial strip(s) of a given width divided into units of equal lengths. Systematic adaptive cluster sampling may be used when the initial sampling procedure is based on a systematic sample of secondary plots within a primary plot. Stratified adaptive cluster sampling may be useful when the population is highly aggregated with different expectations of densities between strata. In this case, follow-up adaptive sampling may cross strata boundaries (Thompson 2002b). Thompson and Seber (1996) also discuss sample size determination based on initial observations within primary units, strata, or observed values in neighboring primary units or strata. Adaptive sampling has considerable potential in ecological research, particularly in studies of rare organisms and organisms occurring in clumped distributions.

In Fig. 4.7, the initial sample consists of five randomly selected strips or primary units. The secondary units are small, square plots. Whenever a target element is located, adjacent plots are added to the sample, which effectively expands the width of the primary strip. As depicted in the figure, because this is a probabilistic sampling procedure not all target elements are located (in fact, you might not know they exist). For systematic adaptive cluster sampling (Fig. 4.8) the initial sample is a spatial systematic sample with two randomly selected starting points. Adjacent plots are added to the initial sample whenever a target element is located. The choice of the systematic or strip adaptive cluster design depends primarily on the a priori decision to use a specific conventional sampling design to gather the initial sample, such as the preceding example using aerial or line transects.

Stratified adaptive cluster sampling essentially works like the previous adaptive designs, and is most often implemented when some existing information on how an initial stratification is available. In conventional (nonadaptive) stratified sampling, units that are thought to be similar are grouped a priori into stratum based on prior information. For example, in Fig. 4.9a, the initial stratified random sample of five units in two strata is established. Then, whenever a sampling unit containing the

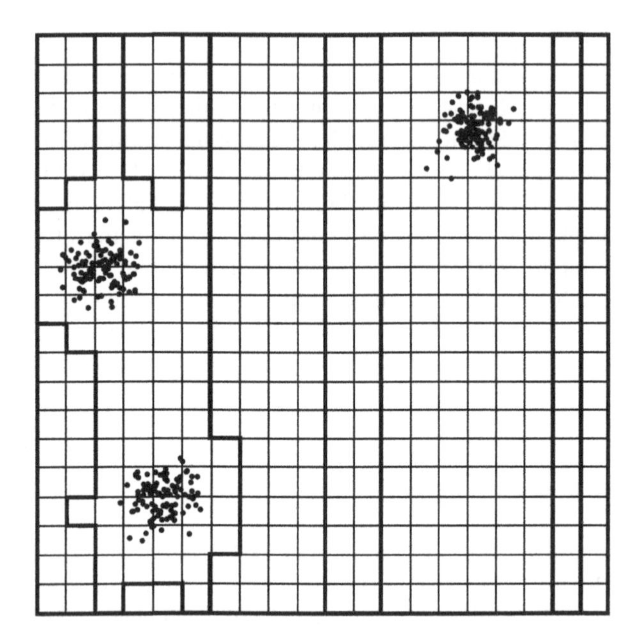

Fig. 4.7 An example of an adaptive cluster sample with initial random selection of five strip plots with the final sample outlined. Reproduced from Thompson (1991a) with kind permission from the International Biometric Society

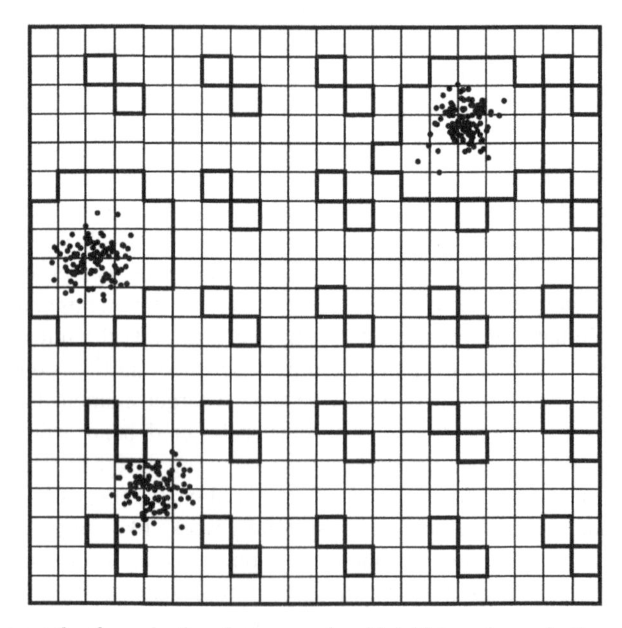

Fig. 4.8 An example of an adaptive cluster sample with initial random selection of two systematic samples with the final sample outlined. Reproduced from Thompson (1991a) with kind permission from the International Biometric Society

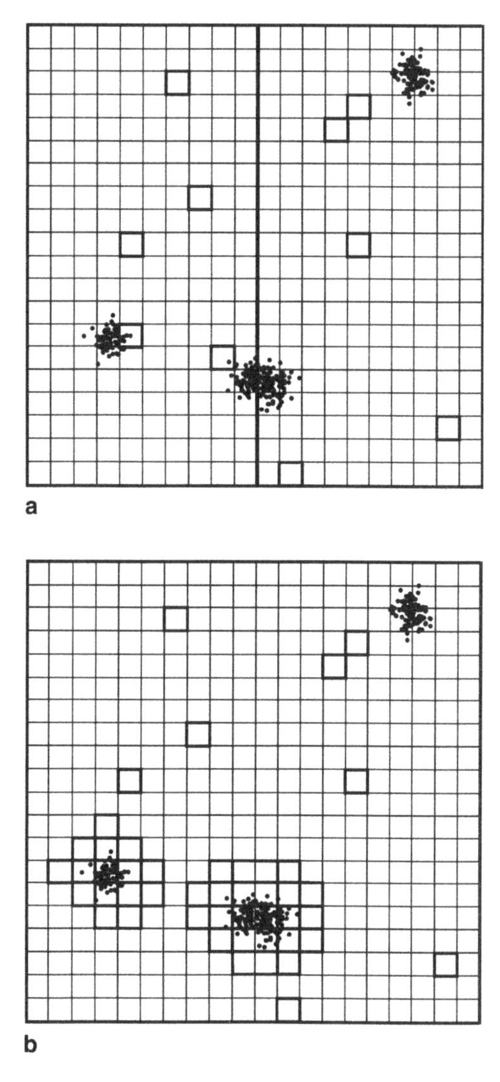

Fig. 4.9 (**a**) Stratified random sample of five units per strata. (**b**) The final sample, which results from the initial sample shown in (**a**). Reproduced from Thompson et al. (1991b) with kind permission from Oxford University Press

desired element is encountered, the adjacent units are added. The final sample in this example (Fig. 4.9b) shows how elements from one strata can be included in a cluster initiated in the other stratum (some units in the right-side stratum were included in the cluster [sample] as a result of an initial selection in the left-side stratum). Thompson (2002b, pp. 332–334) provides a comparison of this example with conventional stratified sampling.

There are four challenges you will encounter when considering implementing an adaptive cluster design (Smith et al. 2004, pp. 86–87):

1. Should I apply adaptive cluster sampling to this population?
2. How large should I expect the final sample size to be?
3. How do I implement adaptive sampling under my field conditions?
4. How can I modify adaptive sampling to account for the biology, behavior, and habitat use of the elements?

Although most biological populations are clustered, adaptive cluster sampling is not necessarily the most appropriate method for all populations. Estimators for adaptive cluster sampling are more complicated and less well understood than those associated with nonadaptive (classical) methods. Adaptive methods should only be used when the benefits of their use clearly outweigh the additional complications.

A difficulty with adaptive sampling is that the final sample size is not known when initiating the study. Although "stopping rules" are available, they can potentially bias results. If the study area is reasonable and well defined, then sampling will "stop" on its own when no additional elements are located. However, if the study area is extremely large and not readily defined (e.g., "the Sierra Nevada," "eastern Texas"), then adaptive sampling becomes difficult to stop once initiated. Perhaps one of the simplest means of stopping, given an appropriate initial sample, is to frequently examine the mean and variances associated with the parameter(s) being sampled and watch for stability of those parameter estimates. These parameter estimates can also be used in conventional power analysis to help guide the stopping decision. Conventional systematic sampling is an effective sampling design for clustered populations, and can be used as a surrogate for designing adaptive sampling.

The suitability of the adaptive design vs. nonadaptive designs depends on the characteristics of the population being sampled. Adaptive cluster sampling is most practical when units are easily located, the condition for selecting units is relatively constant and set at a reasonable level, and the distribution of elements of interest is truly clumped. Adaptive designs are most practical for ground-based surveys for things such as contaminants, plant(s) growing in scattered clumps, wildlife species that exhibit a general seasonal phenology which cause aggregations (e.g., breeding grounds). Adaptive sampling is not recommended for aerial surveys where locating sampling units is difficult at best and locating borders of a neighborhood of units would be extraordinarily difficult and time consuming. Adaptive sampling is also not recommended for situations where the condition is temporary. If the condition causing a unit to be included in a network is likely to change, e.g., presence or absence of a bird in a study plot, then a survey once started would need to be completed as quickly as possible, making planning for surveys difficult. If the conditions were too sensitive or the distribution of the elements of interest not sufficiently clumped (e.g., broadly distributed species like deer) the survey requirements would quickly become overwhelming.

4.3.5 Double Sampling

In double sampling, easy-to-measure or economical indicators are measured on a relatively large subset or census of sampling units in the treatment and reference areas and expensive or time-consuming indicators are measured on a subset of units from each area. As always, easily obtainable ancillary data should be collected. Analysis formulas are available in Cochran (1977). The principles for double sampling are straightforward and the method is easy to implement.

Consider the following examples where y is the primary variable of interest that is relatively expensive to measure on each experimental unit compared with an indicator variable x:

1. y = the number of pairs of breeding ducks present in a certain strip transect measured by ground crews, X = the number of breeding pairs seen during an aerial survey of the same strip
2. y = number of moose seen in a strip transect during an intensive aerial survey, X= number of moose seen in the same strip during a regular aerial survey (e.g., Gasaway et al. 1986)
3. y = the amount of vegetative biomass present on a sample plot, X = ocular estimate of the vegetative biomass on the same plot

In some cases the total (or mean) of the indicator variable may be known for the entire study area while the more expensive variable is known for only a portion of the area. If x and y are positively correlated then double sampling may be useful for improving the precision of estimates over the precision achieved from an initial, small, and expensive sample of both x and y (Eberhardt and Simmons 1987).

4.3.5.1 Double Sampling with Independent Samples

Double sampling would normally be used with independent samples where an initial (relatively small) sample of size n_1 is taken where both y and x are measured. The means for the two variables are calculated or, if the mean is known, the value of the variable is estimated as

$$\bar{y}_l = \sum_i y_i/n_i \quad \text{or} \quad \hat{Y} \; N\bar{y}_l \text{ and}$$

$$\bar{x}_1 = \sum_i x_i/n_i \quad \text{or} \quad \hat{X}_i = N\bar{x}_i.$$

In a relatively large sample of size n_2 (or a census) only the variable x is measured. Its mean is

$$\bar{x}_2 = \sum_i x_i/n_2 \quad \text{or} \quad \hat{X}_2 = N\bar{x}_2.$$

In some situations, the mean for X_2 or $(\bar{X}_2)$ is known from census data, thus the standard error is zero ($X_2 = 0.0$). As an example, suppose X_2 = total production for a farmer's native hay field and Y = potential production without deer as measured in n_1 = 10 deer proof exclosures randomly located in a field. Two variables (X_i, y_i) are measured on the ith enclosure, where y_i is the biomass present on a plot inside the enclosure and X_i is the biomass present on a paired plot outside the enclosure.

The *ratio* of production inside the exclosures to production outside the exclosures is

$$R = \frac{\bar{y}_i}{\bar{x}_1} = \frac{\hat{y}_1}{\hat{x}_1} = \frac{\sum y_i}{\sum x_i}.$$

The ratio estimator for the total production without deer is

$$\hat{Y}_R = \left[\frac{\bar{y}_1}{\bar{x}_1}\right] \cdot \hat{x}_2 = \hat{R}\hat{X}_2,$$

and the estimate of the mean production per plot ($\bar{Y}$) without deer is

$$\bar{y}_R = \left[\frac{\bar{y}_1}{\bar{x}_1}\right] \cdot \bar{x}_2 = \hat{R}\bar{x}_2.$$

There is the tendency to obtain as small a sample as possible of the first more expensive sample. As with any form of probability sampling, the smaller the sample size the greater the likelihood of bias. However, using the ratio estimator, the effect of this bias is reduced. Consider the following example. Suppose the size of the field (N) is 100,000 m^2, the mean production outside the exclosure is 60 gm m^{-2}, the mean production inside the exclosure is 75 gm m^{-2}, and the total production for the field is (X_2) = 100,000 m^2 (50 gm m^{-2}) = 5,000,000 gm outside exclosures. The ratio of the estimates of production is 60 gm m^{-2}/75 gm m^{-2} = 1.25. Thus, there is an additional 0.25 gm of production per m^2 of production inside exclosures for every gm of production outside the exclosures. The estimated production without deer is (50 gm m^{-2}) (1.25) = 62.5 gm m^{-2} and total production of the entire field (Y_2) = 100,000 m^2 (62.5 gm m^{-2}) = 6,250,000 gm if the field could have been protected from deer. Note that the estimate of 75 gm m^{-2} for sample plots inside exclosures is adjusted *down* since the total production ($\bar{X}_2$ = 50 gm m^{-2}) is below the average of paired sample plots outside the exclosures ($\bar{X}_1$ = 60 gm m^{-2}). In our example, the small sample of exclosures apparently landed on higher production areas of the field by *chance*. We assume that the ratio R is adequately estimated by the initial, small but expensive sample. The large, inexpensive, second sample (i.e., total production by the farmer) adjusts for the fact that the initial sample may not truly represent the entire field.

Computation of the variances and standard errors is tedious because a ratio and a product are involved. The variance of the product with independent samples is estimated by the unbiased formula proposed by Reed et al. (1989).

4.3.5.2 Applications of Double Sampling

Smith's (1979) two-stage sampling procedure is a variation of the general double-sampling method for use in environmental impact studies. Smith suggests oversampling in an initial survey, when knowledge concerning impacts is most limited, and recording economical easy-to-measure indicators. For example, animal use (an index to abundance sampled according to a probability sample) might be taken during a pilot study, allowing one to identify species most likely affected by a treatment or impact. In the second stage and with pilot information gained, the more expensive and time-consuming indicators (e.g., the actual number of individuals) might be measured on a subset of the units. If the correlation between the indicators measured on the double-sampled units is sufficiently high, precision of statistical analyses of the expensive and/or time-consuming indicator is improved.

Application of double sampling has grown in recent years, particularly for correcting for visibility bias. Eberhardt and Simmons (1987) suggested double sampling as a way to calibrate aerial observations. Pollock and Kendall (1987) included double sampling in their review of the methods for estimating visibility bias in aerial surveys. Graham and Bell (1969) reported an analysis of double counts made during aerial surveys of feral livestock in the Northern Territory of Australia using a similar method to Caughley and Grice (1982) and Cook and Jacobson (1979). Several studies have used radiotelemetered animals to measure visibility bias, including Packard et al. (1985) for manatees (*Trichechus manatus*), Samuel et al. (1987) for elk, and Flowy et al. (1979) for white-tailed deer (*Odocoileus virginianus*). McDonald et al. (1990) estimated the visibility bias of sheep groups in an aerial survey of Dall sheep (*Ovus dalli*) in the Arctic National Wildlife Refuge (ANWR), Alaska using this technique. Strickland et al. (1994) compared population estimates of Dall sheep in the Kenai Wildlife Refuge in Alaska using double sampling following the Gasaway et al. (1986) ratio technique and double sampling combined with logistic regression. Recently, Bart and Earnst (2002) outlined applications of double sampling to estimate bird population trends. Double sampling shows great promise in field sampling where visibility bias is considered a major issue.

4.3.6 Additional Designs

First, wildlife studies are usually plagued with the need for a large sample size in the face of budgetary and logistical constraints. Ranked set sampling provides an opportunity to make the best of available resources through what Patil et al. (1994) referred to as observational economy. Ranked set sampling can be used with any sampling scheme resulting in a probabilistic sample. A relatively large probabilistic sample of units (N) is selected containing one or more elements (n_i) of interest. The elements then are ranked within each unit based on some obvious and easily discernible characteristic (e.g., patch size, % cover type). The ranked elements are then selected in ascending or descending order of rank – one per unit – for further analysis. The resulting rank-ordered sample provides an unbiased estimator of the population mean superior in efficiency to a simple random sample of the same size (Dell and Clutter 1972).

Ranked set sampling is a technique originally developed for estimating vegetation biomass during studies of terrestrial vegetation; however, the procedure deserves much broader application (Muttlak and McDonald 1992). The technique is best explained by a simple illustration. Assume 60 uniformly spaced sampling units are arranged in a rectangular grid on a big game winter range. Measure a quick, economical indicator of plant forage production (e.g., plant crown diameter) on each of the first three units, rank order the three units according to this indicator, and measure an expensive indicator (e.g., weight of current annual growth from a sample of twigs) on the highest ranked unit. Continue by measuring shrub crown diameter on the next three units (numbers 4, 5, and 6), rank order them, and estimate the weight of current annual growth on the second-ranked unit. Finally, rank order units 7, 8, and 9 by plant crown diameter and estimate the weight of current annual growth on the lowest-ranked unit; then start the process over on the next nine units. After, completion of all 60 units, a ranked set sample of 20 units will be available for estimates of the weight of current annual growth. This sample is not as good as a sample of size 60 for estimating the weight of current annual growth, but should have considerably better precision than a simple random sample of size 20. Ranked set sampling is most advantageous when the quick, economical indicator is highly correlated with the expensive indicator, and ranked set sampling can increase precision and lower costs over simple random sampling (Mode et al. 2002). These relationships need to be confirmed through additional research. Also, the methodology for estimation of standard errors and allocation of sampling effort is not straightforward.

One of the primary functions of sampling design it to draw a sample that we hope provides good coverage of the area of interest and allows for precise estimates of the parameter of interest. The simple Latin square sampling +1 design can provide better sample coverage than systematic or simple random sampling, especially when the distribution of the target species exhibits spatial autocorrelation (Munholland and Borkowski 1996). A simple Latin square +1 design is fairly straightforward; a sampling frame is developed first (note that a Latin square +1 is irrespective of plot shape or size), then a random sample of plots is selected from each row–column combination (Fig. 4.10a), and then a single plot (the +1) is selected at random from the remaining plots (6 showing in Fig. 4.10a). Simple Latin square +1 sampling frames need not be square; they could also be linear (Fig. 4.10b) or any other a range of various shapes (Thompson et al. 1998) so long as the sampling frame can be fully specified.

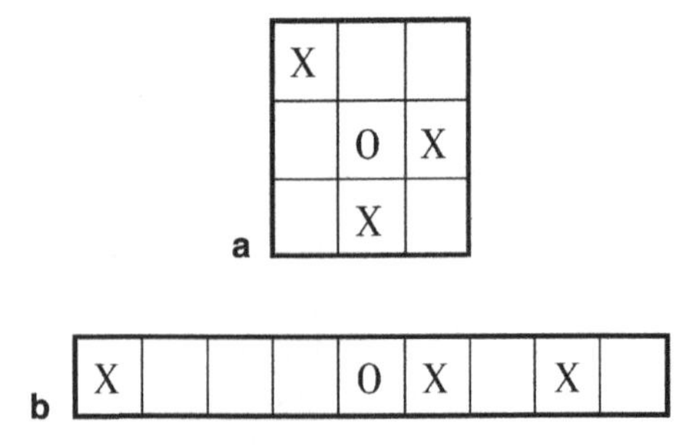

Fig. 4.10 (**a**) A simple Latin square sample of +1 drawn from a sampling frame consisting of nine square plots. Those plots having an "X" were the initial randomly selected plots based for each row–column; the plot having an "O" is the +1 plot, which was randomly selected from the remaining plots. (**b**) The same sampling frame adapted to a population tied to a linear resource. Reproduced from Thompson et al. (1998) with kind permission from Elsevier

Another approach to sampling natural resources, called generalized random-tessellation stratified designs (GRTS; Stevens and Olsen 1999, 2004), was developed to assist with spatial sampling of natural resources and ensure that the samples are evenly dispersed across the resource. Stratified sampling designs tend to spread out sample plots evenly across a landscape, simple random sampling tends to give patterns that are more spatially clumped. Under GRTS designs, the assumption is that segments of a population tend to be more similar the closer they are in space. So, in order to gather a sample of the resource in question, it is desirable to attempt to spread the points fairly evenly across the study frame. For each sampling procedure, a reverse hierarchical ordering is applied and generalized random-tessellation samples are designed such that for a given sample size (N) the first n units will be spatially balanced across a landscape (Stevens and Olsen 2004). GRTS designs have been used for large-scale environmental monitoring studies although they could potentially be used for smaller scale studies.

4.4 Point and Line Sampling

In the application of probability sampling, as seen above, one assumes each unit in the population has equal chance of being selected. Although this assumption may be modified by some sampling schemes (e.g., stratified sampling), the decisions regarding sample selection satisfy this assumption. In the cases where the probability of selection is influenced in some predictable way by some characteristic of the object or organism, this bias must be considered in calculating means and totals. Examples include line intercept sampling of vegetation (McDonald 1980; Kaiser 1983), plotless techniques such as the Bitterlich plotless technique for the estimation of forest cover (Grosenbaugh 1952), aerial transect methods for estimating big game numbers (Steinhorst and Samuel 1989; Trenkel et al. 1997), and the variable circular plot method for estimating bird numbers (Reynolds et al. 1980). If the probability of selection is proportional to some variable, then equations for estimating the magnitude and mean for population characteristics can be modified by an estimate of the bias caused by this variable. *Size bias* estimation procedures are illustrated where appropriate in the following discussion of sample selection methods.

4.4.1 Fixed Area Plot

Sampling a population is usually accomplished through a survey of organisms in a collection of known size sample units. The survey is assumed complete (e.g., a census), so the only concern is plot-to-plot variation. Estimating the variance of these counts uses standard statistical theory (Cochran 1977). Results from the counts of organisms on sample units are extrapolated to area of interest based on the proportion of area sampled. For example, the number of organisms (N) in the area of interest is estimated as

$$\hat{N} = \frac{N'}{\alpha},$$

where the numerator (N') equals the number of organisms counted and the denominator (α) equals the proportion of the area sampled. In the case of a simple random sample, the variance is estimated as

$$\hat{\text{var}}(x_i) = \sum_{i=1}^{n} \frac{(x_i - \bar{x})^2}{(n-1)},$$

where n = the number of plots, x_i = the number of organisms counted on plot i, and $\bar{x}$ = the mean number of organisms counted per sample plot.

Sampling by fixed plot is best done when organisms are sessile (e.g., plants) or when sampling occurs in a short time frame such that movements from plots has no effect (e.g., aerial photography). We assume, under this design, that counts are made without bias and no organisms are missed. If counts have a consistent bias and/or organisms are missed, then estimation of total abundance may be inappropriate (Anderson 2001). Aerial surveys are often completed under the assumption that few animals are missed and counts are made without bias. However, as a rule, total counts of organisms, especially when counts are made remotely such as with aerial surveys, should be considered conservative. Biases are also seldom consistent. For example, aerial counts are likely to vary depending on the observer, the weather, ground cover, pilot, and type of aircraft.

4.4.2 Line Intercept Sampling

The objective in line intercept sampling is estimation of parameters of two-dimensional objects in a two-dimensional study area. The basic sampling unit is a line randomly or systematically located perpendicular to a baseline and extended across the study area. In wildlife studies, the objects (e.g., habitat patches, fecal pellets groups) will vary in size and shape and thus will be encountered with a bias toward larger objects relative to the baseline. This *size bias* does not affect the estimate of aerial coverage of the objects but may bias estimates of other parameters. For example, estimates of age or height of individual plants would be biased toward the larger plants in the study area. Estimates of these parameters for the study area must be corrected for this source of bias.

Parameters in line intercept sampling are illustrated in Fig. 4.11. The study region (R) can be defined by its area (A). Within the study area there is a population (N) of individual objects ($N = 5$ in Fig. 4.11) with each defined by an area (a_i). Each object may also have an attribute (Y_i) (e.g., biomass, height, or production of shrubs) and a total of the attribute (Y) over all objects. A mean of the attribute ($\bar{Y}$) can also be calculated (Y/N). Finally, the aerial coverage (C) of N objects can be calculated where the percentage cover is the total area of individual plants divided by the area of the study area ($C = \Sigma_{ai}/A$).

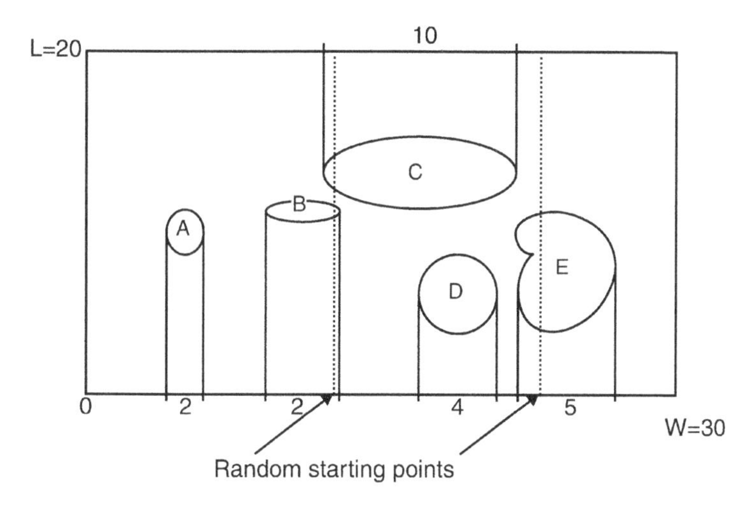

Fig. 4.11 Parameters in line intercept sampling, including the area ($A = L \times W$) of the study area, the objects of interest (1–5), aerial coverage ($a_1,\ldots,a_n$) of the objects, the intercept lines and their random starting point and spacing interval. Reproduced from McDonald (1991) with kind permission from Lyman McDonald

Here we define the following statistics for the line transect:

1. L = length of the randomly located line
2. v_i = length of the intersection of the line and the ith object
3. w_i = width of the projection of the ith object on the baseline
4. m = number of replications of the basic sampling unit (e.g., the number of lines randomly selected)
5. n = number of individual objects intercepted by the line

The primary application of line intercept sampling has been to estimate coverage by the objects of interest (Canfield 1941). The procedure also has been used to record data on attributes of encountered objects (Lucas and Seber 1977; Eberhardt 1978; McDonald 1980; Kaiser 1983), to estimate a variety of parameters including the aerial coverage of clumps of vegetation, coverage and density (number per unit area) of a particular species of plant, number of prairie dog burrows, and the coverage by different habitat types on a map.

4.4.3 *Size-biased Sampling*

Even though biologists often do not recognize that items have been sampled with unequal probability and that these data are size biased, care should be taken to recognize and correct for this source of bias. Size bias can be accounted for by calculating the probability of encountering the ith object with a given length (L) and width (W) with a line perpendicular to the baseline from a single randomly selected point

$$P_i = w_i / W; \; i = 1,2,3,\ldots, N,$$

where w_i is the width of the object in relation to the baseline. The estimate of the number of objects N is

$$\hat{N} = \sum_{i=1}^{n} (1/p_i) = W \sum_{i=1}^{n} (1/w_i),$$

and the density of objects, $D = D/A$, is estimated by

$$\hat{D} = W \sum_{i=1}^{n} (1/w_i)/(LW) = (1/L) \sum_{i=1}^{n} (1/w_i)$$

where n is the number of objects intercepted by the single line of length L.

The total of the attribute, $\hat{Y} = \sum_{i=1}^{n} Y_1$; over all objects in the area sampled is estimated by

$$\hat{Y} = W \sum_{i=1}^{n} (y_i / w_i)$$

and the mean of the attribute per object is estimated by

$$\hat{u}_y = \left(\sum_{i=1}^{n} (y_i / w_i) \right) \Bigg/ \left(\sum_{i=1}^{n} (1/w_i) \right).$$

Means and standard errors for statistical inference can be calculated from independent (m) replications of the line-intercept sample. Lines of unequal length result in means weighted by the lengths of the replicated lines.

4.4.4 Considerations for Study Design

Since the probability of encountering an object is typically a divisor in estimators, it is desirable to design sampling to maximize p_i, minimizing the variance of the estimates. The width of objects ($w_1, w_2,\ldots, w_n$) is in the denominator of the formula for calculating the probability of encountering the objects. Thus, the baseline should be established so that the projections of the objects on the baseline are maximized, increasing the probability that lines extending perpendicular to the baseline will encounter the objects. Lines of unequal length require that weighted means be used for making estimates of parameters when combining the results of independent replicate lines. As an example,

$$\hat{D} = \sum_{j=1}^{m} L_j / D_j / \sum_{j=1}^{m} L_j.$$

4.4.5 Estimation of Coverage by Objects

Estimation of coverage of objects, such as clumps of vegetation, is a common use of line-intercept sampling in wildlife studies (Canfield 1941). The estimate of percent cover of objects is unbiased and can be estimated by the percentage of the line that is intersected by the objects (Lucas and Seber 1977) using the formula

$$\hat{C} = \sum_{i=1}^{n} v_i / L,$$

where v_i is the length of the intersection of the *i*th object with a single replicate line of length L. Again, replication of lines of intercept m times allows the estimate of a standard error for use in making statistical inferences. Equal length lines can be combined in the above formula to equal L. Weighted means are calculated when lines are of unequal length.

4.4.6 Systematic Sampling

Line intercept methodologies often employ systematic sampling designs. In the systematic placement of lines, the correct determination of the replication unit and thus the correct sample size for statistical inferences is an issue. If sufficient distance between lines exists to justify an assumption of independence, then the proper sample size is the number of individual lines and the data are analyzed as if the individual lines are independent replications. However, if the assumption of independence is not justified (i.e., data from individual lines are correlated) then the set of correlated lines is considered the replication unit. The set of m lines could be replicated m' times using a new random starting point each time, yielding an independent estimate of parameters of interest with $L' = m(L)$ as the combined length of the transects to yield m' independent replications. Statistical inferences would follow the standard procedures.

The general objectives in systematic location of lines are to:

1. Provide uniform coverage over the study region, R
2. Generate a relatively large variance within the replication unit vs. a relatively small variance from replication to replication

For example, the total biomass and cover by large invertebrates on tidal influenced beaches may be estimated by line intercept sampling with lines perpendicular to the tidal flow. Standard errors computed for systematically located lines should be conservative (too large) if densities of the invertebrates are more similar at the same tidal elevation on all transects vs. different tidal elevations on the same transect (condition 2 above is satisfied). Even if individual lines cannot be considered independent, when condition 2 is satisfied then standard computational procedures for standard errors can be used (i.e., compute standard errors as if the data were independent) to produce conservative estimates.

4.4.7 One Line with No Replication

Often one or more long lines are possible but the number is not sufficient to provide an acceptable estimate of the standard error. Standard errors can be estimated by breaking the lines into subsets, which are then used in a jackknife or bootstrap procedure. A good example might be surveys along linear features such as rivers or highways. Skinner et al. (1997) used bootstrapping for calculating confidence intervals around estimates of moose density along a long transect zigzagging along the Innoko River in Alaska. Each zigzag is treated as an independent transect. While there may be some lack of independence where the segments join, it is ignored in favor of acquiring an estimate variance for moose density along the line. This works best with a relatively large sample size that fairly represents the area of interest. Skinner et al. (1997) reported satisfactory results with 40–60 segments per stratum.

Generally, the jackknife procedure estimates a population parameter by repeatedly estimating the parameter after one of the sample values is eliminated from the calculation resulting in several pseudoestimates of the parameter. The pseudoestimates of the parameter are treated as a random sample of independent estimates of the parameter, allowing an estimate of variance and confidence intervals. The bootstrap is the selection of a random sample of n values $X_1, X_2, \ldots, X_n$ from a population and using the sample to estimate some population parameter. Then a large number of random samples (usually >1,000) of size n are taken from the original sample. The large number of bootstrap samples is used to estimate the parameter of interest, its variance, and a confidence interval. Both methods require a large number of calculations and require a computer. For details on jack-knife, bootstrap, and other computer-intensive methods, see Manly (1991).

4.5 Line Transects

Line transects are similar to line intercept sampling in that the basic sampling unit is a line randomly or systematically located on a baseline, perpendicular to the baseline, and extended across the study region. Unlike line intercept sampling, objects are recorded on either side of the line according to some rule of inclusion. When a total count of objects is attempted within a fixed distance of the line, transect sampling is analogous to sampling on fixed plot (see Sect. 4.4.1). This form of line transect, also known as a belt (strip) transect, has been used by the US Fish and Wildlife Service (Conroy et al. 1988) in aerial counts of black ducks. As with most attempts at total counts, belt transect surveys usually do not detect 100% of the animals or other objects within the strip. When surveys are completed according to a standard protocol, the counts can be considered an index. Conroy et al. (1988) recognized ducks were missed and suggested that survey results should be considered an index to population size.

Line-transect sampling wherein the counts are considered incomplete has been widely applied for estimation of density of animal populations (Laake et al. 1979, 1993). Burnham et al. (1980) comprehensively reviewed the theory and applications of this form of line-transect sampling. Buckland et al. (1993) updated the developments in line-transect sampling through the decade of the 1980s. Alpizar-Jara and Pollock (1996), Beavers and Ramsey (1998), Manly et al. (1996), Quang and Becker (1996, 1997), and Southwell (1994) developed additional theory and application. The notation in this section follows Burnham et al. (1980).

Line-transect studies have used two basic designs and analytic methods depending on the type of data recorded (1) perpendicular distances (x) or sighting distances (r) and (2) angles (θ) (Fig. 4.12). Studies based on sighting distances and angles are generally subject to more severe biases and are not emphasized in this discussion.

There are several assumptions required in the use of line-transect surveys (Buckland et al. 2001), including:

1. Objects on the line are detected with 100% probability.
2. Objects do not move in response to the observer before detection (e.g., animal movements are independent of observers).
3. Objects are not counted twice.
4. Objects are fixed at the point of initial detection.
5. Distances are measured without errors.
6. Transect lines are probabilistically located in the study area.

4.5.1 Detection Function

The probability of detecting an object at a perpendicular distance of x from the transect line is known as the object's detection function $g(x)$ illustrated in Fig. 4.12. Assumption 1, above, that $g(0) = 1$ (i.e., the probability is 1.0 that an object with $x = 0$ will be detected) is key and allows estimation of the necessary parameter for correcting for visibility bias away from the line (i.e., $g < 1.0$). The detection function can be made up of a mixture of more simple functions which depend on factors

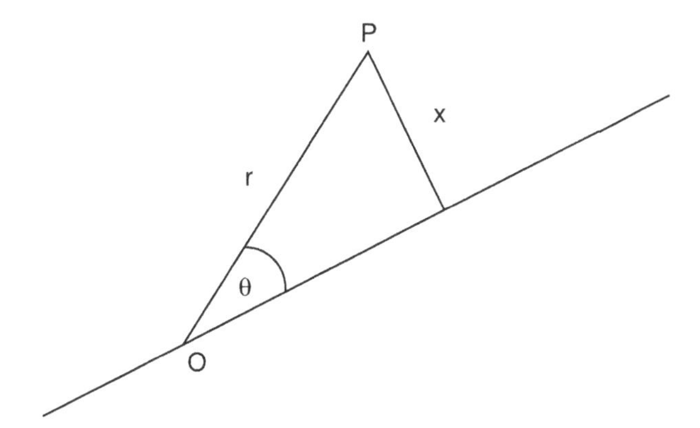

Fig. 4.12 The types of data recorded for the two basic types of line-transect study designs including perpendicular distances (x) or sighting distances (r) and angles (θ). The probability of detecting an object at a perpendicular distance of x from the transect line is known as the object's detection function $g(x)$. Reproduced from Burnham et al. (1980) with kind permission from The Wildlife Society

such as weather, observer training, vegetation type, etc., so long as all such functions satisfy the condition that probability of detection is 100% at the origin $x = 0$ (Burnham et al. 1980).

The average probability of detection for an object in the strip of width $2w$ is estimated by

$$\hat{P}_w = 1/\hat{w}f(0)$$

where $f(x)$ denotes the relative probability density function of the observed right angle distances, x_i, $i = 1, 2,\ldots,n$. The function $f(x)$ is estimated by a curve fitted to the (relative) frequency histogram of the right angle distances to the observed objects and $\hat{f}(0)$ is estimated by the intersection of $f(x)$ with the vertical axis at $x = 0$. Given $\hat{P}_w = 1/\hat{w}f(0)$, and detection of n objects in the strip of width $2w$ and length L, the observed density is computed by

$$\hat{D} = n/2Lw.$$

The observed density is corrected for visibility bias by dividing by the average probability of detection of objects to obtain

$$\begin{aligned}\hat{D} &= (n/2Lw)/(1/w\hat{f}(0)) \\ &= n\hat{f}(0)/2L\end{aligned}$$

The width of the strip drops out of the formula for estimation of density of objects allowing line-transect surveys with no bound on w (i.e., $w = \infty$). However, at large distances from the line, the probability of detection becomes very low and it is desirable to set an upper limit on w such that 1–3% of the most extreme observations are truncated as *outliers*. Decisions on dropping outliers from the data set can be made after data are collected.

4.5.2 Replication

Estimates of the variances and standard errors associated with line-transect sampling are usually made under the assumption that the sightings are independent events and the number of objects detected is a Poisson random variable. If there are enough data (i.e., ≥40 detected objects) on independent replications of transect lines or systematic sets of lines, then a better estimate of these statistics can be made. Replications must be physically distinct and be located in the study area according to a true probability sampling procedure providing equal chance of detection for all individuals. Given independent lines, the density should be estimated on each line and the standard error of density estimated by the usual standard error of the mean density (weighted by line length if lines vary appreciably in length).

If there are not enough detections on independent replications, then jackknifing the lines should be considered (Manly 1991). For example, to jackknife the lines, repeatedly leave one line out of the data set and obtain the pseudoestimate of density by biasing estimates on the remaining lines. The mean of the pseudoestimates and the standard error of the pseudoestimates would then be computed. While jackknifing small samples will allow the estimation of variance, sample sizes are not increased and the pseudovalues are likely to be correlated to some extent, resulting in a biased estimate of variance. The significance of this bias is hard to predict and should be evaluated by conducting numerous studies of a range of situations before reliance is placed on the variance estimator (Manly 1991).

4.5.3 Line-transect Theory and Application

Size bias is an issue when the probability of detecting subjects is influenced by size (e.g., the subject's width, area, etc.). In particular, animals typically occur in groups, and the probability of detecting an individual increases with group size. Estimates of group density and mean group size are required to estimate the density of individuals and an overestimate of mean group size will lead to an overestimate of true density. Drummer and McDonald (1987) proposed bivariate detection functions incorporating both perpendicular distance and group size. Drummer (1991) offered the software package SIZETRAN for fitting size-biased data. Quang (1989) presented nonparametric estimation procedures for size-biased line-transect surveys.

Distance-based methods have been combined with aerial surveys (Guenzel 1997) to become a staple for some big game biologists in estimating animal abundance. As pointed our earlier (Sect. 4.5.1), the probability of detecting objects during line-transect surveys can influence parameter estimates. Quang and Becker (1996) offered an approach for incorporating any appropriate covariate influencing detection into aerial surveys using line-transect methodology by modeling scale parameters as log-linear functions of covariates. Manly et al. (1996) used a double-sample protocol during aerial transect surveys of polar bear. Observations by two observers were analyzed using maximum likelihood methods combined with an information criterion (AIC) to provide estimates of the abundance of polar bears. Beavers and Ramsey (1998) illustrated the use of ordinary least-squares regression analyses to adjust line-transect data for the influence of variables (covariates).

The line-transect method is also proposed for use with aerial surveys and other methods of estimating animal abundance such as a form of capture–recapture (Alpizar-Jara and Pollock 1996) and double sampling (Quang and Becker 1997; Manly et al. 1996). Lukacs et al. (2005) investigated the efficiency of trapping web designs, which can be combined with distance sampling to estimate density or abundance (Lukacs et al. 2004) and provided software for survey design (Lukacs 2002). In addition, line-transect methods have been developed which incorporate covariates (Marques and Buckland 2004), combine capture–mark–recapture data (Burnham et al. 2004), and a host of other potential topics (Buckland et al. 2004). The field of abundance and density estimation from transect-based sampling schemes is active, so additional methodologies are sure to be forthcoming.

4.6 Plotless Point Sampling

The concept of plotless or distance methods was introduced earlier in our discussion of the line intercept method (see Sect. 4.4.2). Plotless methods from sample points using some probability sampling procedure are considered more efficient than fixed area plots when organisms of interest are sparse and counting of individuals within plots is time consuming (Ludwig and Reynolds 1988).

4.6.1 *T-square Procedure*

In the *T*-square procedure, sampling points are at random, or systematically selected locations, and two distances are taken at each point (Fig. 4.13). For example, this method has been used in the selection individual plants and animals for study. McDonald et al. (1995) used the method for selection of invertebrates in the study of the impacts of the Exxon Valdez oil spill. The two measurements include:

1. The distance (x_i) from the random point (*O*) to the nearest organism (*P*)
2. The distance (z_i) from the organism (*P*) to its nearest neighbor (*Q*) with the restriction that the angle *OPQ* must be more than 90° (the *T*-square distance).

The most robust population density estimator from *T*-square data is the compound estimate using both measures x_i and z_i (Byth 1982), computed as

$$\hat{N}_{\mathrm{T}} = \frac{n^2}{2\sum(x_i)\sqrt{2}\sum(z_i)},$$

where *n* represents the number of random points (the sample size). The somewhat complicated standard error is calculated on the reciprocal of the compound density given by Diggle (1983) as

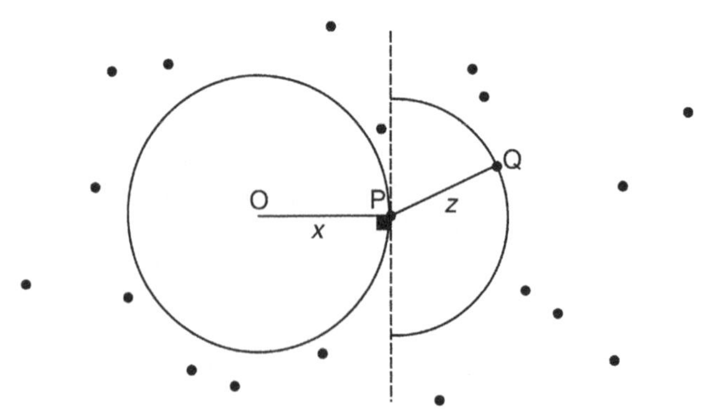

Fig. 4.13 The two *T*-square sampling points and the two distances measured at each point. Reproduced from Morrison et al. (2001) with kind permission from Springer Science + Business Media

$$\text{Standard error of } (1/\hat{N}_T) = \sqrt{\frac{8(\bar{z}^2 S_x^2 + 2\bar{x}\bar{z}s_{xz} + \bar{x}^2 s_z^2)}{n}},$$

where S_x^2 is the variance of point-to-organism distances, S_z^2 is the variance of *T*-square organism-to-neighbor distances, and S_{xz} is the covariance of *x* and *z* distances.

4.6.2 Variable Area Circular Plots

The variable circular plot is often applied as a variation of line-transect sampling for estimating the number of birds in an area (Reynolds et al. 1980). Counts of organisms along a transect is a standard sampling procedure, particularly when the organisms of interest are relatively rare. The variable circular plot is recommended, however, in dense vegetation and rough terrain where attention may be diverted from the survey and toward simply negotiating the transect line. An added advantage of the circular plot is that the observer can allow the surveyed animals to settle down. For example, in breeding bird surveys, observers wait several minutes to allow the songbirds disturbed by their arrival to settle down before counts begin and sound as well as visual observation can be used to detect birds.

While the plot is perceived as circular, the procedure is plotless since all observations made from a point, in any direction, are recorded. Plot size is a function of the observer's ability to detect the organism of interest and not the design (Ramsey and Scott 1979). As with a line transect, estimation of the number of organisms within the area surveyed is based on a detection function $g(x)$ that represents the distance at which the observer can detect organisms of interest. Density is estimated as

$$\hat{D} = \frac{n}{\pi P^2},$$

where *n* is the number of birds detected and the denominator is the area of a circle with a radius of ρ, the distance from the plot center within which we would expect *n* birds to be located (Lancia et al. 2005).

Program DISTANCE (Buckland et al. 1993, 2001) can be used to estimate bird densities from variable circular plot data. The theoretical models and estimation methods used in DISTANCE work best when at least 40 independent observations exist for the area of interest. Data may be pooled across time periods or species to estimate detection functions resulting in an average detection probability.

Distance estimates are usually recorded as continuous data. Buckland (1987) proposed binomial models for variable circular plots where subjects are categorized as being within or beyond a specified distance. Estimates of distances to detected subjects may also be assigned to intervals with the frequency of observations analyzed in the intervals. Placing detected subjects in intervals of distance should be more accurate for subjects close to the observer so we recommend that intervals near the center of the plot be smaller than intervals farthest from the observer. The critical point estimate is the intersection of the detection function at the origin. Burnham et al. (1980) suggested trimming data so that roughly 95% of the observed distances are used in the analysis. The assumption is that the outer 5% of observations are outliers that may negatively affect density estimates.

The assumption that counts are independent may be difficult, as subjects being counted are seldom marked or obviously unique. Biologists may consider estimating use per unit area per unit time as an index to abundance. When subjects are relatively uncommon, the amount of time spent within distance intervals can be recorded. In areas with a relatively high density of subjects, surveys can be conducted as instantaneous counts of animals at predetermined intervals of time during survey periods.

4.7 Model-based Sampling

The major advantage of classic probability sampling is that assumptions regarding the underlying population are unnecessary. Using this approach, the population of interest is considered fixed in time and space. Randomness is present only because of the sample-selection process and variation within the population must be dealt with in the course of data analysis. Model-based sampling uses assumptions to account for patterns of variability within the population and uses these patterns in sampling schemes.

As a note of caution, literature dealing with strictly model-based studies often focuses on the analysis of data. Model-based approaches are often promoted as a less costly and logistically easier alternative to large design-based field studies. The assumption sometimes seems to be that design deficiencies in model-based studies can be overcome by modeling. Data analysis can improve the quality of the information produced by these studies; however, fundamentally flawed design issues should not be ignored. It is worth repeating the philosophy of model selection and data analysis advice on modeling in science as offered by McCullagh and Nelder (1983, p. 6) and Lebreton et al. (1992):

> Modeling in science remains, partly at least, an art. Some principles exist, however, to guide the modeler. The first is that all models are wrong; some, though, are better than others and we can search for the better ones. At the same time we must recognize that eternal truth is not within our grasp. The second principle (which applies also to artists!) is not to fall in love with one model, to the exclusion of alternatives. Data will often point with almost equal emphasis at several possible models and it is important that the analyst accepts this. A third principle involves checking thoroughly the fit of the model to the data, for example by using residuals and other quantities derived from the fit to look for outlying observations, and so on. Such procedures are not yet fully formalized (and perhaps never will be), so that imagination is required of the analyst here, as well as in the original choice of models to fit.

Our distinction between model-based and design-based sampling is somewhat artificial. Modeling is defined as the mathematical and statistical processes involved in fitting mathematical functions to data. Given this definition, models are included in all study designs. The importance of models and assumptions in the analysis of empirical data ranges from little effect in design-based studies to being a critical part of data analysis in model-based studies. Design-based studies result in predicted values and estimates of precision as a function of the study design. Model-based studies lead to predicted values and estimates of precision based on a combination of study design and model assumptions often open to criticism. The following discussion focuses on the most prevalent model-based studies that are heavily dependent on assumptions and estimation procedures involving linear and logistic regression for data analysis. These study methods are only slightly more model-based than some previously discussed (e.g., plotless and line intercept) involving estimates of *nuisance parameters* such as detection probabilities, probabilities of inclusion, and encounter probabilities.

4.7.1 Capture–Recapture Studies

When observational characteristics make a census of organisms difficult, capture–recapture methods may be more appropriate for estimating population abundance, survival, recruitment, and other demographic parameters (e.g., breeding probabilities, local extinction, and recolonization rates). In capture–recapture studies, the population of interest is sampled two or more times and each captured animal is uniquely marked. Depending upon study objectives, captures may be by live trapping, harvest, passive integrated transponder (PIT) tags, radioactive markers, radio-telemetry, observing marks such as neck or leg bands, or repeated counts. Some individual animals may carry unique markings such as color patterns (e.g., stripes on a tiger), vocal patterns (e.g., unique bird sonograms), and even genetic markers. With capture–recapture studies, there is a concern with variation from both the sampling procedure and detectability (capture probability) issues related to the individuals under study (Lancia et al. 2005; Williams et al. 2002). Some detectability issues can be solved through study design, as described by our discussion of line intercept and double sampling (see Sects. 4.3.5 and 4.5). Capture–recapture studies, and the extensive theory dealing with models for the analysis of these data, combine issues related to the sampling process and those issues related to the uncertainty regarding the appropriate model to be used to explain the data (Williams et al. 2002).

In general, sample plans should allow the study to meet the assumptions of the model being used to analyze the resulting data and allow the desired statistical inference. We consider a range of models including the relatively simple Petersen–Lincoln model (Lincoln 1930), the closed and open population capture–recapture Cormack–Jolly–Seber and Jolly–Seber model (Otis et al. 1978; Seber 1982; Pollock et al. 1990; Williams et al. 2002), models for survival of radio-tagged individuals (Pollock et al. 1989; Venables and Ripley 2002), and models for presence–absence data (MacKenzie et al. 2002). For a general review of modeling of capture–recapture statistics we refer you to Pollock (1991) and Williams et al. (2002).

4.7.2 Petersen–Lincoln Model

The Petersen–Lincoln model has been used for years by wildlife biologists to estimate animal abundance and is considered a closed population model. The Petersen–Lincoln model should be considered an index to abundance when a systematic bias prevents of one or more of the assumptions described below from being satisfied. In a Petersen–Lincoln study, a sample n_1 of the population is taken at time t_1 and all organisms are uniquely marked. A second sample n_2 is taken at time t_2 and the organisms captured are examined for a mark and a count is made of the recaptures (m_2). Population size (N) is estimated as

$$\hat{N} = n_1 n_2 / m_2.$$

The assumptions for the Petersen–Lincoln model are:

1. The population is closed (i.e., N does not change between time t_1 and time t_2).
2. All animals have equal probability of capture in each sample.
3. There are no errors of measurement.

The assumption of closure is fundamental to the Petersen–Lincoln and other closed population models. Populations can increase or decrease through reproduction or immigration and mortality or emigration, respectively. The elimination of immigration and emigration is difficult in large and relatively mobile species. The success of mark–recapture studies with mobile populations often depends on the selection of study area boundaries grounded in this assumption. Lebreton et al. (1992 [from Gaillard 1988]) provided an example of studies of roe deer (*Capreolus capreolus*) in a large fenced enclosure, essentially creating an artificially closed population. Numerous studies of larger and more mobile species have attempted to document and account for immigration and emigration through the use of radiotelemetry (e.g., Miller et al. 1997). The assumption can best be met for small and relatively immobile species by keeping the interval between samples short. Lancia et al. (2005) reported 5–10 days as the typical interval, although the appropriate period between samples will be taxon-specific.

The assumption of closure can be relaxed in some situations (Seber 1982). Losses from the population are allowed if the rate of loss is the same for marked and unmarked individuals, which is a difficult assumption to justify. If there are deaths at the time of marking the first sample, then the Petersen–Lincoln estimate applies to the number of animals alive in the population after time t_1. If there is natural mortality of animals between the two samples and it applies equally to marked and unmarked animals, then the estimate applies to the population size at the time of the release of the first sample. Kendall (1999) suggested that if animals are moving in and out of the study area in a completely random fashion, then the Petersen–Lincoln estimator (and closed population methods in general) is unbiased for the larger estimate of abundance. The jackknife estimator of Burnham and Overton (1978) is a good general tool for dealing with heterogeneity of capture probabilities. When heterogeneity is not severe, turning multiple samples into two, as in Menkins and Anderson (1988), works reasonably well. Kendall (1999) also discussed the implications of these and other types of closure violations for studies involving greater than two samples of the population.

The second assumption is related to the first and implies that each sample is a simple random sample from a closed population and that marked individuals have the same probability of capture as the unmarked animals. If the probability of capture is different for different classes of animals (say young vs. adults) or for different locations, then the sampling could follow the stratified random sampling plan. It is common in studies of large populations that a portion of the animal's range may be inaccessible due to topography or land ownership. The estimate of abundance is thus limited to the area of accessibility. This can be a problem for animals that have large ranges, as there is no provision for animals being unavailable during either of the sampling periods. The probability of capture can also be influenced by the conduct of the study such that animals become trap happy (attracted to traps) or trap shy (repulsed from traps). The fact that study design seldom completely satisfies this assumption has led to the development of models (discussed below) that allow the relaxation of this requirement.

The third assumption depends on an appropriate marking technique. Marks must be recognizable without influencing the probability of resighting or recapture. Thus, marks must not make the animal more or less visible to the observer or more or less susceptible to mortality. Marks should not be easily lost. If the loss of marks is a problem, double marking (Caughley 1977; Seber 1982) can be used for corrections to the recapture data. New methods of marking animals are likely to help refine the design of mark–recapture observational studies and experiments (Lebreton et al. 1992). This assumption illustrates the need for standardized methods and good standard operating procedures so that study plans are easy to follow and data are properly recorded.

An appropriate study design can help meet the assumptions of the Petersen–Lincoln model, but the two trapping occasions do not allow a test of the assumptions upon which the estimates are based. Lancia et al. (2005) suggested that in two-sample studies, the recapture method be different and independent of the initial sample method. For example, one might trap and neckband mule deer and then use observation as the *recapture* method. This recommendation seems reasonable and should eliminate the concern over trap response and heterogeneous capture probabilities.

4.7.3 Closed Population Mark–Recapture

Otis et al. (1978) and White et al. (1982) offered a modeling strategy for making density and population size estimates using capture data on closed animal populations. With a complete capture history of every animal caught, these models allow relaxation of the equal catchability assumption (Pollock 1974; Otis et al. 1978; Burnham and Overton 1978; White et al. 1982; Pollock and Otto 1983; Chao 1987, 1988, 1989; Menkins and Anderson 1998; Huggins 1989, 1991; Brownie et al. 1993;

Lee and Chao 1994). A set of eight models is selected to provide the appropriate estimator of the population size. The models are M_0, M_t, M_b, M_h, M_{tb}, M_{th}, M_{bh}, and M_{tbh}, where the subscript "0" indicates the null case, and t, b, and h, are as follows:

- 0 – All individuals have the same probability of capture throughout the entire study
- t – Time-specific changes in capture probabilities (i.e., the Darroch 1958 model where probability of capture is the same for all individuals on a given occasion)
- b – Capture probabilities change due to behavioral response from first capture (i.e., probability of capture remains constant until first capture, can change once, and then remains constant for the remainder of the study)
- h – Heterogeneity of capture probabilities in the population (i.e., different subsets of the individuals have different probability of capture but, probability of capture does not change during the course of the study)

This series of eight models includes all possible combinations of the three factors, including none and all of them (Table 4.1 and Fig. 4.14). Population estimates from removal data can also be obtained because the estimators for the removal model of Zippen (1958) are the same as the estimators under the behavioral model M_b.

Estimators for the eight models can be found in Rexstad and Burnham (1991). We suggest you also check the US Geological Survey Patuxent Wildlife Research Center's software archive (http://www.pwrc.usgs.gov) for additional information and updated software for mark–recapture data. Since explicit formulas do not exist for the estimators, they must be solved by iterative procedures requiring a computer. The design issues are essentially identical to the two-sample Petersen–Lincoln study with the condition of assumption 2 met through the repeated trapping events and modeling.

4.7.4 Population Parameter Estimation

When studying animal populations, survival and recruitment may be of equal or greater interest than density or absolute abundance. Capture–recapture models

Table 4.1 The eight models summarized by symbol, sources of variation in capture probability, and the associated estimator, if any

Model	Sources of variation in capture possibilities	Appropriate estimator
M_0	None	Null
M_t	Time	Darroch
M_b	Behavior	Zippin
M_h	Heterogeneity	Jacknife
M_{tb}	Time, behavior	None
M_{th}	Time, heterogeneity	None
M_{bh}	Behavior, heterogeneity	Generalized removal
M_{tbh}	Time, behavior, heterogeneity	None

The names provided are those used by program Capture and MARK for these estimators

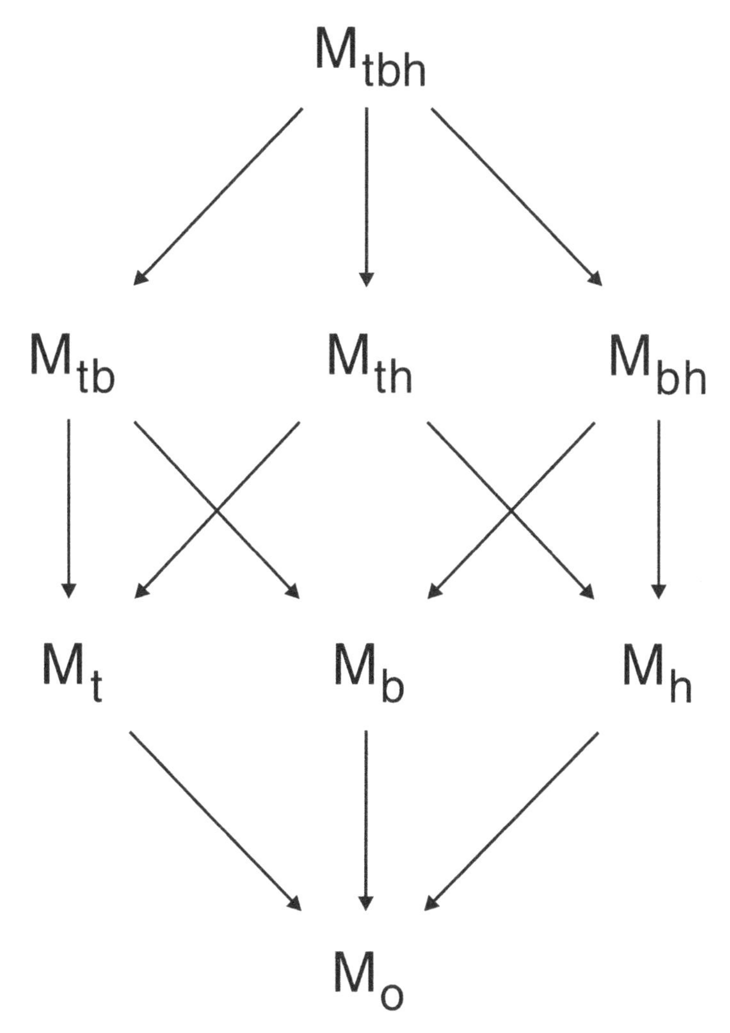

Fig. 4.14 The series of eight closed population models proposed includes all possible combinations of three factors, including none and all of them. Reproduced from Otis et al. (1978) with kind permission from The Wildlife Society

focused on estimation of survival originally treated survival as a nuisance parameter to estimation of abundance (Williams et al. 2002). Beginning around the 1980s, however, survival estimation became a primary state variable of interest in wildlife population ecology (Lebreton et al. 1992). Here we provide a brief overview of several related topics with respect to parameter estimation, but refer readers to Seber (1982), Williams et al. (2002), and Amstrup et al. (2005) for a detailed discussion as the literature and methods for estimating population parameters are continually being reevaluated and refined. Additionally, see Chap. 2 for a list of statistical programs that can be used for estimation procedures.

4.7.4.1 Open Population Mark–Recapture

The Cormack–Jolly–Seber and Jolly–Seber methods (Seber 1982; Williams et al. 2002) allow for estimates of abundance, survival, recruitment, and account for nuisance parameters (e.g., detectability). These models are referred to as open population models because they allow for gain or loss in animal numbers during the study. Note that the rate of gain, sometimes called the birth rate, could be recruitment and immigration and the rate of loss, sometimes called the death rate, could be death and permanent emigration. Estimates of population size follow the Petersen–Lincoln estimator previously discussed (see Sect. 4.7.2). The estimate of survival is the ratio of the number of marked animals in the $i+1$ sample to the number of marked animals in the ith sample. Recruitment from time period i to time period $i+1$ is estimated as the difference between the size of the population at time i and the expected number of survivors from i to $i+1$. Formulas for these estimators are presented with examples in Williams et al. (2002) and Lancia et al. (2005).

Assumptions required by the models and possible sampling implications include:

1. The probability of capture in the ith sample is the same for all animals (marked and unmarked).
2. The probability of survival from t_i to t_{i-1} is the same for all marked animals.
3. The probability that an animal in the ith sample is returned alive is the same for all animals.
4. Marks are not lost or overlooked.
5. The sampling is instantaneous and fates of marked individuals are independent.
6. Emigration from the sample area is permanent.

If the probability of capture varies by characteristics such as age and sex of animal then the data can be stratified during analysis. Similarly, if capture probabilities vary depending on habitat or other environmental variables, then stratification of the study area during sampling may be attempted with separate estimates made for each habitat. The assumption of equal probability of survival (and its reciprocal, the probability of death) of marked animals is not required for either method. For example, young and adult marked animals can have different survival probabilities, a common occurrence in wildlife populations. Using a classic design of one capture occasion per period, survival–immigration and death–permanent emigration are completely confounded in capture–recapture data. However, for the models to be useful, one must accept the assumption that survival probability is the same for marked and unmarked animals. In many situations, one can assume that immigration and emigration offset and thus have little impact on the assumption that estimates of the rate of gain and loss equal birth rate and death rate. If a segregation of these rates is desired, however, then study boundaries should minimize this interchange or interchange must be estimated (e.g., Miller et al. 1997). Emigration and immigration could be significant problems in short-term studies of highly mobile animals with large home ranges (e.g., bears) or in source populations where emigration far exceeds immigration (e.g., dispersal of young mountain lions as described by Hornocker 1970). The confounding mentioned above can be partially avoided by using more complex applications of these models. If the study is being done at multiple sites then multistate models (e.g., Brownie et al. 1993; Williams et al. 2002) can be used to estimate probabilities of movement between areas. Supplemental telemetry could be used to estimate some of the movement. Band recoveries can be combined with recapture information to separate philopatry from survival (Burnham 1993). In age-dependent models, recruitment from a lower age class can be separate from immigration (Nichols and Pollock 1983). There are many different types of capture–recapture models including approaches outlined by Burnham (1993), the super-population approach of Schwarz and Arnason (1996), a host of models by Pradel (1996) which focus on survival and recruitment, as well as the Link and Barker (2005) reparameterization of the Pradel (1996) model to better estimate those recruitment parameters.

4.7.4.2 Pollock's Robust Design

Lancia et al. (2005) pointed out that the distinction between open and closed populations is made to simplify models used to estimate population parameters of interest. The simplifications are expressed as assumptions and study design must take these simplifying assumptions into account. Pollock (1982) noted that long-term studies often consist of multiple capture occasions for each period of interest. He showed that the extra information from the individual capture occasions could be exploited to reduce bias in Jolly–Seber estimates of abundance and recruitment when there is heterogeneity in detection probabilities.

Under Pollock's robust design, each sampling period consists of at least two subsamples, ideally spaced closely together so that the population can be considered closed to additions and deletions during that period. Kendall and Pollock (1992) summarized other advantages of this basic design, in that abundance, survival rate, and recruitment can be estimated for all time periods in the study, whereas with the classic design one cannot estimate abundance for the first and last periods, survival rate to the last period, and the first and last recruitment values; recruitment can be separated into immigration and recruitment from a lower age class within the population when there are at least two age classes, whereas the classic design requires three age classes (Nichols and Pollock 1990); abundance and survival can be estimated with less dependence, thereby lessening some of the statistical problems with density-dependent modeling (Pollock et al. 1990); and study designs for individual periods can be customized to meet specific objectives, due to the second level of sampling. For instance, adding more sampling effort in period i (e.g., more trapping days) should increase precision of the abundance estimate for period i. However, adding more sampling effort after period i should increase precision of survival rate from i to $i + 1$.

The additional information from the subsamples in the robust design allows one to estimate the probability that a member of the population is unavailable for detection (i.e., a temporary emigrant) in a given period (Kendall et al. 1997). Depending on the context of the analysis, this could be equivalent to an animal being a nonbreeder or an animal in torpor. Based on the advantages listed above, we recommend that Pollock's robust design be used for most capture–recapture studies. There are no apparent disadvantages in doing so. Even the assumption of closure across subsamples within a period is not necessarily a hindrance (Schwarz and Stobo 1997; Kendall 1999). Even where it turns out that it is not possible to apply sufficient effort to each subsample to employ the robust design, the data still can be pooled and traditional methods used. The advantages of the robust design derive from the second source of capture information provided by the subsamples. Obviously, the overall study design must recognize the desired comparisons using the open models, even though the distribution of the samples for the closed model (as long as it is a probabilistic sample) is of relatively little consequence.

4.7.4.3 Time-to-event Models

Survival analysis is a set of statistical procedures for which the outcome variable is the time until an event occurs (Kleinbaum 1996). As such, survival analysis is concerned with the distribution of lifetimes (Venables and Ripley 2002). In wildlife research, survival analysis is used to estimate survival ($\hat{S}$), or the probability that an individual survives a specified period (days, weeks, years). Because estimates of survival are used in population models, evaluations of changing population demography, and as justification for altering management practices, approaches to survival analysis have becoming increasingly common in wildlife research. Probably the most common approach to survival analysis in wildlife science is estimation using known fate data based on radio-telemetry where individuals are relocated on some regular basis. Another common application of time to event models has been recent work focused on estimating survival of nests where the event of interest is the success or failure of a nest (Stanley 2000; Dinsmore et al. 2002; Rotella et al. 2004; Shaffer 2004).

Generally, estimation of survival is focused on the amount of time until some event occurs. Time-to-event models are not constrained to evaluating only survival, as the event of interest could include not only death, but also recovery (e.g., regrowth after a burn), return to a location (e.g., site fidelity), incidence (e.g., disease transmission or relapse), or any experience of interest that happens to an individual (Kleinbaum 1996). Typically, the time in time-to-event models refers to an anthropomorphic specification set by the researchers (e.g., days, months, seasons) based on knowledge of the species of interest. In wildlife studies, the event of interest is usually death (failure).

One key point that must be addressed is censoring, both right and interval censoring and left truncation. Censoring occurs when the information on the individual(s) survival is incomplete, thus we do not know the survival times exactly. There are three types of censoring which influence survival modeling:

- Right censoring – when the dataset becomes incomplete on the right size of the follow-up period
- Middle censoring – when during the study, the probability of detecting an individual is <1
- Left truncation – when the dataset is incomplete at the left side of the follow-up period

Censoring of individuals in wildlife studies can be caused by several factors, including loss or failure of the radio-tag, detection probabilities <1, topography, or observer search image as well as emigration, which we break into two classes, temporary, or when a radio tagged individual leaves the study area for 1 sampling occasion and then returns (e.g., middle censoring), and permanent emigration, or when an individual leaves the study area and does not return. One primary difficulty in radio-telemetry studies is distinguishing radio failure from permanent emigration. Additionally, nonrelocation due to temporary emigration during an encounter (sampling) occasion also causes censoring. For example, temporary emigration can be a problem when attempting to evaluate effects of some factor on survival, but this factor differs on and off the study area (e.g., hunting on public lands [study area], but no hunting off the study area on adjacent private lands).

There are three basic survivorship functions using for analysis of time to event data. First, consider that T is a random variable that indicated the length of time before a specific event occurs, e.g., the event typically is "failure", i.e., death of a study individual, but it could be "success," such as returning to an area.

The three potential survivorship functions (from Venables and Ripley 2002; Hosmer and Lemeshow 1999) are

- $S(t) = \Pr(T>t)$, which is the survivorship function which described the probability than an individual animal survives longer than time T. This is frequently estimated as the proportion of animals surviving longer than t, $\hat{S}(t)$
- $f(t) = 1 - S(t)$ or $f(t) = \mathrm{d}F(t)/\mathrm{d}t = -\,\mathrm{d}S(t)/\mathrm{d}t$, which is the probability density function for the time until event. $f(t)$ is most often called the life distribution or the failure time distribution
- $h(t) = f(t)/S(t)$ is the hazard function and is interpreted as a conditional probability of failure rate

Perhaps the most common estimator for survival is the Kaplan–Meier product limit estimator (Kaplan and Meier 1958; Pollock et al. 1989). The Kaplan–Meier estimator does not make any underlying assumptions about the function being estimated and is basically an extension of the binomial estimator (Williams et al. 2002). In its simplest form, the Kaplan–Meier estimator is

$$\hat{S}(t) = \Pi_{a=1}^{J}\left(1 - \frac{d_j}{r_j}\right),$$

where the product of all j terms for which $a_j <$ time t given that a_j are the discrete time points (j) when death occurs, d_j is the number of deaths at the jth time point, and r_j is the number of animals at risk at the jth time point. Thus, the probability of surviving from time 0 to a_1 (interval during which first death occurs) is estimated as

$$\hat{S}(a_1) = 1 - \frac{d_1}{r_1}$$

and the probability of surviving from time a_1 to a_2 is $1 - d_2/r_2$, so $\hat{S}(a_2)$ is the product of the first 2 is given by

$$\hat{S}(a_2) = \left(1 - \frac{d_1}{r_1}\right)\left(1 - \frac{d_2}{r_2}\right), \quad \text{and so on.}$$

There are several general assumptions for time to event studies (see Pollock et al. 1989; Williams et al. 2002). First, we assume that radio-tagged individuals are a random sample from the population of interest. This assumption can be satisfied by using random location of trapping sites or perhaps stratifying trapping effort by perceived density of the population. We also assume that survival times are independent among different animals; violating this assumption leads to *overdispersion*. For example, you catch a brood of quail (say 6 young) and radio-tag each, but a predator finds the brood and predates the hen and all the young – thus survival time between individuals was not independent. Additionally, we assume that radio transmitters (or other marks) do not affect the survival of marked individuals and that the censoring mechanism in random or that censoring is not related to fate of the individual (e.g., a radio destroyed during predation or harvest event). For staggered entry studies, newly marked individuals have the same survival function as previously marked individuals.

4.7.4 Occupancy Modeling

Occupancy modeling is a recent entry into the field of capture–recapture analysis (MacKenzie et al. 2002; MacKenzie 2005). This approach stems from historical work done to confirm presence of a species in a particular location at a particular time, and as such relates data on site-specific features (e.g., canopy cover) to the presence of a species. Thus, the presence or absence of the feature can be used as a surrogate for abundance in monitoring temporal and spatial changes in species distributions (MacKenzie et al. 2006). Research on animal detectability has focused primarily on density or abundance estimation (e.g., Buckland et al. 2001; Borchers et al. 2002; Williams et al. 2002), but more limited efforts have been expended on presence–absence approaches (Vojta 2005). Occupancy modeling focuses on estimating the proportion of an area of suitable habitat that is occupied by an individual of the species of interest (MacKenzie et al. 2004).

Occupancy surveys make the same general assumptions as most capture–mark–recapture studies and several specific assumptions (MacKenzie et al. 2006) including (1) survey sites are closed to changes in occupancy over the survey season, (2) occupancy probabilities and detection probabilities are either constant across sites or a function of survey covariates, and (3) detections at each location are independent. Surveys for occupancy are usually less labor intensive than surveys for estimation of abundance in that both active (e.g., point counts during breeding season) and passive approaches (e.g., track counts or hair snares) can be used to survey for presence. However, the difficulty becomes determining when a species is truly absent from the study plot, because failing to locate an individual during a survey does not imply absence (MacKenzie et al. 2006).

From a survey design standpoint, the percentage of sampling units occupied by a species of interest across a landscape is important for population management and monitoring (MacKenzie and Royle 2005). Occupancy surveys are developed to estimate this quantity (fraction of sampling units occupied), while accounting for incomplete detectability and those factors, which influence detectability. Consider the hypothetical situation where we conduct bird point counts to evaluate presence–absence of an endangered passerine across a physiographic region. We are unable to sample every potential area the birds might inhabit, but we know that the birds select a specific habitat type (e.g., closed canopy forest). Thus, our sampling frame will be all potential bird habitat within this ecoregion, and we will sample, according to some probabilistic design, a subset of the total number of sampling units (sites). For each site, we will conduct several visits; the number of visits depends upon bird phenology and survey effort necessary, although MacKenzie and Royle (2005) suggest 3 visits. On each visit, presence or absence is noted, with our intent being to estimate occupancy (ψ_i) as well as detection probability (p_i) for the ith sampling unit for the species of interest (MacKenzie and Royle 2005; MacKenzie et al. 2006). Currently, occupancy models are available for single or multiple season surveys, and considerable research is continuing on combining occupancy surveys with count data or marked individuals to estimate population size (MacKenzie et al. 2006).

Occupancy modeling provides an alternative to managers for monitoring species trends (proportion of plots with the species) as well as evaluating colonization and extinction from study sites. Occupancy approaches require less data and effort. More precise abundance estimation for a rare species across a landscape may not be implementable due to costs associated with capture and marking animals over a broad spatial and temporal frame, while collection of presence–absence data can demonstrate whether the population is expanding or contracting over time. This might be all the information required for sound management.

Species detectability frequently hinders the ability of managers to make appropriate management decisions. Detectability becomes extremely important when dealing with species that are rare either functionally or operationally (McDonald 2004). Work by Royle and Nichols (2003), Royle (2004a,b), Kery et al. (2005), Royle and Link (2005), and Royle et al. (2005) focused on estimating species abundance by combining repeated survey counts and mixture models (beta-binomial mixtures) to estimate both detectability and abundance. We see occupancy modeling as a considerable improvement over uncorrected surveys (e.g., bird point counts) and these approaches should be evaluated for applicability across wildlife science.

4.8 Resource Selection

A primary concern of the biologist is the identification, availability, and relative importance of resources (e.g., food, cover, or water) used by animals (i.e., habitat). Habitat or *resource selection* by animals is of interest when evaluating habitat management and the impact of perturbations on wildlife populations. These studies have far reaching importance to wildlife management, particularly as they relate to federally protected species. For example, results of habitat use studies were central to the debate over the importance of old-growth timber to the spotted owl (*Strix occidentalis*) and instream flows in the central Platte River in Nebraska to the whooping crane (*Grus americana*).

In resource selection studies, the availability of a resource is the quantity accessible to the animal (or population of animals) and the use of a resource is that quantity utilized during the time period of interest (Manly et al. 1993). When use of a resource is disproportionate to availability, then the use is selective (i.e., the animal is showing a preference or avoidance for the resource). Manly et al. (1993) provide a unified statistical theory for the analysis of selection studies. The theory and application of resource selection studies were updated (Johnson 1998). We recommend a thorough review of both of these references for anyone considering this type of study.

Biologists often identify resources used by animals and document their availability (usually expressed as abundance or presence/absence). Resource selection models can be developed using most of the designs previously discussed. In most observational studies, it will be impossible to identify unique animals. However, using observations of animals seen from randomly or systematically chosen points, it is possible to use resource variables with known availability (e.g., vegetation) as predictor variables. For example, assume that a certain vegetation type is preferentially selected as feeding sites for elk within a certain distance of conifer cover (Thomas 1979). For example, if the distance was 0.5 km, then one could predict that the impact of timber harvest on elk forage would increase if logging occurs <0.5 km from this vegetation type. Alternatively, the study area could be classified into available units characterized on the basis of a set of predictor variables, such as vegetation type, distance to water, distance to cover, and distance to roads. If use is defined as the presence or absence of feeding elk, resource selection could be used to evaluate the effect of a set of predictor variables on available forage.

4.8.1 Sampling Designs

Alldredge et al. (1998) reviewed the multitude of methods used in the study of resource selection. Resource selection occurs in a hierarchical fashion from the geographic range of a species, to individual animal ranges within a geographic range, to use of general features (habitats) within the individual's range, to the selection of particular elements (food items) within the feeding site (Manly et al. 1993). The first design decision in a resource selection study is the scale of study (Johnson 1980). Manly et al. (1993) suggested conducting studies at multiple scales. Additional important decisions affecting the outcome of these studies include the selection of the study area boundary and study techniques (Manly et al. 1993).

Resource selection probability functions give probabilities of use for resource units of different types. This approach may be used when the resource being studied can be classified as a universe of *N* available units, some of which are used and the remainder not used. Also, every unit can be classified by the values that it possesses for certain important variables ($X = X_1, X_2, \ldots, X_p$) thought to affect use. Examples include prey items selected by predators based on color, size, and age, or plots of land selected by ungulates based on distance to water, vegetation type, distance to disturbance, and so on. Sampling of used and unused units must consider the same issues as discussed previously for any probability sample.

Thomas and Taylor (1990) described three general study designs for evaluating resource selection. In design I, measurements are made at the population level. Units available to all animals in the population are sampled or censused and classified into used and unused. Individual animals are not identified. In design II, individual animals are identified and the use of resources is measured for each while availability is measured at the level available to the entire population. In design III, individuals are identified or collected as in design II and at least two of the sets of resource units (used resource units, unused resource units, available resource units) are sampled or censused for each animal.

Manly et al. (1993) also offered three sampling protocols for resource selection studies. First, one outlines random sampling or complete counts on available units and randomly samples used resource units. Next, one = outlines randomly samples or census subjects within available units and randomly samples unused units. Finally, one takes an independent sample of both used and unused units. Also, it is possible in some situations to census both used and unused units. Erickson et al. (1998) described a moose (*Alces alces*) study on the Innoko National Wildlife Refuge in Alaska that evaluated habitat selection following Design I and sampling protocol A (Fig. 4.15).

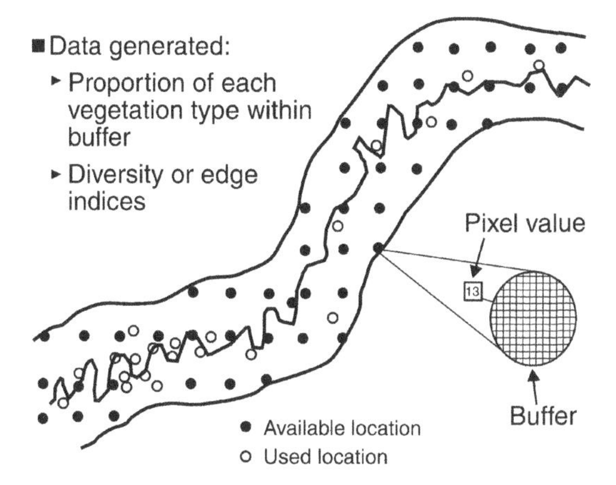

Fig. 4.15 Schematic of design I and sampling protocol A (from Manly et al. 1993) as used in a survey of potential moose use sites in a river corridor in Innoko National Wildlife Refuge in Alaska. Reproduced from Erickson et al. (1998) with kind permission from American Statistical Society

The selection of a particular design and sampling protocol must consider the study area, the habitats or characteristics of interest, the practical sample size, and the anticipated method of analysis. The design of studies should also consider the relationship between resource selection and the anticipated benefits of the selection of good resources, such as increased survival rate, increased productivity, and/or increased fitness (Alldredge et al. 1998).

4.9 Spatial Statistics

Wildlife studies frequently are interested in describing the spatial pattern of resources or contaminants. The application of spatial statistics offers an opportunity to evaluate the precision of spatial data as well as improve the efficiency of spatial sampling efforts. Spatial statistics combine the geostatistical prediction techniques of kriging (Krige 1951) and simulation procedures such as conditional and unconditional simulation (Borgman et al. 1984, 1994). Both kriging and simulation procedures are used to estimate random variables at unsampled locations. Kriging produces best linear unbiased predictions using available known data, while the simulation procedures give a variety of estimates usually based on the data's statistical distribution. Kriging results in a smoothed version of the distribution of estimates, while simulation procedures result in predicted variance and correlation structure, and natural variability of the original process are preserved (Kern 1997). If the spatial characterization of the mean of the variable in the mean in each cell of a grid, for example, then kriging procedures are satisfactory. However, if the spatial variability of the process is of importance, simulation procedures are more appropriate. For a more complete treatment of simulation techniques see Borgman et al. (1994) or Deutsch and Journel (1992). Cressie (1991) gave a complete theoretical development of kriging procedures, while Isaaks and Srivastava (1989) provided a more applied treatment appropriate for the practitioner. For the original developments in geostatistics, we refer you to Krige (1951), Matheron (1962, 1971), and Journel and Huigbregts (1978).

In a study using spatial statistics, data generally are gathered from a grid of points and the spatial covariance structure of variables is used to estimate the variable of interest at points not sampled. The data on the variable of interest at the sample locations could be used to predict the distribution of the variable for management or conservation purposes. For example, suppose a wind plant is planned for a particular area and there is concern regarding the potential for the development to create risk to birds. If bird counts are used as an index of local use, then estimates of local mean bird use could be used to design the wind plant to avoid high bird use areas. Preservation of local variability would not be necessary, and kriging would provide a reasonable method to predict locations where bird use is low and hence wind turbines should be located. Using this sort of linear prediction requires sampling in all areas of interest.

Geostatistical modeling, which considers both linear trends and correlated random variables, can be more valuable in predicting the spatial distribution of a variable of interest. These geostatistical simulation models are stochastic, and

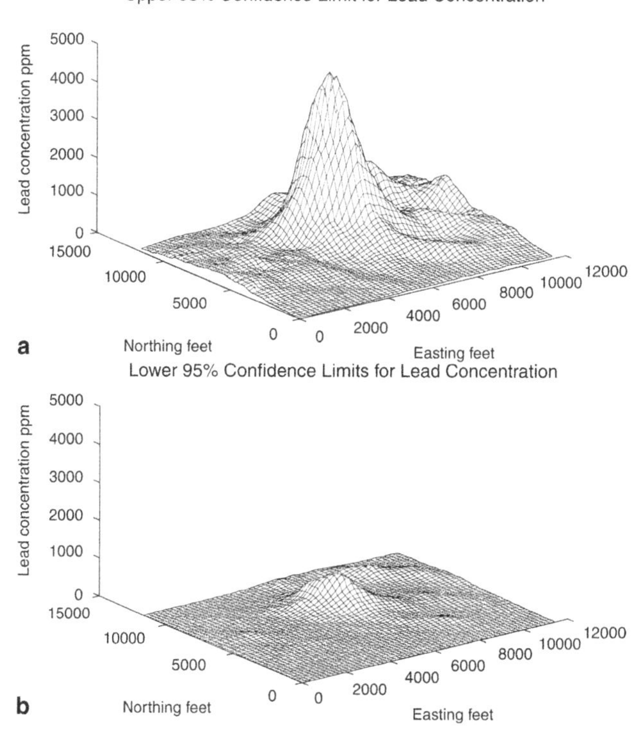

Fig. 4.16 A hypothetical three-dimensional map generated by geostatistical modeling illustrating an upper (**a**) and lower (**b**) 95% confidence limit for lead concentration (from Kern 1997)

predict a degree of randomness in spatial perception of the parameter (Borgman et al. 1994). For example, if one is interested in the spatial distribution of a contaminant for the purposes of cleanup, then a fairly high degree of interest would exist in the location of high concentrations of the contaminant as well as the degree of confidence one could place in the predicted distribution. This confidence in the predicted distribution of contaminates would lead to decisions about where cleanup is required and where more sampling effort is needed. Figure 4.16 illustrates a hypothetical upper and lower confidence limit for lead concentration. These kinds of maps could be valuable in impact evaluation, as well as management situations such as contaminant clean up.

4.10 Summary

The goal of wildlife ecology research is to learn about wildlife populations and the habitats that they use. Thus, the objective of Chap. 4 was to provide a description of the fundamental methods for sampling and making inferences in wildlife studies.

We began with a discussion of the basics of sample survey design, plot shape and size, random and nonrandom sample survey selection as well as a description of common definitions used in wildlife sample survey design. Within Sect. 4.1, we detail the necessity to define clearly study objectives, the area of inference, and the sampling unit(s) of importance. Additionally, we discuss the need for clear definition of the parameters to measure. In Sects. 4.2 and 4.3, we discussed numerous methods for probability sampling, ranging from simple random sampling to strip adaptive cluster sampling. Under this framework, we outline the need for probabilistic sampling procedures and how their use lead to strong inference. We outlined several methods to sample populations, ranging from simple fixed area plots to more complicated distance-based estimators under design-based inference.

Next, we focused on model-based sampling (Sect. 4.7). We outlined the rationale for using model-based techniques and discussed the differences between model-based and design-based studies (also see Chap 2). Often, as each wildlife study is unique, decisions regarding the sampling plan will require use of a variety of methods. With this in mind, we discussed several variant of capture–mark–recapture techniques, analysis of presence–absence data, and time to event models; all of which are used for model-based inferences. We conclude this chapter with a discussion on resource selection and spatial statistics and their application to wildlife conservation.

References

Alldredge, J. R., D. L. Thomas, and L. McDonald. 1998. Survey and comparison of methods for study of resource selection. J. Agric. Biol. Environ. Stat. 3: 237–253.

Alpizar-Jara, R., and K. H. Pollock. 1996. A combination line transect and capture re-capture sampling model for multiple observers in aerial surveys. J. Environ. Stat. 3: 311–327.

Amstrup, S. C., T. L. McDonald, and B. F. J. Manly. 2005. Handbook of Capture–Recapture Analysis. Princeton University Press, Princeton.

Anderson, D. R. 2001. The need to get the basics right in wildlife field studies. Wildl. Soc. Bull. 29: 1294–1297.

Bart, J. and S. Earnst. 2002. Double sampling to estimate density and population trends in birds. Auk 119: 36–45.

Beavers, S. C., and F. L. Ramsey. 1998. Detectability analysis in transect surveys. J. Wildl. Manage. 62(3): 948–957.

Block, W. M., and L. A. Brennan. 1992. The habitat concept in ornithology: Theory and applications. Curr. Ornithol. 11: 35–91.

Borchers, D. L., S. T. Buckland, and W. Zucchini. 2002. Estimating Animal Abundance. Springer, Berlin Heidelberg New York.

Borgman, L. E., M. Taheri, and R. Hagan. 1984. Three-dimensional, frequency-domain simulations of geologic variables, in G. Verly, M. David, A. G. Journel, and A. Marechal, Eds. Geostatistics for Natural Resources Characterization, Part I. Reidel, Dordrecht.

Borgman, L. E., C. D. Miler, S. R. Signerini, and R. C. Faucette. 1994. Stochastic interpolation as a means to estimate oceanic fields. Atmos.-Ocean. 32(2): 395–419.

Bormann, F. H. 1953. The statistical efficiency of sample plot size and shape in forest ecology. Ecology 34: 474–487.

Brownie, C., J. E. Hines, J. D. Nichols, K. H. Pollock, and J. B. Hestbeck. 1993. Capture–recapture studies for multiple strata including non-Markovian transitions. Biometrics 49: 1173–1187.

Buckland, S. T. 1987. On the variable circular plot method of estimating animal density. Biometrics 43: 363–384.
Buckland, S. T., D. R. Anderson, K. P. Burnham, and J. L. Laake. 1993. Distance Sampling: Estimating Abundance of Biological Populations. Chapman and Hall, London.
Buckland, S. T., D. R. Anderson, K. P. Burnham, J. L. Laake, D. L. Borchers, and L. Thomas. 2001. Introduction to Distance Sampling. Oxford University Press, Oxford.
Buckland, S. T., D. R. Anderson, K. P. Burnham, J. L. Laake, D. L. Borchers, and L. Thomas. 2004. Advanced Distance Sampling. Oxford University Press, Oxford.
Burnham, K. P. 1993. A theory for combined analysis of ring recovery and recapture data, in J. D. Lebreton and P. M. North, Eds. Marked Individuals in the Study of Bird Population, pp. 199–214. Birkhäuser-Verlag, Basel, Switzerland.
Burnham, K. P., and W. S. Overton. 1978. Estimation of the size of a closed population when capture probabilities vary among animals. Bometrika 65: 625–633.
Burnham, K. P., D. R. Anderson, and J. L. Laake. 1980. Estimation of density from line transect sampling of biological populations. Wildl. Monogr. 72: 1–202.
Burnham, K. P., S. T. Buckland, J. L. Laake, D.L. Borchers, T. A. Marques, J. R. B. Bishop, and L. Thomas. 2004. In S. T. Buckland, D. R. Anderson, K. P. Burnham, J. L. Laake, D. L. Borchers, and L. Thomas, Eds. Advanced Distance Sampling, pp. 307–392. Oxford University Press, Oxford.
Byth, K. 1982. On robust distance-based intensity estimators. Biometrics 38: 127–135.
Canfield, R. H. 1941. Application of the line intercept method in sampling range vegetation. J. Forest. 39: 388–394.
Caughley, G. 1977. Sampling in aerial survey. J. Wildl. Manage. 41: 605–615.
Caughley, G., and D. Grice. 1982. A correction factor for counting emus from the air and its application to counts in western Australia. Aust. Wildl. Res. 9: 253–259.
Chao, A. 1987. Estimating the population size for capture–recapture data with unequal catchability. Biometrics 43: 783–791.
Chao, A. 1988. Estimating animal abundance with capture frequency data. J. Wildl. Manage. 52: 295–300.
Chao, A. 1989. Estimating population size for sparse data in capture–recapture experiments. Biometrics 45: 427–438.
Christman, M. C. 2000. A review of quadrat-based sampling of rare, geographically clustered populations. J. Agric. Biol. Environ. Stat. 5: 168–201.
Cochran, W. G. 1977. Sampling Techniques, 3rd Edition. Wiley, New York.
Collier, B. A., S. S. Ditchkoff, J. B. Raglin, and J. M. Smith. 2007. Detection probability and sources of variation in white-tailed deer spotlight surveys. J. Wildl. Manage. 71: 277–281.
Conroy, M. J., J. R. Goldsberry, J. E. Hines, and D. B. Stotts. 1988. Evaluation of aerial transect surveys for wintering American black ducks. J. Wildl. Manage. 52: 694–703.
Cook, R. D., and J. O. Jacobson. 1979. A design for estimating visibility bias in aerial surveys. Biometrics 35: 735–742.
Cressie, N. A. C. 1991. Statistics for Spatial Data. Wiley, New York.
Darroch, J. N. 1958. The multiple recapture census: I. Estimation of a closed population. Biometrika 45: 343–359.
Dell, T. R., and J. L. Clutter. 1972. Ranked set sampling theory with order statistics background. Biometrics 28: 545–553.
Deutsch, C. V., and A. G. Journel. 1992. GSLIB Geostatistical Software Library and User's Guide. Oxford University Press, New York.
Diggle, P. J. 1983. Statistical Analysis of Spatial Point Patterns. Academic, London.
Dinsmore, S. J., G. C. White, and F. L. Knopf. 2002. Advanced techniques for modeling avian nest survival. Ecology 83: 3476–3488.
Drummer, T. D. 1991. SIZETRAN: Analysis of size-biased line transect data. Wildl. Soc. Bull. 19(1): 117–118.
Drummer, T. D., and L. L. McDonald. 1987. Size bias in line transect sampling. Biometrics 43: 13–21.
Eberhardt, L. L. 1978. Transect methods for populations studies. J. Wildl. Manage. 42: 1–31.

Eberhardt, L. L., and M. A. Simmons. 1987. Calibrating population indices by double sampling. J. Wildl. Manage. 51: 665–675.

Erickson, W. P., T. L. McDonald, and R. Skinner. 1998. Habitat selection using GIS data: A case study. J. Agric. Biol. Environ. Stat. 3: 296–310.

Flowy, T. J., L. D. Mech, and M. E. Nelson. 1979. An improved method of censusing deer in deciduous–coniferous forests. J. Wildl. Manage. 43: 258–261.

Foreman. K. 1991. Survey Sampling Principles. Marcel Dekker, New York.

Fretwell, S. D., and H. L. Lucas. 1970. On territorial behavior and other factors influencing habitat distribution in birds. I. Acta Biotheoret. 19: 16–36.

Gaillard, J. M. 1988. Contribution a la Dynamique des Populations de Grands Mammiferes: l'Exemple du Chevreuil (*Capreolus capreolus*). Dissertation. Universite Lyon I, Villeurbanne, France.

Gasaway, W. C., S. D. DuBois, D. J. Reed, and S. J. Harbo. 1986. Estimating Moose Population Parameters from Aerial Surveys. Institute of Arctic Biology, Biological Papers of the University of Alaska, Fairbanks, AK 99775, No. 22.

Gilbert, R. O. 1987. Statistical Methods for Environmental Pollution Monitoring. Van Nostrand Reinhold, New York.

Gilbert, R. O., and J. C. Simpson. 1992. Statistical methods for evaluating the attainment of cleanup standards. Vol. 3, Reference-Based Standards for Soils and Solid Media. Prepared by Pacific Northwest Laboratory, Battelle Memorial Institute, Richland, WA, for U.S. Environmental Protection Agency under a Related Services Agreement with U.S. Department of Energy, Washington, DC. PNL-7409 Vol. 3, Rev. 1/UC-600.

Graham, A., and R. Bell. 1969. Factors influencing the countability of animals. East Afr. Agric. For. J. 34: 38–43.

Grosenbaugh, L. R. 1952. Plotless timber estimates – new, fast, easy. J. For. 50: 532–537.

Guenzel, R. J. 1997. Estimating Pronghorn Abundance Using Aerial Line Transect Sampling. Wyoming Game and Fish Department, Cheyenne, WY.

Hasel, A. A. 1938. Sampling error in timber surveys. J. Agric. Res. 57: 713–736.

Hendricks, W. A. 1956. The Mathematical Theory of Sampling. The Scarecrow Press, New Brunswick.

Hornocker, M. G. 1970. An analysis of mountain lion predation upon mule deor and elk in the Idaho Primitive Area. Wildl. Monogr. 21.

Horvitz, D. G., and D. J. Thompson. 1952. A generalization of sampling without replacement from a finite universe. J. Am. Stat. Assoc. 47: 663–685.

Hosmer Jr., D. W., and S. Lemeshow. 1999. Applied Survival Analysis. Wiley, New York.

Huggins, R. M. 1989. On the statistical analysis of capture experiments. Biometrika 76: 133–140.

Huggins, R. M. 1991. Some practical aspects of a conditional likelihood approach to capture experiments. Biometrics 47: 725–732.

Hurlbert, S. H. 1984. Pseudoreplication and the design of ecological field experiments. Ecol. Monogr. 54: 187–211.

Isaaks, E. H., and R. M. Srivastava. 1989. An Introduction to Applied Geostatistics, Oxford University Press, New York.

Johnson, D. H. 1980. The comparison of usage and availability measurements for evaluating resource preference. Ecology 61: 65–71.

Johnson, D. H., Ed. 1998. J. Agric. Biol. Environ. Stat. 3(3) Special Issue: Resource Selection Using Data from Geographical Information Systems (GIS).

Johnson, D. H. 2002. The importance of replication in wildlife research. J. Wildl. Manage. 66: 919–932.

Journel, A. G., and C. J. Huijbregts. 1978. Mining Geostatistics. Academic, London.

Kaiser, L. 1983. Unbiased estimation in line-intercept sampling. Biometrics 39: 965–976.

Kalamkar, R. J. 1932. Experimental error and the field plot technique with potatoes. J. Agric. Sci. 22: 373–383.

Kaplan, E. L., and P. Meier. 1958. Nonparametric estimation from incomplete observations. J. Am. Stat. Assoc. 53: 457–481.

Kendall, W. L. 1999. Robustness of closed capture–recapture methods to violations of the closure assumption. Ecology 80: 2517–2525.
Kendall, W. L., and K. H. Pollock. 1992. The robust design in capture–recapture studies: A review and evaluation by Monte Carlo simulation, in D. R. McCullough, and R. H. Barrett, Eds. Wildlife 2001: Populations, pp. 31–43. Elsevier, London.
Kendall, W. L., J. D. Nichols, and J. E. Hines, 1997. Estimating temporary emigration using capture–recapture data with Pollock's robust design. Ecology 78: 563–578.
Kern, J. W. 1997. Data analysis techniques for point prediction and estimation of spatial means. Technical Research Work Order Dan Goodman, Department of Biology, Montana State University, Bozeman, MT. Contract 290801.
Kery, M., J. A. Royle, and H. Schmid. 2005. Modeling avian abundance from replicated counts using binomial mixture models. Ecol. Appl. 15: 1450–1461.
Kleinbaum, D. G. 1996. Survival Analysis: A self-learning text. Springer, New York.
Krebs, C. J. 1989. Ecological Methodology. Harper Collins, New York.
Krebs, C. J. 1999. Ecological Methodology, 2nd Edition. Addison-Welsey, Menlo Park.
Krige, D. G. 1951. A Statistical Approach to Some Mine Valuation and Allied Problems at the Witwaterstrand. Unpublished Masters Thesis, University of Witwaterstrand.
Kuehl, R. O. 2000. Design of Experiments: Statistical Principles of Research Design and Analysis, 2nd Edition. Brooks/Cole, Pacific Grove.
Laake, J. L., K. P. Burnham, and D. R. Anderson. 1979. User's Manual for Program TRANSECT, 26 pp. Utah State University Press, Logan, Utah.
Laake, J. L., S. T. Buckland, D. R. Anderson, and K. P. Burnham. 1993. DISTANCE User's Guide. Version 2.0. Colorado Cooperative Fish and Wildlife Research Unit, Colorado State University, Fort Collins, CO.
Lancia, R. A., W. L. Kendall, K. H. Pollock, and J. D. Nichols. 2005. Estimating the number of animals in wildlife populations, in C. E. Braun, Ed. Research and Management Techniques for Wildlife and Habitats, pp. 106–153. Wildlife Society, Bethesda, MD.
Lebreton, J. -D., K. P. Burnham, J. Clobert, and D. R. Anderson. 1992. Modeling survival and testing biological hypotheses using marked animals: A unified approach with case studies. Ecol. Monogr. 62: 67–118.
Lee, S. -M., and A. Chao. 1994. Estimating population size via sample coverage for closed capture–recapture models. Biometrics 50: 88–97.
Lincoln, F. C. 1930. Calculating waterfowl abundance on the basis of banding returns. US Dept. Agric. Circ. No. 118: 1–4.
Link, W. A., and R. J. Barker. 2005. Modeling association among demographic parameters in analysis of open population capture–recapture data. Biometrics 61: 46–54.
Lucas, H. A., and G. A. F. Seber. 1977. Estimating coverage and particle density using the line intercept method. Biometrika 64: 618–622.
Ludwig, J. A., and J. F. Reynolds. 1988. Statistical Ecology. Wiley, New York.
Lukacs, P. M. 2002. WebSim: Simulation software to assist in trapping web design. Wildl. Soc. Bull. 30: 1259–1261.
Lukacs, P. M., A. B. Franklin, D. R. Anderson. 2004. In S. T. Buckland, D. R. Anderson, K. P. Burnham, J. L. Laake, D. L. Borchers, and L. Thomas, Eds. Advanced Distance Sampling, pp. 260–278. Oxford University Press, Oxford.
Lukacs, P. M., D. R. Anderson, and K. P. Burnham. 2005. Evaluation of trapping-web design. Wildl. Res. 32: 103–110.
MacKenzie, D. I. 2005. What are the issues with presence–absence data for wildlife managers? J. Wildl. Manage. 69: 849–860.
MacKenzie, D. I., and J. A. Royle. 2005. Designing occupancy studies: General advice and allocating survey effort. J. Appl. Ecol. 42: 1105–1114.
MacKenzie, D. I., J. D. Nichols, G. B. Lachman, S. Droege, J. A. Royle, and C. A. Langtimm. 2002. Estimating site occupancy rates when detection probabilities are less than one. Ecology 83: 2248–2255.

MacKenzie, D. I., J. A. Royle, J. A. Brown, and J. D. Nichols. 2004. Occupancy estimation and modeling for rare and elusive species. In W. L. Thompson, Ed. Sampling Rare or Elusive Species, pp. 149–172. Island Press,Washington.

MacKenzie, D. I., J. D. Nichols, J. Andrew Royle, K. H. Pollock, L. L. Bailey, and J. E. Hines. 2006. Occupancy Estimation and Modeling. Academic Press, London.

Manly, B. F. J. 1991. Randomization and Monte Carlo Methods in Biology. Chapman and Hall, London.

Manly, B. F. J., L. McDonald, and D. Thomas. 1993. Resource Selection by Animals: Statistical Design and Analysis for Field Studies. Chapman and Hall, London.

Manly, B. F. J., L. McDonald, and G. W. Garner. 1996. Maximum likelihood estimation for the double-count method with independent observers. J. Agric. Biol. Environ. Stat. 1(2): 170–189.

Marques, F. C. C., and S. T. Buckland. 2004. In S. T. Buckland, D. R. Anderson, K. P. Burnham, J. L. Laake, D. L. Borchers, and L. Thomas, Eds. Advanced Distance Sampling, pp. 31–47. Oxford University Press, Oxford.

Matheron, G. 1962. Traite de Geostatistique Appliquee, Tome I. Memoires du Bureau de Recherches Geologiques et Minieres, No. 14. Editions Technip, Paris.

Matheron, G. 1971. The theory of regionized variables and its applications. Cahiers du Centre de Morphologie Mathematique, No. 5. Fontaine-bleau, France.

McCullagh, P., and J. A. Nelder. 1983. Generalized Linear Models. Chapman and Hall, London.

McDonald, L. L. 1980. Line-intercept sampling for attributes other than cover and density. J. Wildl. Manage. 44: 530–533.

McDonald, L. L. 1991. Workshop Notes on Statistics for Field Ecology. Western Ecosystems Technology, Inc. Cheyenne, WY.

McDonald, L. L. 2004. Sampling rare populations, in W. L. Thompson, Ed. Sampling Rare or Elusive Species, pp. 11–42. Island Press, Washington.

McDonald, L. L., H. B. Harvey, F. J. Mauer, and A. W. Brackney. 1990. Design of aerial surveys for Dall sheep in the Arctic National Wildlife Refuge, Alaska. Seventh Biennial Northern Wild Sheep and Goat Symposium. May 14–17, 1990, Clarkston, Washington.

McDonald, L. L., W. P. Erickson, and M. D. Strickland. 1995. Survey design, statistical analysis, and basis for statistical inferences in Coastal Habitat Injury Assessment: *Exxon Valdez* Oil Spill, in P. G. Wells, J. N. Buther, and J. S. Hughes, Eds. *Exxon Valdez* Oil Spill: Fate and Effects in Alaskan Waters. ASTM STP 1219. American Society for Testing and Materials, Philadelphia, PA.

Menkins Jr., G. E., and S. H. Anderson. 1988. Estimation of small-mammal population size. Ecology 69: 1952–1959.

Miller, S. D., G. C. White, R. A. Sellers, H. V. Reynolds, J. W. Schoen, K. Titus, V. G. Barnes, R. B. Smith, R. R. Nelson, W. B. Ballard, and C. C. Schwartz. 1997. Brown and black bear density estimation in Alsaka using radiotelemetry and replicated mark-resight techniques. Wildl. Monogr. 133.

Mode, N. A., L. L. Conquest, and D. A. Marker. 2002. Incorporating prior knowledge in environmental sampling: ranked set sampling and other double sampling procedures. Environmetrics 13: 513–521.

Mood, A. M., F. A. Graybill, and D. C. Boes. 1974. Introduction to the Theory of Statistics, 3rd Edition. McGraw-Hill, Boston.

Morrison, M. L., W. M. Block, M. D. Strickland, and W. L. Kendall. 2001. Wildlife Study Design. Springer.

Munholland, P. L., and J. J. Borkowski. 1996. Simple latin square sampling +1: A spatial design using quadrats. Biometrics 52: 125–136.

Muttlak, H. A., and L. L.McDonald. 1992. Ranked set sampling and the line intercept method: a more efficient procedure Biometrical J. 34: 329–346.

Nichols, J. D., and K. H. Pollock. 1983. Estimation methodology in contemporary small mammal capture–recapture studies. J. Mammal. 64: 253–260.

Nichols, J. D., and K. H. Pollock. 1990. Estimation of recruitment from immigration versus in situ reproduction using Pollocks robust design. Ecology 71: 21–26.

Noon, B. R., N. M. Ishwar, and K. Vasudevan. 2006. Efficiency of adaptive cluster sampling and random sampling in detecting terrestrial herptofauna in a tropical rainforest. J. Wildl. Manage. 34: 59–68.

Otis, D. L., K. P. Burnham, G. C. White, and D. R. Anderson. 1978. Statistical inference from capture data on closed animal populations. Wildl. Monogr. 62.

Otis, D. L., L. L. McDonald, and M. Evans. 1993. Parameter estimation in encounter sampling surveys. J. Wildl. Manage. 57: 543–548.

Overton, W. S., D. White, and D. L. Stevens. 1991. Design Report for EMAP: Environmental Monitoring and Assessment Program. Environmental Research Laboratory, U.S. Environmental Protection Agency, Corvallis, OR. EPA/600/3-91/053.

Packard, J. M., R. C. Summers, and L. B. Barnes. 1985. Variation of visibility bias during aerial surveys of manatees. J. Wildl. Manage. 49: 347–351.

Patil, G. P., A. K. Sinha, and C. Taillie. 1994. Ranked set sampling, in G. P. Patil and C. R. Rao, Eds. Handbook of Statistics, Environmental Statistics, Vol. 12. North-Holland, Amsterdam.

Pechanec, J. F., and G. Stewart. 1940. Sage brush-grass range sampling studies: size and structure of sampling unit. Am. Soc. Agron. J. 32: 669–682.

Peterjohn, B. G., J. R. Sauer, and W. A. Link. 1996. The 1994 and 1995 summary of the North American Breeding Bird Survey. Bird Popul. 3: 48–66.

Pollock, K. H. 1974. The Assumption of Equal Catchability of Animals in Tag-Recapture Experiments. Ph.D. Thesis, Cornell University, Ithaca, NY.

Pollock, K. H. 1982. A capture–recapture design robust to unequal probability of capture. J. Wildl. Manage. 46: 752–757.

Pollock, K. H. 1991. Modeling capture, recapture, and removal statistics for estimation of demographic parameters for fish and wildlife populations: past, present, and future. J. Am. Stat. Assoc. 86: 225–238.

Pollock, K. H., and M. C. Otto. 1983. Robust estimation of population size in closed animal populations from capture–recapture experiments. Biometrics 39: 1035–1049.

Pollock, K. H., and W. L. Kendall. 1987. Visibility bias in aerial surveys: A review of estimation procedures. J. Wildl. Manage. 51: 502–520.

Pollock, K. H., S. R. Winterstein, C. M. Bunck, and P. D. Curtis. 1989. Survival analysis in telemetry studies: The staggered entry design. J. Wildl. Manage. 53: 7–15.

Pollock, K. H., J. D. Nichols, C. Brownie, and J. E. Hines. 1990. Statistical inferences for capture–recapture experiments. Wildl. Monogr. 107.

Pradel, R. 1996. Utilization of capture–mark–recapture for the study of recruitment and population growth rate. Biometrics 52: 703–709.

Quang, P. X. 1989. A nonparametric approach to size-biased line transect sampling. Draft Report. Department of Mathematical Sciences, University of Alaska, Fairbanks, AK 99775. Biometrics 47(1): 269–279.

Quang, P. X., and E. F. Becker. 1996. Line transect sampling under varying conditions with application to aerial surveys. Ecology 77: 1297–1302.

Quang, P. X., and E. F. Becker. 1997. Combining line transect and double count sampling techniques for aerial surveys. J. Agric. Biol. Environ. Stat. 2: 230–242.

Ramsey, F. L., and J. M. Scott. 1979. Estimating population densities from variable circular plot surveys, in R. M. Cormack, G. P. Patil and D. S. Robson, Eds. Sampling Biological Populations, pp. 155–181. International Co-operative Publishing House, Fairland, MD.

Reed, D. J., L. L. McDonald, and J. R. Gilbert. 1989. Variance of the product of estimates. Draft report. Alaska Department of Fish and Game, 1300 College Road, Fairbanks, AK 99701.

Rexstad, E., and K. Burnham. 1991. Users Guide for Interactive Program CAPTURE, Abundance Estimation for Closed Populations. Colorado Cooperative Fish and Wildlife Research Unit, Fort Collins, CO.

Reynolds, R. T., J. M. Scott, and R. A. Nussbaum. 1980. A variable circular-plot method for estimating bird numbers. Condor 82(3): 309–313.

Rotella, J. J., S. J. Dinsmore, and T. L. Shaffer. 2004. Modeling nest-survival data: A comparison of recently developed methods that can be implemented in MARK and SAS. Anim. Biodivers. Conserv. 27: 187–205.

Royle, J. A. 2004a. Generalized estimators of avian abundance from count survey data. Anim. Biodivers. Conserv. 27: 375–386.

Royle, J. A. 2004b. Modeling abundance index data from anuran calling surveys. Conserv. Biol. 18: 1378–1385.

Royle, J. A., and W. A. Link. 2005. A general class of multinomial mixture models for anuran calling survey data. Ecology 86: 2505–2512.

Royle, J. A., and J. D. Nichols. 2003. Estimating abundance from repeated presence–absence data or point counts. Ecology 84: 777–790.

Royle, J. A., J. D. Nichols, and M. Kery. 2005. Modelling occurrence and abundance of species when detection is imperfect. Oikos 110: 353–359.

Samuel, M. D., E. O. Garton, M. W. Schlegel, and R. G. Carson. 1987. Visibility bias during aerial surveys of elk in northcentral Idaho. J. Wildl. Manage. 51: 622–630.

Scheaffer, R. L., W. Mendenhall, and L. Ott. 1990. Elementary Survey Sampling. PWS-Kent, Boston, MA.

Schoenly, K. G., I. T. Domingo, and A. T. Barrion. 2003. Determining optimal quadrat sizes for invertebrate communities in agrobiodiversity studies: A case study from tropical irrigated rice. Environ. Entomol. 32: 929–938.

Schwarz, C. J., and A. N. Arnason. 1996. A general methodology for the analysis of capture–recapture experiments in open populations. Biometrics 52: 860–873.

Schwarz, C. J., and W. T. Stobo. 1997. Estimating temporary migration using the robust design. Biometrics 53: 178–194.

Seber, G. A. F. 1982. The Estimation of Animal Abundance and Related Parameters, 2nd Edition. Griffin, London.

Shaffer, T. L. 2004. A unified approach to analyzing nest success. Auk 121: 526–540.

Skinner, R., W. Erickson, G. Minick, and L. L. McDonald. 1997. Estimating Moose Populations and Trends Using Line Transect Sampling. Technical Report Prepared for the USFWS, Innoko National Wildlife Refuge, McGrath, AK.

Smith, W. 1979. An oil spill sampling strategy, in R. M. Cormack, G. P. Patil, and D. S. Robson, Eds. Sampling Biological Populations, pp. 355–363. International Co-operative Publishing House, Fairland, MD.

Smith, D. R., M. J. Conroy, and D. H. Brakhage. 1995. Efficiency of adaptive cluster sampling for estimating density of wintering waterfowl. Biometrics 51: 777–788.

Smith, D. R., J. A. Brown, and N. C. H. Lo. 2004. Application of adaptive sampling to biological populations, in W. L. Thompson, Ed. Sampling Rare or Elusive Species, pp. 77–122. Island Press, Washington.

Southwell, C. 1994. Evaluation of walked line transect counts for estimating macropod density. J. Wildl. Manage. 58: 348–356.

Stanley, T. R. 2000. Modeling and estimation of stage-specific daily survival probabilities of nests. Ecology 81: 2048–2053.

Steinhorst, R. K., and M. D. Samuel. 1989. Sightability adjustment methods for aerial surveys of wildlife populations. Biometrics 45: 415–425.

Stevens, D. L., and A. R. Olsen. 1999. Spatially restricted surveys over time for aquatic resources. J. Agric. Biol. Environ. Stat. 4: 415–428.

Stevens, D. L., and A. R. Olsen. 2004. Spatially balanced sampling of natural resources. J. Am. Stat. Assoc. 99: 262–278.

Strickland, M. D., L. McDonald, J. W. Kern, T. Spraker, and A. Loranger. 1994. Analysis of 1992 Dall's sheep and mountain goal survey data, Kenai National Wildlife Refuge. Bienn. Symp. Northern Wild Sheep and Mountain Goat Council.

Strickland, M. D., W. P. Erickson, and L. L. McDonald. 1996. Avian Monitoring Studies: Buffalo Ridge Wind Resource Area, Minnesota. Prepared for Northern States Power, Minneapolis, MN.

Thomas, J. W. Ed. 1979. Wildlife Habitats in Managed Forests: The Blue Mountains of Oregon and Washington. US Forest Service, Agriculture Handbook 553.

Thompson, S. K. 1990. Adaptive cluster sampling. J. Am Stat. Assoc. 85: 1050–1059.

Thompson, S. K. 1991a. Adaptive cluster sampling: designs with primary and secondary units. Biometrics 47: 1103–1115.

Thompson, S. K. 1991b. Stratified adaptive cluster sampling. Biometrika 78: 389–397.

Thompson, S. K. 1992. Sampling. Wiley, New York.

Thompson, W. L. 2002a. Towards reliable bird surveys: Accounting for individuals present but not detected. Auk 119: 18–25.

Thompson, S. K. 2002b. Sampling, 2nd Edition. Wiley, New York.

Thompson, S. K., and G. A. F. Seber. 1996. Adaptive Sampling. Wiley, New York.

Thomas, D., and E. Taylor. 1990. Study designs and tests for comparing resource use and availability. J. Wildl. Manage. 54: 322–330.

Thompson, W. L., G. C. White, and C. Gowan. 1998. Monitoring vertebrate populations. Academic, London.

Trenkel, V. M., S. T. Buckland, C. McLean, and D. A. Elston. 1997. Evaluation of aerial line transect methodology for estimating red deer (*Cervus elaphus*) abundance in Scotland. J. Environ. Manage. 50: 39–50.

Venables, W. N., and B. D. Ripley. 2002. Modern Applied Statistics with S. Springer, New York.

Vojta, C. D. 2005. Old dog, new tricks: Innovations with presence–absence information. J. Wildl. Manage. 69: 845–848.

White, G. C., D. R. Anderson, K. P. Burnham, and D. L. Otis, 1982. Capture–Recapture and Removal Methods for Sampling, Closed Populations. Rpt. LA-8787-NERP. Los Alamos National Laboratory, Los Alamos, NM.

Wiegert, R. G. 1962. The selection of an optimum quadrat size for sampling the standing crop of grasses and forbs. Ecology 43: 125–129.

Williams, B. K., J. D. Nichols, and M. J. Conroy. 2002. Analysis and Management of Animal Population. Academic, London.

Wolter, K. M. 1984. Investigation of some estimators of variance for systematic sampling. J. Am. Stat. Assoc. 79: 781–790.

Zippen, C. 1958. The removal method of population estimation. J. Wildl. Manage. 22: 82–90.

Chapter 5
Sampling Strategies: Applications

5.1 Introduction

We have now presented the philosophy and basic concepts of study design, experimental design, and sampling. These concepts provide the foundation for design and execution of studies. Once a general design is conceptualized, it needs to be applied. A conceptual study design, however, is not always an executable design. During the application of the planned study design, additional steps and considerations are often necessary. These include ensuring appropriate sampling across space and time, addressing sampling errors and missing data, identifying appropriate response variables, applying appropriate sampling methodology, and establishing sampling points. Jeffers (1980) provides very useful guidance in the form of a checklist of factors to consider in developing and applying a sampling strategy. These include (1) stating the objectives, (2) defining the population to which inferences are to be made, (3) defining sampling units, (4) identifying preliminary information to assist development and execution of the sampling design, (5) choosing the appropriate sampling design, (6) determining the appropriate sample size, and (7) recording and analyzing data. We have discussed many of these topics in Chaps. 1–4; we further elaborate on some of these considerations here.

5.2 Spatial and Temporal Sampling

Wildlife populations exhibit variations in population dynamics and patterns of resource use across both time and space. Patterns often vary with the scale at which a study is conducted (Wiens 1989). Given that populations and resource-use patterns vary in time and space, studies should be designed and conducted at the scale that encompasses the appropriate variation (Bissonette 1997) (Fig. 5.1). Scale, therefore, denotes the appropriate resolution one should employ to measure or study an ecological system or process (Schneider 1994). For some species, a single scale may be appropriate, whereas other species should be studied at multiple scales. Even then, ecological relationships observed at one scale may differ from

M.L. Morrison et al., *Wildlife Study Design.*

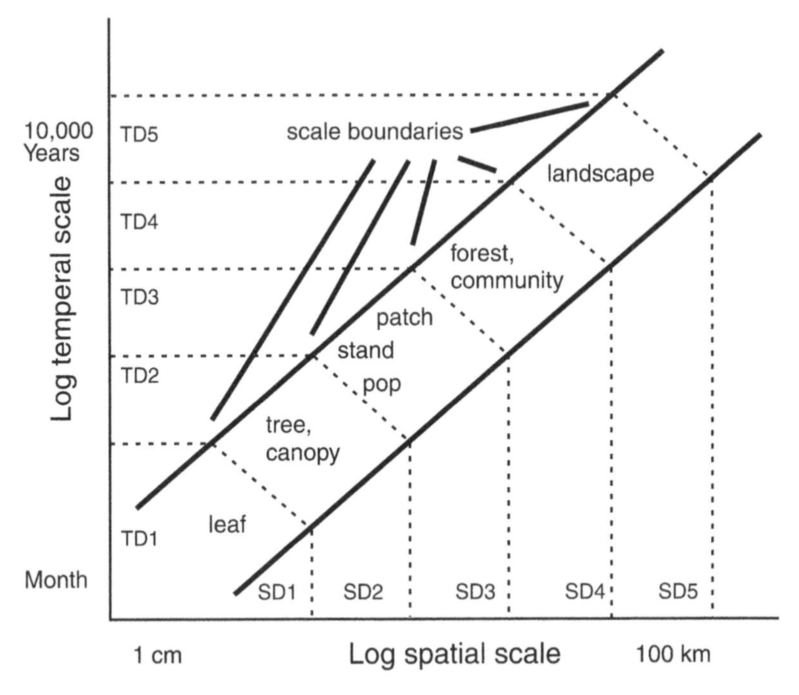

Fig. 5.1 Approximate matching of spatial and temporal scales in ecological studies. Domains of scale are represented by the *dotted lines*. Reproduced from Bissonette (1997) with kind permission from Springer Science+Business Media

those observed at others (Wiens et al. 1986). For example, consider populations of small mammals from ponderosa pine/Gambel oak (*Pinus ponderosa/Quercus gambeli*) forests of north-central Arizona. These small mammals are typically reproductively inactive during winter, initiating reproduction during spring and continuing through summer and fall with the population gradually increasing over this period (Fig. 5.2). As a result, estimates of population size depend largely on the time of year when data were collected. Population estimates based on sampling during fall would be greater than those derived using winter data or using a pooled data set collected throughout the year. Not only do populations fluctuate within years, but also they exhibit annual variation (see Fig. 5.2) as is often the case with *r*-selected species. Thus, a study conducted during a given year may not be representative of the population dynamics for the species studied. Depending on whether the objective of a study is to characterize "average or typical" population size, patterns of population periodicity, or population trend, sampling should occur over the appropriate time period for unbiased point estimates or estimates of trend.

Wildlife populations and patterns of resource use vary spatially as well. Spatial variations occur at different scales, ranging from within the home range of an individual to the geographic range of the species. For example, Mexican spotted owls (*Strix occidentalis lucida*) often concentrate activities within a small portion of

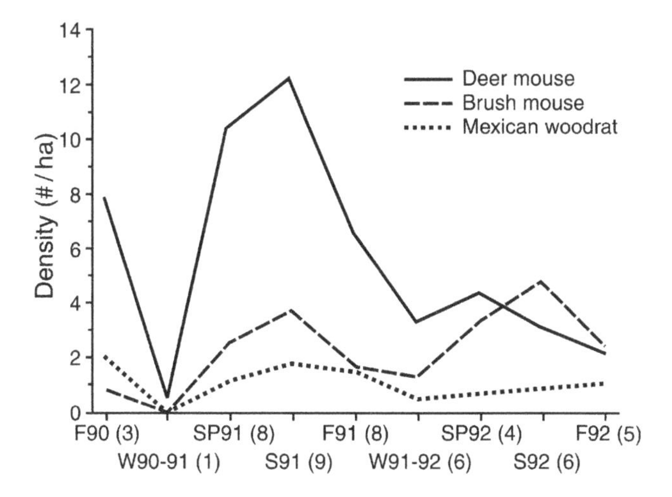

Fig. 5.2 Crude density estimates for the deer mouse (*Peromyscus maniculatus*), brush mouse (*P. boylii*), and Mexican woodrat (*Neotoma mexicana*) from Arizona pine/oak forests illustrating season and yearly variation in abundance. Reproduced from Block et al. (2005) with kind permission from The Wildlife Society

their estimated home ranges, thereby differentially using potentially available resources (Ganey et al. 1999). A study to understand resource use must take into account patchy activity patterns. When evaluating foraging ecologies of birds in oak (*Quercus* spp.) woodlands, Block (1990) found geographic variation in tree use and foraging behavior of most birds studied. A broad-based study to understand foraging ecology in oak woodlands would require a design that ensured variation across oak woodlands was sampled adequately.

5.2.1 Designs for Time Spans Related to Populations

Wildlife populations are subjected to variations in environmental conditions and ecosystem processes over time. Processes can be rare or common, occur predictably or unpredictably, act slowly or quickly, be stochastic or deterministic, and can be local or widespread. Given the variety of factors that might influence a population, sampling should be structured to incorporate as much of the variation resulting from those factors as possible. The most obvious solution is to conduct studies over a long enough time to incorporate most influences on a population (Wiens 1984; Strayer et al. 1986; Morrison 1987). Strayer et al. (1986) concluded that long-term studies were appropriate to study slow processes, rare events, subtle processes, and complex phenomena. Examples of slow processes potentially influencing wildlife populations include plant succession, invasion by exotic species, and long-term population cycles (for example, prey population cycles [e.g., snowshoe hares] that influence population dynamics of the predator [e.g., northern goshawk]). Rare events include "disasters" such as fire, floods, population irruptions of a food item (e.g., insect epidemics resulting in a numerical response by birds), and various environmental crunches (Morrison 1987). Subtle processes are those that may show little change over a short period, but whose effects are greater when viewed within a longer time frame. An example might be effects of global warming on wildlife population distributions. Complex phenomena are typically the result of multiple interacting factors. Contemporary southwestern ponderosa pine forests are the result of natural forest processes as influenced by past grazing, logging, and fire suppression (Moir et al. 1997). Given that these factors have occurred at various times and locations, and for various durations, long-term study would be required to unravel their effects on the species under study.

This begs the question, how long is long-term? Strayer et al. (1986) defined *long term* as a study "if it continues for as long as the generation time of the dominant organism or long enough to include examples of the important processes that structure the ecosystem under study…the length of study is measured against the dynamic speed of the system being studied." Clearly, long-term depends greatly on the species and system under study.

Although not necessarily equivalent to long-term studies, alternatives exist to provide some insight into long-term phenomena (1) substituting space for time, (2) retrospective studies, (3) use of a system with fast dynamics for slow dynamics, and (4) modeling. Each of these approaches offers certain benefits and disadvantages that are discussed in more detail in Chap. 7.

5.2.2 Designs for Spatial Replication Among Populations

Hurlbert's (1984) treatise on pseudoreplication provided an elegant argument for the inclusion of replicate plots to increase the scope of inferences possible from a research study. Frequently, investigators take numerous samples from a defined study area and regard those as replicates, when in actuality they are more appropriately regarded as subsamples within a sample. Inferences drawn from results of such a study would be restricted to that place of study. If the objective of the study is to understand aspects of a population over a larger geographic area, a study is needed that samples the population across that area.

5.3 Sampling Biases

Thompson et al. (1998) distinguish two types of errors: sampling and nonsampling errors. Sampling error is random in nature, often resulting from the selection of sampling units. Sampling error is most appropriately addressed by developing and implementing a sampling design that provides adequate precision. We refer you to the preceding chapters for guidance on study design development. In contrast, nonsampling error or *sampling bias* is typically a systematic bias where a parameter is consistently under- or overestimated. Although considerable thought and planning goes into development of a sampling design, execution of that design may result in unaccounted biases. These include differences among observers, observer bias, measurement error, missing data, and selection biases. Realization that bias can and does occur and knowledge of the types of bias that will arise are crucial to any wildlife study. By understanding potential biases or flaws in execution of a study design, an investigator can try to avoid or minimize their effects (see Sect. 2.2) on data quality, results, and conclusions. Aspects of sampling bias are discussed below.

5.3.1 Observer Biases

Many types of wildlife data are collected. Although some collection methods have little sensitivity to interobserver variation, others can result in systematically biased data dependent on observer. Differences among observers can result from the technique used to collect the data, differences in observer skill, or human error in collecting, recording, or transcribing data.

Interobserver differences are unavoidable when collecting many types of wildlife data. Not only will observer results differ among each other, but also deviations from true values (i.e., bias) often occur. For example, Block et al. (1987) found that ocular estimates or guesses of habitat variables differed among the observers in their study. Not only did these observers provide significantly different estimates for many variables, but also their estimates differed unpredictably from measured or baseline values (e.g., some overestimated while others underestimated). Given the vast interobserver variation, Block et al. (1987) found that number of samples for precise estimates were greater for ocular estimates than for systematic measurements (Fig. 5.3). Verner (1987) and Nichols et al. (2000) found similar observer variation when comparing bird count data among observers. Verner (1987) found that one person might record fewer total detections than other observers, but would detect more bird species. Observer variation is not limited to differences among individuals; Collier et al. (2007) found that evidence detection rates by the same observers varied between sampling occasions. Without an understanding of the magnitude and direction of observer variation, one is restricted in the ability to correct or adjust observations and minimize bias.

5.3.2 Sampling and Measurement Biases

If measurement errors are independent among observers such that they average to be 0 (i.e., some observers overestimate and others underestimate a parameter), they are reflected within standard estimates of precision. These errors can, however,

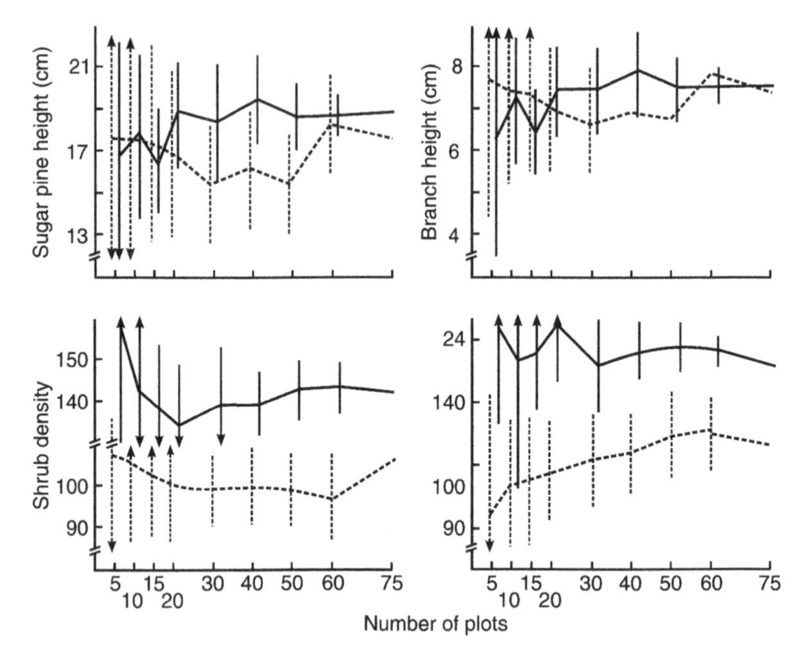

Fig. 5.3 Influence of sample size on the stability of estimates (*dotted lines*) and measurements (*solid lines*). *Vertical lines* represent 1 SD for point estimates. Reproduced from Block et al. (1987) with kind permission from the Cooper Ornithological Society

result in precision estimates beyond an acceptable level, in which case actions are necessary to reduce the frequency and magnitude of such errors. This can be accomplished by better training and quality control or by increasing sample sizes to increase precision of the estimates.

If measurement errors are correlated among observations, then estimates of precision may be biased low (Cochran 1977). Examples of correlated measurement errors are when an observer consistently detects fewer or more individuals of a species than what exists, a tool that is calibrated incorrectly (e.g., compass, altimeter, rangefinder), or when systematic errors are made during data transcription or data entry.

5.3.3 Selection Biases

Habitat- and resource-selection studies typically entail identifying a general resource such as a specific habitat type or tree species, and then determining whether a population uses part of that resource disproportionately to its occurrence. Studies can examine one variable at a time or multiple variables simultaneously. Manly et al. (1993, pp. 29–30) listed seven assumptions that underlie resource selection studies (1) the distributions of the variables measured to index available resources remain constant during the course of a study, (2) the probability function of resource selection remains constant within a season, (3) available resources have been correctly identified, (4) used and unused resources are correctly identified, (5) the variables that influence the probability of selection have been identified, (6) individuals have free access to all available resource units, and (7) when sampling resources, units are sampled randomly and independently. The degree to which these assumptions are violated introduces bias into the study.

The assumptions of constant resource availability and a constant probability function of resource selection are violated in most field studies. Resources change over the course of a study, both in quantity and quality. Further, populations' patterns of resource use and resource selection may also change with time. The severity of the potential bias depends on the length of study and the amount of variation in the system. Many studies are conducted within a season, most typically the breeding season. Consider the breeding chronology of many temperate passerine birds (Fig. 5.4). The breeding season begins with establishment of territories and courtship behavior

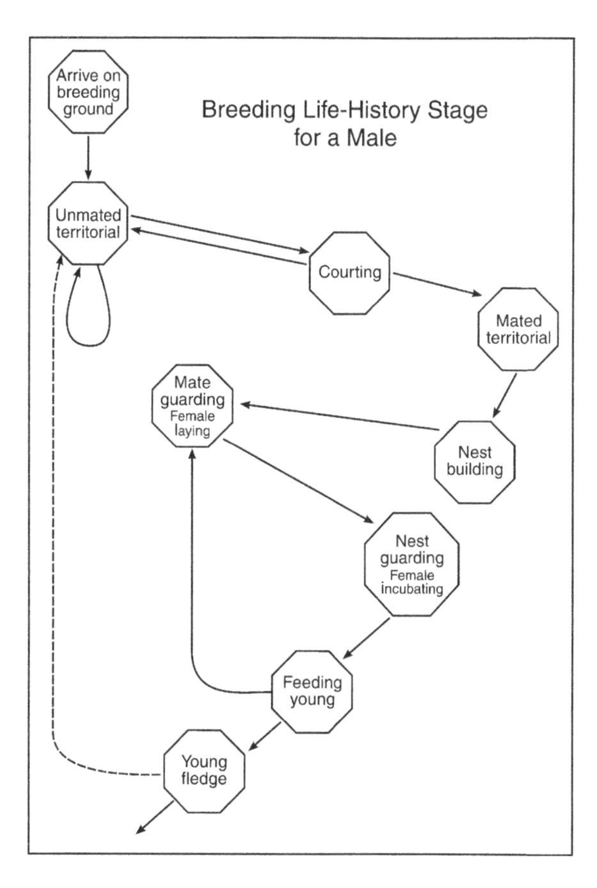

Fig. 5.4 Transition diagram of the breeding history stage for male white-crowned sparrows (*Zonotrichia leucophrys*) (from Wingfield and Jacobs 1999)

during early spring. Once pair bonds are established, they select or construct a nest and then go through various stages of reproduction, such as egg laying, incubation, and nestling and fledgling rearing before the juveniles disperse. Some species may nest multiple times during a breeding season. Initiation of breeding behavior to dispersal of the young may encompass 3–4 months, a period that likely includes fluctuations in food resources, vegetation structure, and in individual resource needs. Thus, an investigator studying birds during the breeding season has two alternatives: either examine "general patterns" of selection across the breeding season or partition the breeding period into smaller units possibly corresponding to different stages in the reproductive cycle (e.g., courtship, nest formation, egg laying, incubation, fledgling, dispersal). Examining general patterns across the breeding season may provide results that have no relevance to actual selection patterns because the average or mean value for a given variable may not represent the condition used by the species at any time during the season. Breaking the breeding season into smaller subunits may reduce the variation in resource availability and resource selection to meet this assumption more closely. A tradeoff here, however, is that if the period is too short, the investigator may have insufficient time to obtain enough samples for analysis. Thus, one must carefully weigh the gain obtained by shortening the study period with the ability to get a large enough sample size.

A key aspect of any study is to determine the appropriate variables to measure. In habitat studies, for example, investigators select variables they think are relevant to the species studied. Ideally, the process of selecting variables should be based on careful evaluation of the species studied including a thorough review of past work. Whether the species is responding directly to the variable measured is generally unknown, and could probably more closely be determined by a carefully controlled experiment such as those Klopfer (1963) conducted on chipping sparrows (*Spizella passerina*). Rarely are such experiments conducted, however.

Even if the appropriate variable is chosen for study, an additional problem is determining if that resource is available to the species. That is, even though the investigator has shown that a species uses a particular resource, this does not mean that all units of that resource are equally available to the species. Numerous biotic and abiotic factors may render otherwise suitable resources unavailable to the species (Wiens 1985) (Fig. 5.5). Identifying which resources are and are not available is a daunting task. Most frequently, the investigator uses some measure of occurrence or relative occurrence of a resource to index availability. Implicit to such an index is a linear relation between occurrence and availability. The degree to which this assumption is violated is usually unknown and rarely tested; effects of violating the assumption on study results are perhaps less well understood.

5.3.4 Minimization of Bias

An underlying goal of any study is to produce reliable data that provide a basis for valid inferences. Thus, the investigator should strive to obtain the most accurate data possible, given the organism or system studied, the methods available for

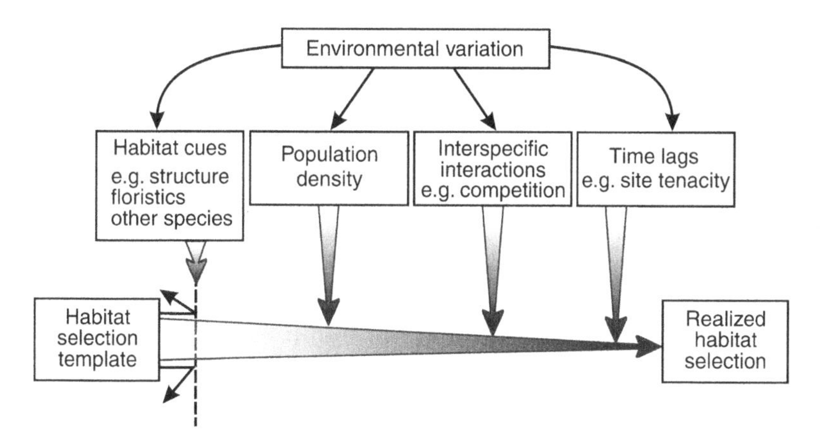

Fig. 5.5 Factors influencing habitat selection in birds. Reproduced from Wiens (1985) with kind permission from Elsevier

collecting data, and constraints imposed by the people, time, and money at hand. Many forms of bias can be minimized through careful planning and execution of a study. Specific aspects that can and should be addressed are discussed below.

5.3.4.1 Quality Assurance and Quality Control

Measurement bias can be eliminated or minimized by developing and following a rigorous quality assurance/quality control (QA/QC) program. The research branch of the United States Department of Agriculture (USDA) Forest Service has developed such a program. They define quality assurance as a process to produce reliable research data with respect to its precision, completeness, comparability, and accuracy. *Quality assurance* includes the steps involved with study planning, design, and implementation, as well as with the steps entailed for analysis and reporting of data. *Quality control* is the routine application of procedures (such as calibration or maintenance of instruments) to reduce random and systematic errors, and to ensure that data are generated, analyzed, interpreted, synthesized, communicated, and used within acceptable limits. Quality control also requires the use of qualified personnel, use of appropriate methods, and adherence to specific sampling procedures.

Critical, yet often ignored steps in QA are development of a written study plan, and then having that study plan reviewed by peers and a biometrician (see Sect. 1.3.1 regarding feedback from peers). Study plans should be regarded as a living document. Studies are often adjusted during implementation for various reasons: restricted access, loss of personnel, changes in methods to collect data, unforeseen logistical obstacles, and lost data. These changes should be documented in the study plan through an amendment, addendum, or some similar change.

Both QA and QC come into play during all phases of a study. Perhaps the very first step in QA/QC is to "obtain some advice from a statistician with experience in your field of research before continuing with the sampling investigation" (Jeffers 1980, p. 5). Thereafter, QA/QC should be invoked continuously during all stages of a study (data collection, data entry, data archiving) to ensure an acceptable quality of collecting data, high standards for project personnel, a rigorous training program, and periodic checks for consistency in data collection. The QA/QC process should also be documented as part of the study files, and QC data retained to provide estimates of data quality.

5.3.4.2 Observer Bias

The first step in addressing observer bias is to recognize that it exists. Once identified, the investigator has various options to limit bias. Potential ways to control observer bias are to (1) use methods that are repeatable with little room for judgment error, (2) use skilled, qualified, and motivated observers, (3) provide adequate training and periodic retraining, and (4) institute QC protocols to detect and address observer bias.

If the bias is uncorrelated such that it averages to be 0 but inflates estimates of sampling variance, it can be addressed by tightening the way data are collected to minimize increases in precision estimates, or by increasing sample sizes to reduce precision estimates. If the bias is systematic and correlated, it may be possible to develop a correction factor to reduce or possibly eliminate bias. Such a correction factor could be developed through double sampling (see Sect. 4.3.5) and as discussed later in Sect. 5.5.5. As an example, consider two ways to measure a tree canopy cover: one using a spherical densiometer and the other by taking ocular estimates. The densiometer is a tool commonly used to measure canopy cover and provides a less biased estimate than ocular estimates. When field readings are compared between the two methods, a relationship can be developed based on a curve-fitting algorithm, and that algorithm can be used to adjust observations using the less reliable technique.

A step often ignored or simply conducted in a cursory fashion is training. Kepler and Scott (1981) developed a training program for field biologists conducting bird counts and found that training reduced interobserver variation. Training is needed not only to ensure that observers have adequate skills to collect data, but also to ensure that observers follow established protocols, to ensure that data forms are completed correctly, and to ensure that data are summarized and stored correctly after they have been collected. Training should not be regarded as a single event, but as a continuous process. Periodic training should be done to verify that observers have maintained necessary skill levels, and that they continue to follow established protocols.

5.4 Missing Data

Investigators use a variety of ways to collect and record data. These include writing observations on data sheets, recording them into a tape recorder, use of hand-held data loggers, use of video cameras, sound equipment, or traps to capture animals.

Occasionally, data forms are lost or equipment malfunctions, is vandalized, or broken resulting in lost data. Little and Rubin (1987) present a comprehensive description of analysis methods that reflect missing data. However, the need for such complex methods should be avoided by attentive implementation of study protocols.

If possible, observers should record data when observations are made and while still fresh in their mind. This prevents the observer from forgetting or recording inaccurate information, commonly referred to as transcription error. Transcription errors occur after data have been collected and they are being transferred from one medium to another: for example, from a tape recorder to a data sheet or from a data sheet to an electronic medium (also known as data entry error). Transcription error may be minimized if two or more people transcribe the data independently, and then compare the work of both people. Discrepancies can be identified and then checked against the original data to correct errors. Granted, this step increases up-front costs of the study, but avoids huge potential problems in data analysis and interpretation.

5.4.1 Nonresponses

Nonresponse error occurs when one fails to record or observe an individual or unit that is part of the selected sample. For example, when conducting point or line-transect counts of birds, an observer will undoubtedly fail to detect all individuals present. This occurs for a variety reasons, including the behavior or conspicuousness of the individual, observer knowledge and skill levels, physical abilities of the observer, environmental conditions (e.g., vegetation structure, noise levels, weather), and type of method used. Likely, detectability of a given species will vary in time and space. As the proportion of nonresponses increases, so does the confidence interval of the parameter estimated (Cochran 1977). Thus, it is important to minimize nonresponses by using appropriate methods for collecting data and by using only qualified personnel. Once appropriate steps have been made to minimize nonresponses, techniques are available to adjust observations collected in the field by employing a detectability correction factor (see Sect. 4.5.1 for an example from line-transect sampling). Thompson et al. (1998, p. 83) caution that methods that use correction factors are (1) more costly and time consuming than use of simple indices and (2) should only be used when the underlying assumptions of the method are "reasonably" satisfied.

5.4.2 Effects of Deviations from Sampling Designs

Deviations from sampling designs can add an unknown bias to the data set. For example, if an observer decides to estimate rather than measure tree heights, it could result in an underestimate of tree heights (see Fig. 5.3) (Block et al. 1987). Also, if

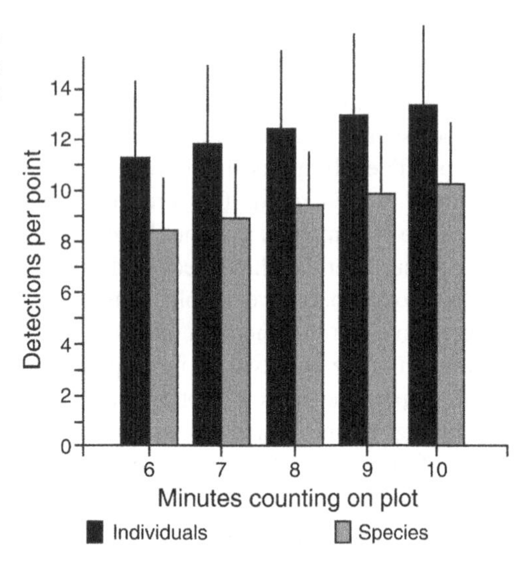

Fig. 5.6 Mean (±1 SD) number of individuals detected during 6- to 10-min point counts (from Thompson and Schwalbach 1995)

an observer records birds during point counts for only 6 min rather than 10 min as established in the study design, the observer may record fewer species and fewer individuals of each species (Fig. 5.6) (Thompson and Schwalbach 1995).

5.5 Selection of the Sampling Protocol

As detailed in Chap. 4, every wildlife study is unique and the selection of a specific sampling protocol depends on the experience and expertise of the investigator. For many studies, there may be more than one valid approach to conduct a given study. The investigator has choices in specific sampling methods used to collect various types of data. Sometimes the choice is obvious; other times the choice may be more obscure. Critical factors that weigh into selection of a sampling protocol include (1) the biology of the species studied, (2) its relative population size and spatial distribution, (3) methods used to detect the species, (4) the study objectives, (5) resources available to the study, (6) the size of the study area(s), (7) time of year, and (8) the skill levels of the observers.

Consider, for example, different methods to sample population numbers of passerine birds (Ralph and Scott 1981; Verner 1985). Four common methods include spot mapping, point counts, line transects, and capture–recapture. Each has merits and limitations in terms of feasibility, efficiency, and accuracy, and variations exist even within a method. With point counts, the investigator must decide to use fixed- or variable-radius plots, radius length for fixed plots, how much time to spend at a point (usually varies between 4 and 10 min), how many times to count each point, and how to array points within a study area. Many of the nuances that influence these decisions are discussed more fully in the papers contained within Ralph et al. (1995). Our point is that numerous details must be considered when selecting the specific sampling protocol to be used in collecting data. Simply choosing a basic procedure does not provide the level of specificity that must be defined in the sampling protocol.

5.5.1 Sampling Intensity

Sampling intensity refers to how many, how long, and how often units should be sampled. Obviously, this depends on the study objectives and the information needed to address them. In addition, the biology of the organism and attributes of the process or system being studied will influence sampling intensity. In considering sampling intensity, one must realize that tradeoffs will be involved. For example, the length of time spent at each sampling unit or number of repeat visits needed may limit the overall number of plots visited. Deriving the optimal allocation of sampling intensity should be a goal of a pilot study. In the absence of a pilot study, the investigator might try to analyze a similar data set or consult with experts knowledgeable in the species studied and the types of statistical analyses involved. A third possibility is to conduct the study in a staged fashion, whereby preliminary data can be evaluated to determine the appropriate sampling intensity. This procedure can be repeated over a series of sampling sessions until the sampling intensity meets the study needs.

How many plots are needed for a study is an issue of sample size. Adequate sample sizes are needed for precise point estimates of the variable being measured, to ensure adequate statistical power to detect a difference or trend should it indeed occur, and to meet specific requirements of specialized analytical tools such as program DISTANCE (Buckland et al. 2001, 2004), program MARK (Cooch and White 2007), and others. As a rule, more plots are needed if a species is rare or has a clumped distribution (Thompson et al. 1998). Sample size considerations have been discussed in more detail in Sect. 2.6.7.

Temporal considerations for sampling intensity involve both the length of time to spend collecting observations during each visit to a sampling point, number of visits to each point, and the length of time needed to conduct a study. The amount of time spent at a sampling point depends on the species studied and the probability of detecting them during a certain period of time. Dawson et al. (1995) found that the probability of detecting a species increased with time spent at a count station (Table 5.1). However, the more time spent at each point limits the number of points that can be sampled, consequently both factors must be considered in the study design (Petit et al. 1995).

One visit to a plot may be inadequate to detect all individuals using an area because of missed observations, behavioral differences that influence detectability,

Table 5.1 Probability of detecting 14 species of neotropical migratory birds within 5, 10, 15, and 20 min at points where they were known to be present

		Probability of detecting within			
Species name	Number of points*	5 min	10 min	15 min	20 min
Yellow-billed cuckoo	258	0.465	0.655	0.812	0.922
Great-crested flycatcher	270	0.540	0.704	0.839	0.926
Eastern wood-pewee	294	0.611	0.752	0.854	0.920
Acadian flycatcher	176	0.747	0.820	0.896	0.936
Blue-gray gnatcatcher	112	0.580	0.728	0.862	0.931
Wood thrush	323	0.784	0.882	0.939	0.971
Gray catbird	171	0.615	0.779	0.893	0.936
Red-eyed vireo	377	0.857	0.922	0.964	0.980
Worm-eating warbler	79	0.507	0.671	0.877	0.929
Ovenbird	244	0.765	0.885	0.940	0.977
Kentucky warbler	82	0.580	0.773	0.827	0.945
Common yellowthroat	125	0.606	0.740	0.852	0.950
Scarlet tanager	295	0.718	0.833	0.910	0.948
Indigo bunting	184	0.582	0.726	0.845	0.912

* Number of points is the number at which the species was detected
Source: Dawson et al. (1995)

or animal movements. As noted above, observers vary in skills, motivation, and competency. As a result, some may fail to detect an animal present during a given sampling period. To minimize this bias, Verner (1987) suggested that multiple visits to each plot, each visit by a different observer, would decrease the probability of missing observations. Because animal behavior can change based on seasonal phenology according to different needs and activities, sampling intensity may need to vary accordingly. For example, fewer visits might be needed at bird counting points during the breeding than during the nonbreeding season. Birds tend to exhibit greater site fidelity while conducting breeding activities than during winter when they are not defending nests and territories. During winter, birds are often found in mobile flocks patchily distributed across the landscape. To increase the probability of detecting birds in these flocks requires increased sampling effort usually accomplished by conducting more visits at each point.

5.5.2 Line Intercepts

Line intercept sampling was first introduced as a method for sampling vegetation cover (Canfield 1941; Bonham 1989). It is widely used in wildlife habitat studies when vegetation cover is anticipated as a strong correlate of use by the species of interest. Effectively, one calculates the proportion or percentage of a line that has vegetation directly above the line. Say, for example, a line intercept was 20-m long, and 15 m of the line were covered by canyon live oak (*Quercus chrysolepis*). Thus, percentage canopy cover would be 15/20 × 100 = 75%. There is some flexibility in how line intercept methods are used. They are used to estimate overall canopy cover, cover by plant species, cover at different height strata, or cover by different plant forms (e.g., herbs, shrubs, trees), providing opportunities to examine both structural and floristic habitat correlates.

A derivation of the line intercept technique is the point intercept method. This methodology was developed primarily to sample grasses and forbs as part of range studies (Heady et al. 1959). Generally, points are systematically arrayed along a transect that is oriented in a random direction. The method entails noting whether the object being sampled is on the point (commonly termed a hit) or not. An estimate of cover would be the percentage of points where a "hit" was recorded. Similar to line intercepts, point intercepts can be used a various ways to address different study objectives.

As noted in Sect. 4.4.2, intercepts can also be a used as a method to sample individuals for collecting additional information or attributes. For example, suppose a study on secondary cavity-nesting birds wanted to estimate the relative abundance of cavities on a study area. An intercept could be used to select trees to sample, and then the numbers of cavities could be counted on each tree that was hit to provide an abundance estimate for cavities.

5.5.3 Plot Shape and Size

Wildlife tends to be distributed nonrandomly across the landscape in patterns that typically correspond to the distribution of their habitat; these patterns are further influenced by intraspecific and interspecific interactions (Fretwell and Lucas 1970; Block and Brennan 1993). Further, the relative abundance of animals also varies both within and among species. Given that distributions and abundance vary, sample units or plots should vary in shape and size depending on the species studied. As Thompson et al. (1998, p. 48) conclude, no single plot design applies to all situations. Consequently, each study should be considered independently.

5.5.3.1 Plot Shape

Four primary factors influence the selection of the plot shape: detectability of individuals, distribution of individuals, edge effects (i.e., ratio of plot perimeter to plot area), and the methods used to collect data (Thompson 2002; Thompson et al. 1998). A plot with a large edge effect may lead to greater commission or omission of individuals during counts, that is, including individuals existing off the plot or excluding individuals occurring on the plot. Given plots of equal area, long, narrow, rectangular plots will have greater edge effect than square plots, which have more edge than circular plots. However, differences in edge effect can be outweighed by the sampling method, dispersion pattern of individuals, and their detectability. For example, if sampling occurs quickly, then animals may not move in and out of the plot, thereby minimizing the errors of commission and omission. Thompson (2002) concluded that rectangular plots were more efficient than square or round plots for detecting individuals, which would reduce the magnitude of adjustments needed for simple counts and provide less biased estimates of animal numbers. Given that most species of wildlife are not randomly distributed, most exhibit some degree of clumping. Thompson et al. (1998) found that precision of estimates increased if few plots had no detection of the species sampled. Thus, the shape of a plot must be such to maximize the probability of including the species. Generally, a long and narrow rectangular plot would have greater chance of encountering a species with a clumped distribution.

5.5.3.2 Plot Size

Numerous factors influence plot size, including the biology of the species, its spatial distribution, study objectives, edge effects, logistical considerations, and cost constraints. Certainly, larger species or top-level predators with large home ranges require larger plots to include adequate numbers. For example, a 10-ha plot might include only 1% of the home range of a spotted owl, whereas it could include the entire home ranges of multiple deer mice (*Peromyscus maniculatus*). The spatial arrangement of a species also influences plot size. If the distribution is strongly aggregated, small plots may fail to include any in a large proportion, whereas the number of empty plots would be reduced with large plots (Fig. 5.7). Another advantage of larger plots, especially when studying highly vagile animals, is that larger plots are more likely to include the entire home range of the species than smaller plots. Effectively, larger plots have a lower ratio of edge to interior, thereby limiting potential edge effects. Disadvantages of larger plots are costs and limitations in the number of plots that can be sampled. Larger plots require more effort to sample than smaller plots. Given a fixed budget, one is faced with a tradeoff between plot size and the number that can be sampled. This tradeoff must be carefully weighed during the planning process by carefully stating the study objectives or hypothesis, specifying the differences, trends, or parameters to be estimated, and determining the sample size needed for adequate statistical treatment of the data. A study that consists of one or two large plots may be of limited value in drawing inferences regarding the species studied in that results cannot be easily extrapolated beyond the study location. Thus, an investigator must carefully weigh the resources at hand with the objectives or information needs of a study to evaluate critically if enough plots of adequate size can be sampled to ensure a valid study. If not, the study should not be done.

5.5.4 Pilot Studies

The first step in critically evaluating the feasibility of conducting a project is to engage in a pilot study. Green (1979, p. 31) states quite succinctly the importance of a pilot study in his fifth principle for study design noting: "Carry out some preliminary sampling to provide a basis for evaluation of sampling design and statistical analysis

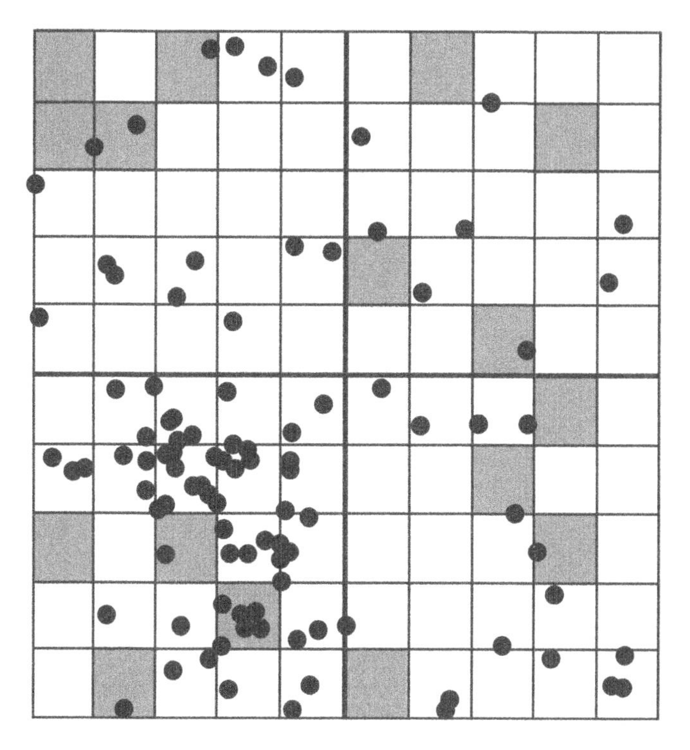

Fig. 5.7 A sampling frame with an underlying clumped spatial distribution of animals is divided into four strata each containing 25 plots. A simple random sample of four plots is selected from each stratum (Figure from *Monitoring Vertebrate Populations* by Thompson, White and Gowan, Copyright © 1998 by Academic Press, reproduced by permission of Elsevier)

options. Those who skip this step because they do not have enough time usually end up losing time." A pilot study allows the investigator a chance to evaluate whether data collection methodologies are effective, estimate sampling variances, establish sample sizes, and adjust the plot shape and size and other aspects of the sampling design. Conducting a pilot study often leads to greater efficiency in the long run because it can ensure that you do not oversample and that the approach that you are taking is the appropriate one. All too often investigators skip the pilot portion of a study, going directly into data collection, only to conclude that their study fails to achieve anticipated goals despite having collected volumes of data. Many biometricians can recount horror stories of students entering their office with reams of field data, only to find out that the data have very limited utility.

5.5.5 Double Sampling

Double sampling entails designs in which two different methods are used to collect data related to the same or a similar variable (Thompson 2002). A primary impetus for double sampling is to find a quicker and cheaper way to get needed information.

For example, one may use line-transect methods (see Chap. 4) as the preferred way to estimate population size of elk. Although previous investigators have demonstrated that line-transect methods provide accurate and precise estimates of elk population size, the costs of conducting the sampling are high, especially when estimates are needed for large areas. An alternative may be to conduct aerial surveys, although aerial surveys by themselves may not be sufficiently accurate to base estimates on them alone. By using both methods for an initial sample, one can use a simple ratio estimate of detectability to calibrate results from the aerial survey. Ratio estimation is effective in situations in which the variable of interest is linearly related to the auxiliary (quick and dirty) variable. The ideal proportion of the sample relative to the subsample depends on the relative costs of obtaining the information and the strength of the relationship between the two (Thompson 2002). For a detailed discussion of double sampling, see Sect. 4.3.5.

Double sampling also can be used for identifying strata and determining allocation of samples in the strata. For example, consider a study to determine the average height of trees used by hairy woodpeckers for foraging. In collecting information on tree height, the investigator also gathers information on age (adult or juvenile) and sex of woodpeckers observed. Assuming a large enough sample, the investigator can estimate the relative proportion of individuals in each age/sex stratum. A stratified random sample is then selected from the initial sample to estimate the variable of interest: in this case, tree height. The advantage of this approach is that it provides an unbiased point estimate of tree height, but with greater precision than provided by a simple random sample.

A third use of double sampling is to correct for nonresponses in surveys. Surveys are often collected to gather information on consumptive or nonconsumptive use of wildlife for a variety of reasons. Surveys done on hunting success are used in population estimation models or to set future bag limits. Surveys conducted on recreational bird-watching provide resource agencies with important information to manage people and minimize potentially deleterious effects on rare or sensitive birds. Thus, surveys are a useful tool in understanding the biology of species of interest and in developing management options to conserve them. A problem arises, however, because only a proportion of those surveyed actually respond for one reason or another. Assuming that the sample of respondents is representative of the nonrespondents is tenuous. A solution is to consider the nonrespondents as a separate stratum and to take a subsample from that. Thus, the double sample is needed to separately sample the strata defined by whether or not people responded to the survey.

5.6 Development of Sampling Categories

Sampling categories can be designated in three ways (1) partitioning observations into blocking groups, (2) redefining a continuous variable as a categorical variable, or (3) dividing a variable into logical categories.

Observations can be collected or partitioned into groups either before or after sampling. Ideally, groups should be identified while developing the study, preferably through a pilot study, rather than afterward to ensure that the sampling design is adequate to obtain the samples needed to meet study objectives. Groups are identified because the investigator may have some reason to suspect that the groups differ in the particular attribute under study. For example, studies of bird foraging have documented differences between the sexes in foraging (Morrison 1982; Grubb and Woodrey 1990). Pooling observations from both sexes, therefore, may result in invalid results and erroneous conclusions (Hanowski and Neimi 1990). By partitioning observations into groups, the investigator can evaluate whether the groups actually differ in the variable of interest. If no differences are found, groups can be pooled and treated as one. Groups can correspond to characteristics of the species studied, such as age (adult, subadult, juvenile), sex (male, female), or size (big, small), or they can correspond to where (study area) or when (year, season) data were collected (Block 1990; Miles 1990).

Occasionally an observer may collect data on a continuous variable and decide to transform it to a categorical variable for analysis. This may occur because the field methodology for collecting data was imprecise, thus limiting confidence in the actual field measurements. An example might be ocular estimates of tree heights in a habitat study. Because such estimates are often inaccurate and quite variable among observers (Block et al. 1987), a more reasonable approach would be to place height estimates into broad categories that would have some biological relevance (e.g., seedlings, saplings, pole-sized trees, and mature tree) and also minimize bias. The investigator may sacrifice some flexibility of analyzing continuous data, but would minimize the chance of reporting erroneous conclusions.

Perhaps the most common use of sampling categories is when collecting categorical data. Examples include age classes, behaviors, offspring numbers, and plant species. Categories are defined at the onset of the study, generally based on the expertise of the investigator or previous knowledge of the species or system under study (i.e., literature review). Typically, frequency data collected for each category, and a variety of nonparametric and Bayesian analyses are available for analyzing the data (Conover 1980; Noon and Block 1990). Because many of these analyses are sensitive to "empty cells" or zero counts, it is important to define categories in such a way as to minimize the occurrence of empty cells. If necessary, the investigator can combine categories during preliminary data screening to avoid low or empty cell counts as long as the resulting categories have biological relevance.

5.6.1 Identifying Resources

The task of identifying resources used by a species of interest is an exercise that requires the investigator to make an informed guess of what resources or classes of resources a species is actually using. Resources can often be identified during a pilot study, based on results of previous studies, information provided by an expert, or through a literature review. Of all of these sources of information, a pilot study is perhaps most ideal given that use of resources can vary in time and space.

5.6.2 Categories

Many studies, particularly of arboreal species, evaluate activity height or investigate relationships between animal activities and habitat elements or relationships of vegetation height to population dynamics or community metrics. For example, studies of foraging birds may be interested in how high they forage in a tree or a study might be done to test MacArthur and MacArthur's (1961) foliage height diversity/bird species diversity model. The investigator could measure height as a continuous variable, or he or she could divide vegetation heights into categories that have some biological relevance. The choice between recording actual heights or height categories depends on the study objective, statistical analysis, and logistics (that is, measuring vegetation height accurately is more time consuming). For a forested system with a multistoried canopy, categories may correspond to the average heights of the vegetation stratum: for example, 0–1 m for shrubs, 1–4 m for saplings, 4–10 m for mid-story trees, and >10 m for the overstory. Statistical analyses specific to study objectives and potential other analyses are very important to consider because while finer resolution categories can always be combined into coarser categories for analysis, the reverse is not true.

5.6.3 Placement and Spacing of Sampling Stations

Once the investigator has selected a study area and designated sampling plots, he or she must place stations in such a way to adequately sample within and among plots. Placement of stations is influenced by a number of factors including study objectives, the sampling design, logistical considerations, and the organism(s) or system(s) studied. Further, different designs for placing stations are needed depending on whether the ultimate goal is to detect all individuals (complete enumeration) or to sample a portion of the population and estimate population size. As noted below, a complete enumeration is rarely achieved, thus the more likely scenario is for a portion of the population to be sampled.

Sampling stations can be placed in an array such as a grid where the array is considered the sampling unit. Grids are commonly used for sampling small mammals, herpetofauna, or passerine birds. In this case, sampling stations are not generally considered independent samples because individuals of the species studied may use multiple stations, and the presence or absence of an animal at one sampling station may influence the presence of animals at others. Using grids or similar arrays allows an investigator to apply capture–recapture (Otis et al. 1978; White et al. 1982) capture–removal (Zippen 1958) methods to study population dynamics. Habitat relations can be studied by comparing stations where a species was captured with those where it was not or by assessing correlations across grids (Block et al. 1998).

Often, study objectives and statistical analyses require that sampling stations be independent. Ad hoc definitions of independence could be (1) when the probability of sampling the same individual at adjacent points is relatively small or (2) when the species or number of a species present at one point does not influence what species and how many are present at nearby points. An empirical way to test for autocorrelation of observations is based on mixed model analysis of variance (Littel et al. 1996) where one can test for correlations between sampling points at various distances. An alternative testing approach might be the analysis provided by Swihart and Slade (1985) for testing for autocorrelations among observations separated in time. Both methods might allow the investigator to estimate the optimal distance between points to ensure that they are independent observations. Once investigators define that distance, they need to physically locate sampling points within their sample plots. An efficient way to place plots is using a systematic random sampling design (Cochran 1977). This entails randomly choosing the initial starting point and a random direction, and then spacing subsequent points at the defined distance along this bearing. Alternatively, one could randomly overlay a grid with spacing between grid points greater than or equal to that needed for independent observations and then collect data at all or a random sample of those grid points. An advantage of systematic random sampling is that it provides for efficient sampling in the field, whereas locating random grid points could require more time to locate and reach each sampling point. However, a disadvantage of systematic random sampling is that it may not sample the variation within a sample plot as well as random points distributed throughout the plot (Thompson et al. 1998). This potential bias, however, is probably ameliorated given large enough (>50) sample sizes (Cochran 1977; Thompson et al. 1998).

5.7 Sampling Small Areas

Many studies are restricted to small areas for a variety of reasons. For example, a study could be conducted to understand population dynamics or habitat relations of species occurring on a small, unique area such as an island or a patch of riparian vegetation surrounded by uplands. Given the uniqueness of the area, the study must be done at that particular location. Another reason a study might be done in a small area is that a study is being done to understand potential effects of a planned activity or impact on the species found there. These types of studies often fall under the umbrella of impact assessment studies that are discussed in detail in Chap. 6. Given a small area, the researcher should strive for a complete enumeration with the realization that he or she will probably miss some individuals (see below). However, given that most of the population within the study area will be sampled, a correction for detectability can be applied to parameter estimates, thereby increasing their precision.

5.8 Sampling vs. Complete Enumeration

Complete enumeration is more popularly referred to as a census. However, censuses are rarely done in field situations for two major reasons. First, logistical constraints often limit which areas, or limit the amount of area, that can be sampled. Some studies occur on rugged terrain or in dense vegetation that is either inaccessible or not safely accessible by field personnel. In addition, many study areas or sample plots are too large to sample completely, thereby leaving some individuals undetected. Second, the probability of detecting species is most typically less than one. This could be a result of the behavior of the species studied, vegetation density, or the field methodology. Whatever the specific reason, it is probably more reasonable to assume that you are not detecting all individuals, thereby making a complete enumeration nearly impossible. In cases where an investigator strives for a complete enumeration and a large proportion of the population is actually sampled, then a finite population correction factor should be applied to increase precision of the point estimates.

5.9 Sampling for Parameter Estimation of Rare Species or Rare Events

5.9.1 Synthesis of Available Methodologies

Thompson's (2004) book on sampling rare and elusive species provides an excellent overview on the topic. Included are well-developed chapters on various methodologies, including adaptive sampling (Smith et al. 2004), two-phase stratified adaptive sampling (Manly 2004), sequential sampling (Christman 2004), and using various techniques such as genetics (Schwartz et al. 2006), photography (Karanth et al. 2004), and a plethora of indices (Becker et al. 2004; Conn et al. 2004).

Perhaps chief among the available techniques is adaptive sampling or adaptive cluster sampling. Thompson (2002) proposed adaptive sampling for studying rare or clustered animals as a way to increase the effectiveness of the sampling effort. In a general sense, it is somewhat akin to a fishing expedition where one randomly trolls trying to locate fish. Once located, fishing (sampling) becomes more concentrated around that area. Adaptive sampling operates in a similar way. A researcher surveys for a rare or clustered species and once encountered, additional neighboring plots are added to the sample. The primary purpose of adaptive sampling is to acquire more precise point estimates for a given sample size or cost. A secondary benefit is that more observations are acquired on the species of interest, thereby increasing the amount of ancillary information that might be gathered.

Smith et al. (2004) provide guidance for when and how to apply adaptive management specifically as it relates to sampling efficiency, sample size requirements, implementation, and adjustment. Adaptive management becomes efficient when "...the ratio of simple random sampling to adaptive sampling variance" (Smith et al. 2004, p. 89) is greater to 1, assuming a sample size needed for adaptive sampling. Determining that sample size is not as easy as it would be for more traditional sampling approaches because the final sample size depends on what you find. This is influenced by the spatial distribution of the target population, the size of the cluster or neighborhood sampled around observations, and a priori stopping rules detailing when sampling should cease. Adaptive sampling must be flexible to account for variations among species behaviors, mobilities, and habitats. For example, modifications are needed to avoid double sampling mobile species that might flush to adjacent sampling units after being disturbed. Adaptive sampling works well for terrestrial sampling, but its use in aerial surveys is limited due to difficulties specifying whether individuals are within the sampling unit. Another consideration is the situations where study species occur in fragile habitats; sampling might need to be modified to minimize or avoid habitat disturbance.

Sequential sampling is similar to adaptive sample, the primary difference being that sample sizes in sequential sampling are determined once a certain criterion is met (Christman 2004). As such, you do not know *a priori* how many units you need to sample and the decision depends on data recorded to that point. A potential limitation of this approach, however, is that it may result in point estimates with lower precision. The advantage is that it assures that you will have a sample that includes at least a certain number of the rare animal under study.

Manley (2004) proposed a variation on adaptive sampling that he termed two-phase adaptive stratified sampling. This involves dividing your study area into two or more strata such that the variation of the variable of interest is constant relative to the variation found throughout the study area. For example, you might identify strata on the basis of population density where one strata includes high density plots and the other low density plots, or strata could be defined on the basis of some habitat feature such as canopy closure. Simple random samples are taken from each stratum and variances are estimated for each. Second-phase sample units are allocated one-by-one to the stratum where it will provide the largest reduction in the variance. The process continues until all samples have been allocated among strata. Then, data are analyzed as if they came from a conventional stratified random sample. This approach works best when you think that your target population will vary by strata based on environmental conditions (e.g., habitat), but when populations are not extremely clumped. If populations are extremely clumped, adaptive sampling is preferred.

5.9.2 Field Applications

Suppose the primary objective of a study is to estimate population density of the yellow-blotched ensatina (*Ensatina eschscholtzii croceater*), and a secondary objective is to collect morphometric data on the animals located. This subspecies of ensatina is patchily distributed within mesic oak woodlands of southern California (Block and

Morrison 1998). Using adaptive sampling, random walk surveys are used to locate the salamander-by searching on and under all suitable substrates. Once a salamander is located, the researcher conducts an area-constrained survey (Heyer et al. 1994) to sample that area intensively for salamanders. An area-constrained survey consists of establishing a plot, and then searching for ensatinas on, under, and in all possible substrates – typically rocks and logs for this species – within that plot. The researcher would also collect morphological information on animals captured. Thus, both a density sample (primary study goal) and morphometric data (ancillary study information) would result from the efforts. Although it appears to have potential for many wildlife studies, adaptive sampling has not been widely used. For a more complete discussion of adaptive sampling and the variations thereof, see Thompson (2002), Thompson and Seber (1996), and Smith et al. (2004).

5.10 When Things Go Wrong

As well as you might plan, things do not always go as you wish. You might encounter problems obtaining samples because of limited access or physical barriers. Weather might limit field collection or influence behavior of the animal under study. Equipment might fail. Field personnel may not follow protocols leading to mistakes or gaps in data collection. And yes, of course, funding might be reduced or even eliminated. When these situations occur, what can you do? Ignore the problems and continue blithely along? End the study and analyze data you have? Evaluate the status of the study and make mid-course corrections? Below we provide some guidance for making the best out of less than desirable sampling situations.

5.10.1 Guidance on Mid-Project Changes

Hopefully, you have the wherewithal to stick with your study and the make mid-project changes. The severity and extent of the changes needed might dictate whether a simple course correction is needed, or if major changes are needed in the scope and direction of the study. Timing of the changes also influences what changes are needed. For example, changes occurring early in a study might require revising study objectives and the basis of the research, whereas changes toward the end might entail minor adjustments to make sure you retain much of your efforts. Changes may be needed for a number of reasons, and those changes may differ depending on whether the study is an experiment, quasiexperiment, or observational study. It is almost impossible to envision all of the problems that can and do occur. In the collective years conducting research by the authors of this book, we continue to encounter new ways for things to go wrong. Our bottom line advice is do not panic. With some perseverance, you can likely salvage a study even if the nature and underlying objectives are a bit different from what you set out to accomplish.

For example, consider an experiment investigating effects of prescribed fire on birds in pine forests. The experimental design calls for paired plots, where one of the two plots is burned with a low severity surface fire and the fire occurs in spring. Treatment units are supposed to be randomly assigned. For discussion, let us evaluate ramifications of various deviations from the experimental design:

- Treatment units not randomly allocated
- Treatments occur during different times of the year
- Two treatment units were not treated

Typically, researchers depend upon managers to execute treatments as part of field experiments. Although involving fire managers, in this case, is essential to getting treatments done, managers bring additional considerations that may compromise aspects of the experiment. For example, logistical considerations of access, fire breaks, ability to control the fire, and the like might preclude some units from being treated. As a result, you compromise randomization to get the treatments done. This might have little effect on your study, but you should acknowledge the deviation from randomization, and incorporate a site factor to evaluate whether or not site explains more variation in your response variable(s) than treatment. Another deviation in the application of treatments is that they may not get done at the same time and are spread across seasons or even years. Again, not much you can do about it, but acknowledge that situation and perhaps include a time factor in your analysis. More problematic, however, is when fewer plots are treated than needed for the experiment. This might necessitate a complete change in study focus from an experiment to an observational study where you focus on correlates related to fire rather than evaluating cause–effect relationships. As a result, you might need to alter your sampling strategies and your data analysis, but you can still gather useful data to address aspects of fire and birds.

5.10.2 How to Make the Best of a Bad Situation

As we noted above, you can probably salvage something from study even if your best laid plans go wrong. Do not panic, but systematically evaluate your options. Some questions to ask: can I still address my original objectives; how much of my existing data can be used; how much more data must I gather; can I follow the same sampling design or does it need to be adjusted; what are my options for data analysis; should I switch from an ANOVA or BACI analysis to more exploratory model selection techniques? By considering these questions, you will understand your options better. It may very well be that your final study has little resemblance to your original plan. It is best to get something out of all of your efforts than waste all of your hard work.

Often you are able to address your main study objective but with less confidence than would have been possible had you implemented the optimal design. For example, losing several study plots due to unforeseen circumstances (e.g., permission for access denied; treatments could not be applied) would result in a more limited inference based on your results; but, you would still have results and be able to make some more qualified inference for application outside your immediate study sites. Likewise, your initial plans to submit your work for publication to a national-level journal might need to be changed to submit to a regional journal. But, in our opinion, your duty as a scientist is to do the best work you can and get the work published; where it is published is a consideration but of secondary importance.

5.11 Summary

The first rule for applying a sampling design is to recognize that the design itself is not self-implementing. By that, we mean that people are needed actually to conduct the study: for example, to locate and mark sample plots, lay out sample points, collect data, and then document, transcribe, and handle the data through analysis and interpretation. During the process of applying a conceived study plan, adjustments will more than likely be necessary. Along this line, we offer a few salient points to keep in mind:

- Wildlife populations and ecologies typically vary in time and space. A study design should account for these variations to ensure accurate and precise estimates of the parameters under study.
- Various factors may lend bias to the data collected and study results. These include observer bias, sampling and measurement bias, and selection bias. Investigators must acknowledge that bias can and does occur, and take measures to minimize or mitigate the effects of that bias.
- A critical aspect of any study is development of and adherence to a rigorous QA/QC program.
- Study plans should be regarded as living documents that detail all facets of a study, including any changes and modifications made during application of the study design.
- Sampling intensity must be sufficient to provide information needed to address the study objectives. Anything less may constitute a waste of resources.
- Plot size and shape are unique to each study.
- Pilot studies are critical as "Those who skip this step because they do not have enough time usually end up losing time" (Green 1979, p. 31).
- Studying rare species or events requires special approaches such as adaptive sampling, adaptive cluster sampling, sequential sampling, and two-phase stratified adaptive sampling.
- Stuff happens. Even the best designed studies require mid-study adjustments.

References

Becker, E. F., H. N. Golden, and C. L. Gardner. 2004. Using probability sampling of animal tracks in snow to estimate population size, in W. L. Thompson, Ed. Sampling Rare or Elusive Species, pp. 248–270. Island Press, Covelo, CA.

Bissonette, J. A. 1997. Scale-sensitive ecological properties: Historical context, current meaning, in J. A. Bissonette, Ed. Wildlife and Landscape Ecology: Effects of Pattern and Scale, pp. 3–31. Springer, New York, NY.

Block, W. M. 1990. Geographic variation in foraging ecologies of breeding and nonbreeding birds in oak woodlands. Stud. Avian Biol. 13: 264–269.

Block, W. M., and L. A. Brennan. 1993. The habitat concept in ornithology: Theory and applications. Curr. Ornithol. 11: 35–91.

Block, W. M., M. L. Morrison, and P. E. Scott. 1998. Development and evaluation of habitat models for herpetofauna and small mammals. For. Sci. 44: 430–437.

Block, W. M., and M. L. Morrison. 1998. Habitat relationships of amphibians and reptiles in California oak woodlands. J. Herpetol. 32: 51–60.

Block, W. M., K. A. With, and M. L. Morrison. 1987. On measuring bird habitat: Influence of observer variability and sample size. Condor 89: 241–251.

Block, W. M., J. L. Ganey, P. E. Scott, and R. King. 2005. Prey ecology of Mexican spotted owls in pine-oak forests of northern Arizona. J. Wildl. Manage. 69: 618–629.

Bonham, C. D. 1989. Measurements of Terrestrial Vegetation. Wiley, New York, NY.

Buckland, S. T., D. R. Anderson, K. P. Burnham, J. L. Laake, D. L. Borchers, and L. Thomas. 2001. Introduction to Distance Sampling. Oxford University Press, New York, NY.

Buckland, S. T., D. R. Anderson, K. P. Burnham, J. L. Laake, D. L. Borchers, and L. Thomas. 2004. Advanced Distance Sampling. Oxford University Press, New York, NY.

Canfield, R. 1941. Application of the line intercept method in sampling range vegetation. J. Forest. 39: 388–394.

Christman, M. C. 2004. Sequential sampling for rare or geographically clustered populations, in W. L. Thompson, Ed. Sampling Rare or Elusive Species, pp. 134–145. Island Press, Covelo, CA.

Cochran, W. G. 1977. Sampling Techniques, 3rd Edition. Wiley, New York, NY.

Collier, B. A., S. S. Ditchkoff, J. B. Raglin, and J. M. Smith. 2007. Detection probability and sources of variation in white-tailed deer spotlight surveys. J. Wildl. Manage. 71: 277–281.

Conn, P. B., L. L. Bailey, and J. R. Sauer. 2004. In W. L. Thompson, Ed. Sampling Rare or Elusive Species, pp. 59–74. Island Press, Covelo, CA.

Conover, W. J. 1980. Practical Nonparametric Statistics, 2nd Edition. Wiley, New York, NY.

Cooch, E., and G. White. 2007. Program MARK: A Gentle Introduction, 6th Edition. http://www.phidot.org/software/mark/docs/book/

Dawson, D. K., D. R. Smith, and C. S. Robbins. 1995. Point count length and detection of forest Neotropical migrant birds, in C. J. Ralph, J. R. Sauer, and S. Drogge, Tech. Eds. Monitoring Bird Populations by Point Counts, pp. 35–43. Gen. Tech. Rpt. PSW-GTR-149. USDA Forest Service, Pacific Southwest Research Station, Albany, CA.

Development and evaluation of habitat models for herpetofauna and small mammals. For. Sci. 44: 430–437.

Fretwell, S. D., and H. L. Lucas. 1970. On territorial behavior and other factors influencing habitat distribution in birds. I. Acta Biotheoret. 19: 16–36.

Ganey, J. L., W. M. Block, J. S. Jenness, and R. A. Wilson. 1999. Mexican spotted owl home range and habitat use in pine-oak forest: Implications for forest management. Forest Sci. 45: 127–135.

Green, R. H. 1979. Sampling Design and Statistical Methods for Environmental Biologists. Wiley, New York, NY.

Grubb Jr., T. C., and M. S. Woodrey. 1990. Sex, age, intraspecific dominance, and use of food by birds wintering in temperate-deciduous and cold-coniferous woodlands: A review. Stud. Avian Biol. 13: 270–279.

Hanowski, J. M., and G. J. Niemi. 1990. Effects of unknown sex in analysis of foraging behavior. Stud. Avian Biol. 13: 280–283.

Heady, H. F., R. P. Gibbons, and R. W. Powell. 1959. A comparison of charting, line intercept, and line point methods of sampling shrub types of vegetation. J. Range Manage. 21: 370–380.

Heyer, W. R., M. A. Donnelly, R. W. McDiarmid, L. C. Hayek, and M. S. Foster, Eds. 1994. Measuring and Monitoring Biological Diversity: Standards and Methods for Amphibians. Smithsonian Institution Press, Washington, DC.

Hurlbert, S. H. 1984. Pseudoreplication and the design of ecological field experiments. Ecol. Monogr. 54: 187–211.
Jeffers, J. N. R. 1980. Sampling. Statistical Checklist 2. Institute of Terrestrial Ecology, National Environment Research Council, Cambridge, United Kingdom.
Karanth, K. U., J. D. Nichols, and N. S. Kumar. 2004. Photographic sampling of elusive mammals in tropical forests, in W. L. Thompson, Ed. Sampling Rare or Elusive Species, pp. 229–247. Island Press, Covelo, CA.
Kepler, C. B., and J. M. Scott. 1981. Reducing bird count variability by training observers. Stud. Avian Biol. 6: 366–371.
Klopfer, P. H. 1963. Behavioral aspects of habitat selection: The role of early experience. Wilson Bull. 75: 15–22.
Littel, R. C., G. A. Milliken, W. W. Stroup, and R. D. Wolfinger. 1996. SAS System for Mixed Models. SAS Institute, Cary, NC.
Little, R. J. A., and D. B. Rubin. 1987. Statistical Analysis with Missing Data. Wiley, New York, NY.
MacArthur, R. H., and J. W. MacArthur. 1961. On bird species diversity. Ecology 42: 594–598.
Manly, B. L. 2004. Two-phase adaptive stratified sampling, in W. L. Thompson, Ed. Sampling Rare or Elusive Species, pp. 123–133. Island Press, Covelo, CA.
Manly, B. L., L. McDonald, and D. Thomas. 1993. Resource Selection by Animals. Chapman and Hall, London.
Miles, D. B. 1990. The importance and consequences of temporal variation in avian foraging behavior. Stud. Avian Biol. 13: 210–217.
Moir, W. H., B. Geils, M. A. Benoit, and D. Scurlock. 1997. Ecology of southwestern ponderosa pine forests, in W. M. Block, and D. M. Finch, Tech. Eds. Songbird Ecology in Southwestern Ponderosa Pine Forests: A Literature Review, pp. 3–27. Gen. Tech. Rpt. RM-GTR-292. USDA Forest Service, Rocky Mountain Forest and Range Experiment Station, Fort Collins, CO.
Morrison, M. L. 1982. The structure of western warbler assemblages: Analysis of foraging behavior and habitat selection in Oregon. Auk 98: 578–588.
Morrison, M. L. 1987. The design and importance of long-term ecological studies: Analysis of vertebrates in the Inyo-White mountains, California, in R. C. Szaro, K. E. Severson, and D. R. Patton, Tech. Coords. Management of Amphibians, Reptiles, and Small Mammals in North America, pp. 267–275. Gen. Tech. Rpt. RM-166. USDA Forest Service, Rocky Mountain Forest and Range Experiment Station, Fort Collins, CO.
Nichols, J. D., J. E. Hines, J. R. Sauer, F. W. Fallon, J. E. Fallon, and P. J. Heglund. 2000. A double observer approach for estimating detection probability and abundance from point counts. Auk 117: 393–408.
Noon, B. R., and W. M. Block. 1990. Analytical considerations for study design. Stud. Avian Biol. 13: 126–133.
Otis, D. L., K. P. Burnham, G. C. White, and D. R. Anderson. 1978. Statistical inference from capture data on closed animal populations. Wildl. Monogr. 62.
Petit, D. R., L. J. Petit, V. A. Saab, and T. E. Martin. 1995. Fixed-radius point counts in forests: Factors influencing effectiveness and efficiency, in C. J. Ralph, F. R. Sauer, and S. Drogge, Tech. Eds. Monitoring Bird Populations by Point Counts, pp. 49–56. Gen. Tech. Rpt. PSW-GTR-149. USDA Forest Service, Pacific Southwest Research Station, Albany, CA.
Ralph, C. J., and J. M. Scott, Eds. 1981. Estimating numbers of terrestrial birds. Stud. Avian Biol. 6.
Ralph, C. J., J. R. Sauer, and S. Drogge, Tech. Eds. 1995. Monitoring Bird Populations by Point Counts. Gen. Tech. Rpt. PSW-GTR-149. USDA Forest Service, Pacific Southwest Research Station, Albany, CA.
Schneider, D. C. 1994. Quantitative Ecology. Academic, San Diego, CA.
Schwartz, M. K., G. Luikart, and R. S. Waples. 2006. Genetic monitoring as a promising tool for conservation and management. Trends Ecol. Evol. 22: 25–33.
Smith, D. R, J. A. Brown, and N. C. H. Lo. 2004. Applications of adaptive sampling to biological populations, in W. L. Thompson, Ed. Sampling Rare or Elusive Species, pp. 77–122. Island Press, Covelo, CA.

Strayer, D., J. S. Glitzenstein, C. G. Jones, J. Kolasa, G. E. Likens, M. J. McDonnell, G. G. Parker, and S. T. A. Pickett. 1986. Long-term ecological studies: An illustrated account of their design, operation, and importance to ecology. Inst. Ecol. Syst. Occas. Pap. 2: 1–38.

Swihart, R. K., and N. A. Slade. 1985. Testing for independence of observations in animal movements. Ecology 66: 1176–1184.

Thompson, S. K. 2002. Sampling. Wiley, New York, NY.

Thompson, W. L., Ed. 2004. Sampling Rare or Elusive Species. Island Press, Covelo, CA.

Thompson, F. R., and M. J. Schwalbach. 1995. Analysis of sample size, counting time, and plot size from an avian point count survey on the Hoosier National Forest, Indiana, in C. J. Ralph, F. R. Sauer, and S. Drogge, Tech. Eds. Monitoring Bird Populations by Point Counts, pp. 45–48. Gen. Tech. Rpt. PSW-GTR-149. USDA Forest Service, Pacific Southwest Research Station, Albany, CA.

Thompson, S. K., and G. A. F. Seber. 1996. Adaptive Sampling. Wiley. New York, NY.

Thompson, W. L., G. C. White, and C. Gowan. 1998. Monitoring Vertebrate Populations. Academic, San Diego, CA.

Verner, J. 1985. An assessment of counting techniques. Curr. Ornithol. 2: 247–302.

Verner, J. 1987. Preliminary results from a system of monitoring population trends in bird populations in oak-pine woodlands, in T. R. Plumb and N. H. Pillsbury, Tech. Coords. Multiple-Use Management of California's Hardwood Resources, pp. 214–222. Gen. Tech. Rpt. PSW-100. USDA Forest Service, Pacific Southwest Research Station, Berkeley, CA.

White, G. C., D. R. Anderson, K. P. Burnham, and D. L. Otis. 1982. Capture–Recapture and Removal Methods for Sampling Closed Populations. Rpt. LA-8787-NERP. Los Alamos National Laboratory, Los Alamos, NM.

Wiens, J. A. 1984. The place for long-term studies in ornithology. Auk 101: 202–203.

Wiens, J. A. 1985. Habitat selection in variable environments: Shrub-steppe birds, in M. L. Cody, Ed. Habitat Selection in Birds, pp. 227–251. Academic, Orlando, FL.

Wiens, J. A. 1989. Spatial scaling in ecology. Funct. Ecol. 3: 385–397.

Wiens, J. A., J. F. Addicott, T. J. Case, and J. Diamond. 1986. Overview: The importance of spatial and temporal scale in ecological investigations, in J. Diamond and T. J. Case, Eds. Community Ecology, pp. 145–153. Harper and Row, New York, NY.

Wingfield, J. C., and J. D. Jacobs. 1999. The interplay of innate and experimental factors regulating the life-history cycle of birds. Proc. Int. Ornithol. Congr. 22: 2417–2443.

Zippen, C. 1958. The removal method of population estimation. J. Wildl. Manage. 22: 82–90.

Chapter 6
Impact Assessment

6.1 Introduction

In this chapter we apply the concepts developed previously in this book to the specific issue of determining the effects of environmental impacts on wildlife. *Impact* is a general term used to describe any change that perturbs the current system, whether it is planned or unplanned, human induced, or an act of nature. Thus, impacts include a 100-year flood that destroys a well-developed riparian woodland, a disease that decimates an elk herd, or the planned or unplanned application of fertilizer. Impacts also include projects that are intended to improve conditions for animals such as ecological restoration. For example, removal of exotic salt cedar from riparian areas to enhance cottonwood regeneration can substantially impact the existing site conditions.

You have likely already encountered many situations that fall into the latter category; namely, studies that are constrained by location and time. Such situations often arise in environmental studies because the interest (e.g., funding agency) is local, such as the response of plants and animals to treatments (e.g., fire, herbicide) applied on a management area of a few 100 to a few 1,000 ha. Often these treatments are applied to small plots to evaluate one resource, such as plants, and you have been funded to study animal responses. In such situations, the initial plots might be too small to adequately sample many animal species. Or, there might be no treatment involved, and the project focus is to quantify the ecology of some species within a small temporal and spatial scale. It is important for students to note that most resource management is applied locally; that is, on a small spatial scale to respond to the needs of local resource managers. The suite of study designs that fall under the general rubric of impact assessment are applicable to studies that are not viewed as having caused an environmental impact per se. Designs that we cover below, such as after-only gradient designs, are but one example.

A distinction should be made between a hypothesis developed within a manipulative experiment framework and a hypothesis developed within an impact framework. By randomizing treatments to experimental units and replicating the experiments, test conditions are basically free from confounding influences of time and space; inferences can be clearly stated. Within an impact framework, however, test conditions are usually outside the control of the investigator, which makes inference problematic (see related discussion in Skalski and Robson 1992, pp. 161–162).

In this chapter we will concentrate on impacts that are seldom planned, and are usually considered to be a negative influence on the environment. But this need not be so, and the designs described herein have wide applicability to the study of wildlife.

6.2 Experimental Designs: Optimal and Suboptimal

In his classic book on study design, Green (1979) outlined the basic distinction between an *optimal and suboptimal study design.* In brief, if you know what type of impact will occur, when and where it will occur, and have the ability to gather pretreatment data, you are in an "optimal" situation to design the study (Fig. 6.1, sequence 1). Main Sequence 1 is, in essence, the classic manipulative study, although Green was developing his outline in an impact assessment framework. That is, you might be establishing control areas and gathering pretreatment data in anticipation of a likely catastrophic impact such as a fire or flood. Thus, the "when" aspect of the optimal design need not be known specifically other than in the future that you can plan for.

As we step through the decision tree developed by Green (1979; Fig. 6.1), your ability to plan aspects of the impact study decline. Unless we are concerned only

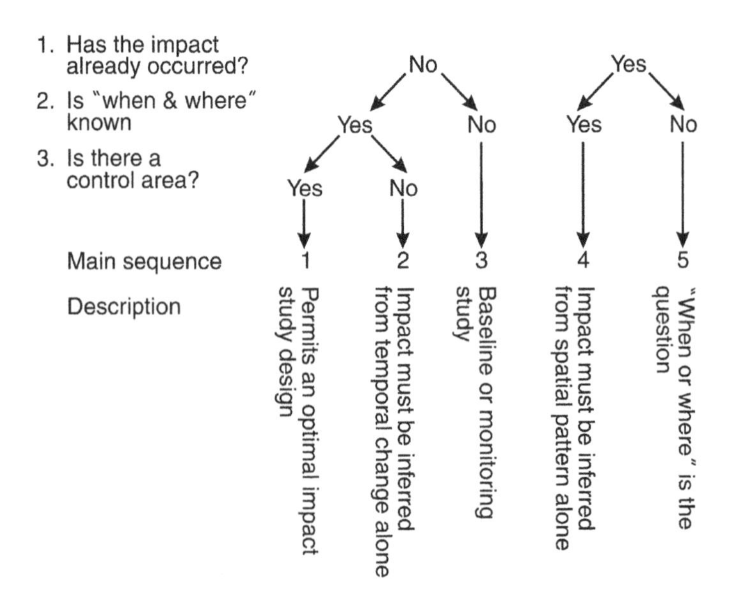

Fig. 6.1 The decision key to the "main sequence" categories of environmental studies. (From *Sampling Design and Statistical Methods for Environmental Biologists* by Green, Copyright © 1979, Wiley. Reprinted by permission of publisher)

with a small geographic area, we can seldom adequately anticipate where an impact will occur, or even if we could, what the type and intensity of the impact would be. This uncertainly is exacerbated by the population structure of many species. Because animals are not distributed uniformly, even within a single vegetation type, we are forced to sample intensively over an area in anticipation of something (the impact) that might never occur – few budgets can allow such luxury (the topic of distribution of plots described below under suboptimal designs applies in this situation).

That we cannot know the specific type of impact that will occur does not always negate, however, implementation of an optimal design. For example, personnel of a wildlife refuge located downstream from a chemical factory could anticipate the chemical, but perhaps not the toxicity, of a spill; or a refuge located in an area of extensive human population growth might anticipate an increase in air pollution and/or a decrease in the water table. Such impacts could likely be modeled using the experience of similar regions in the recent past. Agencies charged with managing game populations could anticipate, perhaps through modeling, some of the major environmental changes resulting from future changes in key limits (e.g., effects on prey, effects on forage quality). Various measures could be prioritized for study. Sensitivity analysis is also a useful tool to aid in understanding the behavior of model parameters and to narrow the number of variables to be monitored. If monitoring sites are distributed across the area of interest (e.g., study site, impacted site), then it is also likely that not all sites would be impacted; thus, some would serve as nonimpacted controls.

As noted by Green, an optimal design is thus an areas-by-times factorial design in which evidence for an impact is a significant areas-by-times interaction. Given that the prerequisites for an optimal design are met, the choice of a specific sampling design and statistical analyses should be based on your ability to (1) test the null hypothesis that any change in the impacted area does not differ statistically or biologically from the control and (2) relate to the impact any demonstrated change unique to the impacted area and to separate effects caused by naturally occurring variation unrelated to the impact (Green 1979, p. 71). Locating appropriate control sites is not a trivial matter. The selection of control sites has been more fully discussed in Chap. 2. It is important to make sure that your control sites are truly unimpacted, even in very subtle ways. For example, the impact could cause animals to vacate the site and move onto the control, which could cause any number of behavioral or density dependent responses by the animals already residing there that you would not perceive.

It is often not possible, however, to meet the criteria for development of an optimal design. Impacts often occur unexpectedly; for example, an unplanned fire substantially reduces the cover of grass or shrub over 35 ha of your management area or a flood destroys most of the remaining native sycamores (*Plantanus wrightii*) along the stream crossing your property. In such cases, a series of *suboptimal study designs* have been described. If no control areas are possible (see Fig. 6.1, sequence 2), then the significance of the impact must be inferred from temporal changes alone (discussed below). If the location and timing of the impact are not known (i.e., it is expected but cannot be planned; e.g., fire, flood, disease), the study becomes a baseline or monitoring study (see Fig. 6.1, sequence 3). If properly planned spatially, then it is likely that nonimpacted areas will be available to serve as controls if and when the impact occurs. This again indicates why "monitoring" studies are certainly research, and might allow the development of a rigorous experimental analysis if properly planned.

Unfortunately, impacts often occur without any preplanning by the land manager. This common situation (see Fig. 6.1, sequence 4) means that impact effects must be inferred from among areas differing in the degree of impact; study design for these situations is discussed below. Finally, situations do occur (see Fig. 6.1, sequence 5) where an impact is known to have occurred, but the time and location are uncertain (e.g., the discovery of a toxin in an animal or plant). This most difficult situation means that all direct evidence of the initial impact could be nonexistent. For example, suppose a pesticide is located in the soil, but at levels below that known to cause death or obvious visual signs of harm to animals. The range of concentration of pesticide known to harm animals varies widely depending on the absolute amount applied, the distribution of application, and the environmental conditions present both during and after application (e.g., soil condition, rainfall). Thus, "backdating" the effect is problematic. Further, it is difficult to know how the pesticide impacted the animal community (e.g., loss of a species) or if recovery has occurred. Ways to evaluate impacts using various suboptimal designs are presented below.

6.3 Disturbances

Three primary types of disturbances occur: pulse, press, and those affecting temporal variance (Bender et al. 1984, Underwood 1994). *Pulse disturbances* are those that are not sustained after the initial disturbance; the *effects* of the disturbance may be long lasting. *Press disturbances* are those that are sustained beyond the initial disturbance. Both pulse and press disturbances can result from the same general impact. For example, a pulse disturbance occurs when a flood washes frog egg masses downstream. A press disturbance occurs subsequently because of the changes in river banks and associated vegetation, which prevent the successful placement of new egg masses. The magnitude of the pulse disturbance will determine our ability to even know that an impact has occurred. For example, Fig. 6.2 depicts mild (B) and relatively severe (C) pulse disturbances; the former would be difficult to detect if sampling was less frequent (i.e., if sampling had not occurred between times 6 and 8) and/or the variance of each disturbance event was high. Figure 6.3 depicts mild (C) and relatively severe (D) press disturbances. The former would be difficult to distinguish from the variation inherent in the control sites.

Disturbances affecting temporal variance are those that do not alter the mean abundance, but change the magnitude of the oscillations between sampling periods. These changes can increase (see Fig. 6.3a) or even decrease (see Fig. 6.3b) the variance relative to predisturbance and/or control sites.

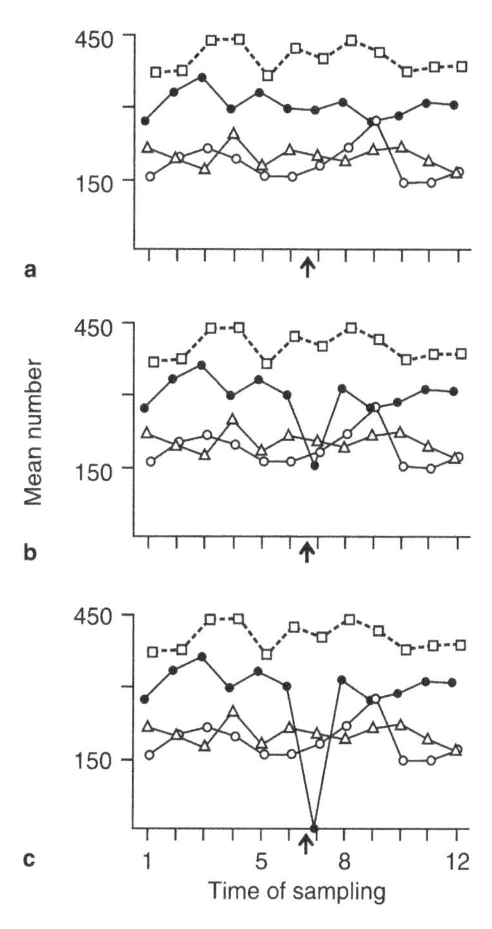

Fig. 6.2 Simulated environmental disturbances in one location (•—•), with three controls, all sampled six times before and after the disturbance (at the times indicated by the *arrow*). (**a**) No effect of the disturbance; (**b**) a pulse reduction of 0.5 of the original mean; and (**c**) a pulse reduction to 0. Reproduced from Underwood (1994), with kind permission from Springer Science + Business Media

Identifying a disturbance is thus problematic because of background variance caused by natural and/or undetected disturbances (i.e., a disturbance other than the one you are interested in). For example, note the similarities between the conclusion of "no effect" in Fig. 6.2a, "temporal variance" in Fig. 6.3b, and a "press disturbance" in Fig. 6.3c; and also the similarities between a "pulse disturbance" in Fig. 6.2b and "temporal variance" in Fig. 6.3a. What we are witnessing here is that the large temporal variance in many populations creates "noise" that obscures more subtle changes (such as environmental impacts) and the lack of concordance in the temporal trajectories of populations in different sites (Underwood 1994).

You need to have some understanding of the type of disturbance that is likely to result from a perturbation to apply the best impact design. For example, spilling of a chemical into a waterway could only result in a pulse disturbance, such as is the case for many herbicide impacts on animals; or the chemical could result in a press disturbance, such as in the case of an insecticide on animals or a herbicide on plants.

The duration of the impact study will be determined by the temporal pattern (length) of the impact. Although by definition a pulse disturbance recovers quickly,

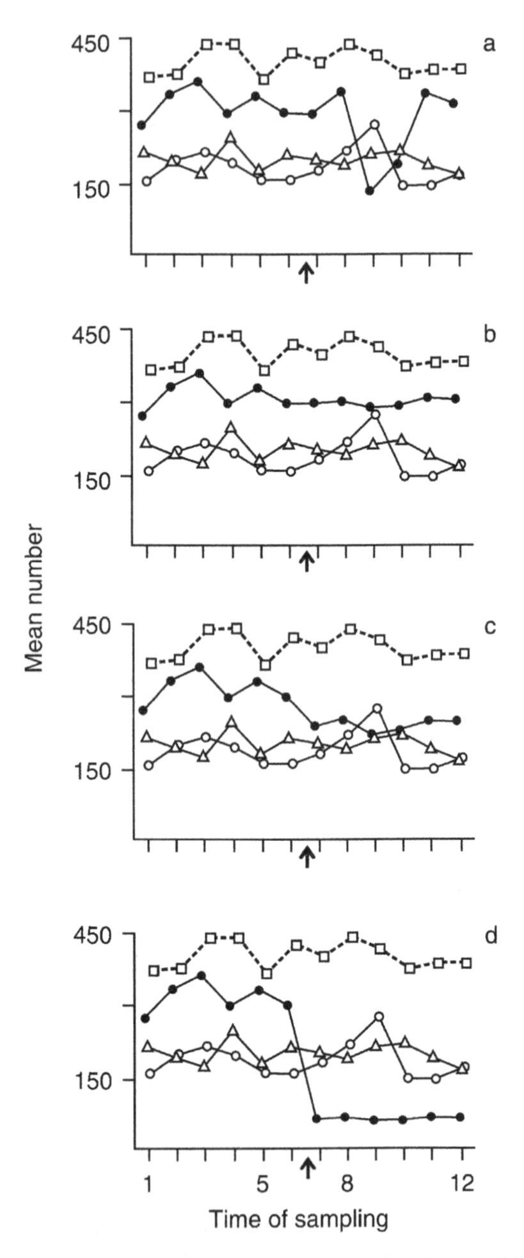

Fig. 6.3 Simulated environmental disturbances in one location (•—•), with three controls, all sampled six times before and after the disturbance (at the times indicated by the *arrow*). (**a**, **b**) The impact is an alteration of temporal variance after the disturbance; temporal standard deviation ×5 in (**a**) and ×0.5 in (**b**). (**c**, **d**) A press reduction of abundance to 0.8 (**c**) and 0.2 (**d**) of the original mean. Reproduced from Underwood (1994), with kind permission from Springer Science + Business Media

a press disturbance usually recovers slowly as either the source of the impact lessens (e.g., a chemical degrades) or the elements impacted slowly recover (e.g., plant growth, animal recolonization).

Also note that an impact can change the temporal pattern of an element, such as the pattern in fluctuation of numbers. A change in temporal patterning could be due to direct effects on the element or through indirect effects that influence the element. Direct effects might result from a change in activity patterns because the impact agent modified sex or age ratios (i.e., differentially impacted survival of different sex–age classes); indirect effects could result because the impact agent severely impacted a predator, competitor, or food source of the element being monitored.

6.3.1 Recovery and Assumptions

Parker and Wiens (2005) provided the following basic definitions that should be used in discussing impact assessment; we have modified some of these slightly to match other concepts and terminology in this book:

- *Biological resources*: Quantifiable components of the systems such as organisms, populations, species, and communities.
- *Levels*: measures of a resource such as abundance, diversity, community structure, and reproductive rates. Hence, levels are quantifiable on an objective scale and can be used to estimate means and variance and to test hypotheses.
- *Natural factors*: Physical and chemical features of the environment that affect the level of a resource at a given time and location, such as temperature, substrate, dissolved oxygen, total organic carbon.
- *Gradient analysis and dose–response regression*: Are often used synonymously; where dose is a measure of exposure to the impact and response is a measure of the biological system.
- *Recovery*: A temporal process in which impacts progressively lessen through natural processes and/or active restoration efforts.
- *Recovered*: When natural factors have regained their influence over the biological resource(s) being assessed.

As summarized by Parker and Wiens (2005), impact assessment requires making assumptions about the nature of temporal and spatial variability of the system under study. Of course, any ecological study makes such assumptions whether or not they are acknowledged; the nature of this variability is critical in designing, analyzing, and interpreting results of a study.

Parker and Wiens (2005; also Wiens and Parker 1995) categorized assumptions about the temporal and spatial variability of a natural (nonimpacted) system as in steady state, spatial, or dynamic equilibrium (Fig. 6.4). As the name implies, a *steady-state system* is typified by levels of resources, and the natural factors controlling them, showing a constant mean through time (a). Hence, the resource at a given location has a single long-term equilibrium to which it will return following perturbation (if it can, indeed, return). Such situations usually only occur in very localized areas. In (A), the arrow denotes when then state of the system (solid line) is perturbed to a lower level (the dashed line). *Spatial equilibrium* occurs when two or more sampling areas, such as impact and reference, have similar natural factors and, thus, similar levels of a resource (B). Thus, in the absence of a perturbation, differences in means are due to sampling error and stochastic variations. Look closely at the dashed line in (A) vs. the dashed line in (B); the primary difference between figures is that multiple areas are considered in (B). *Dynamic equilibrium* incorporates both temporal and spatial variation, where natural factors and levels of resources usually differ between two or more areas being compared, but the differences between mean levels of the resource remain similar over time (C). In such systems recovery occurs when the dynamics of the impacted areas once again parallel those of the reference area. Note in (C) that the reference (solid line) line fluctuates around the mean (also solid line), while the impacted area (dashed line) drops well below the natural (although lower than the reference) condition (lower solid line).

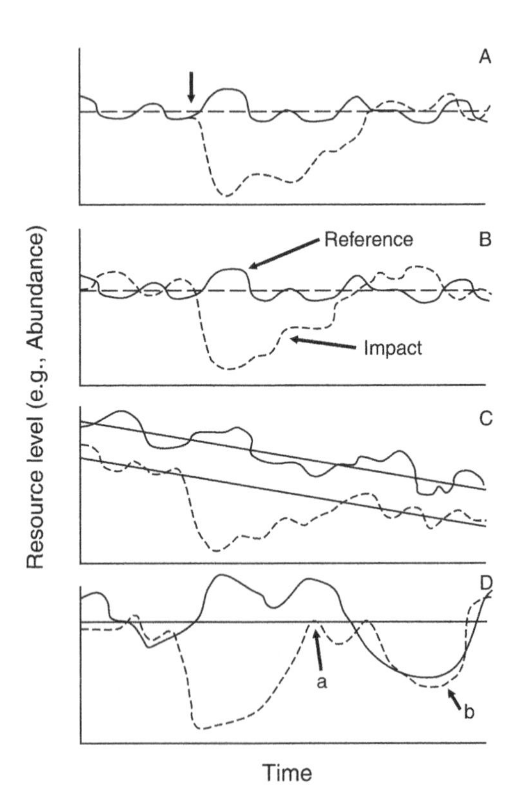

Fig. 6.4 Ecological assumptions affecting the assessment of recovery from an environmental accident. Reproduced from Parker and Wiens (2005), with kind permission from Springer Science + Business Media

Parker and Wiens (2005) also presented an example of when ignorance of the underlying system dynamics can lead to erroneous conclusions on recovery. In Fig. 6.4D, we see considerable natural variation around the long-term, steady-state mean. In this figure the horizontal solid line represents the long-term mean of the fluctuating solid line. If this long-term variation is not known or not considered, the perturbed system might erroneously be deemed recovered when it is not; for example, at point 'a' in the figure. Conversely, the system might be deemed to be impacted when in fact it has recovered (point 'b'; note that the dashed line is now tracking the solid line that represents the natural state).

The assumptions surrounding all three of these scenarios about system equilibrium also require that the perturbation did not cause the resource to pass some threshold beyond which it cannot recover. In such situations a new equilibrium will likely be established. For example, when an event such as fire, over grazing, or flooding permanently changes the soil. Under such situations the system would recover to a different state.

6.4 Before–After Designs

6.4.1 Before–After/Control–Impact Design

As outlined above, Green (1979) developed many of the basic principles of environmental sampling design. Most notably, his before–after/control–impact, or *BACI*, design is the standard upon which many current designs are based. In the BACI design, a sample is taken before and another sample is taken after a disturbance, in each of the putatively disturbed (impacted) sites and an undisturbed (control) site. If there is an environmental disturbance that affects a population, it would appear as a statistical interaction between the difference in mean abundance of the sampled populations in the control and impacted sites before the disturbance, and that difference after the disturbance (Fig. 6.5a).

However, the basic BACI design is confounded because any difference from before to after may occur between two times of sampling as a result of natural variation, and not necessarily by the impact itself (Hurlbert 1984; Underwood 1994). To address this problem, the basic BACI was expanded to include temporal replication, which involves several replicated sampling times before and after the impact (see Fig. 6.5d).

Stewart-Oaten et al. (1986) discussed the advantage that taking samples at nonregular time intervals, rather than on a fixed schedule, had in impact assessment. Sampling at nonregular times will help ensure that no cyclical differences unforeseen by the worker will influence the magnitude of the before–after difference. Taking samples at regular intervals means that temporal variance might not be estimated accurately and that the magnitude of the impact may be overestimated or underestimated. For example, sampling rodents only during the fall postbreeding period, which is a common practice, will obviously underestimate annual variance, and overestimate the annual mean.

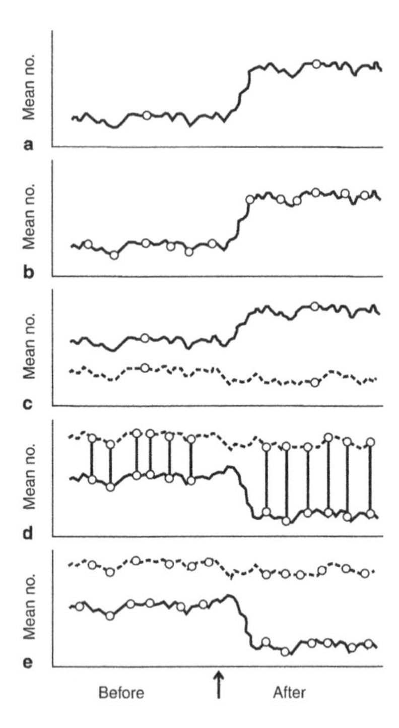

Fig. 6.5 Common sampling designs to detect environmental impacts, with *circles* indicating times of sampling: (**a**) a single sample in one location before and after an impact (at the time of the *arrow*); (**b**) random samples in one location before and after an impact; (**c**) BACI design with a single sample before and after the impact in each of a control (*dashed line*) and the putatively impacted location (*solid line*); (**d**) modified BACI where differences between mean abundances in the control and potentially impacted locations are calculated for random times before and after the disturbance begins (*vertical lines* indicate difference); and (**e**) further modification of (**d**) to allow sampling at different times in each location. Reproduced from Underwood (1991), with kind permission from CSIRO Publishing

Analyses based on temporal replication must assume that each sampling date represents an independent estimate of the true change (see also time-series analysis, below). Osenberg et al. (1994) examined patterns of serial correlation from a long-term study of marine invertebrates to gain insight into the frequency with which samples could be collected without grossly violating the assumption of temporal independence. They found that serial correlation was not a general problem for the parameters they estimated; other analyses have produced similar results (Carpenter et al. 1989; Stewart-Oaten et al. 1992). Many studies in the wildlife literature have examined serial correlation in telemetry studies, and independence of observations should not be assumed without appropriate analysis (White and Garrott 1990).

Underwood (1991, 1994) presented an excellent review of the development of impact analysis, including basic and advanced statistical analyses appropriate to different designs. Analyses of basic BACI designs are summarized, based on analysis of variance (ANOVA), in Table 6.1, and are keyed to the patterns of disturbance described in Fig. 6.5b–d.

The Stewart-Oaten et al. (1986) modification (of taking samples at nonregular intervals) solved some of the problems of lack of temporal replication, but did not address the problem of lack of spatial replication. The comparison of a single impact site and a single control site is still confounded by different factors between the two sites that are not due to the identified impact. Remember that local populations do not necessarily have the same trajectory of abundance and behavior, and temporal interaction among sites is common. Stewart-Oaten et al. (1986) concluded that such a temporal change in the difference between the two sites before the impact would render the variable being used to assess the impact unusable; we can only assume that such a temporal difference would continue after the impact. Because many populations can be expected to vary in their patterns of abundance across times and sites, the basic BACI design – even with temporal replication – is a serious restriction on the usefulness of this design (Underwood 1994).

Table 6.1 Statistical analyses for the detection of environmental impact using various sampling designs[a]

Source of variation	Degrees of freedom	*F*-ratio vs.	Degrees of freedom
(a) Replicated before/after sampling at a single location; samples are taken at t random times before and t times after the putative impact (see Fig. 6.5**b**)			
Before vs. after = B	1	$T(B)$	$1, 2(t-1)$
Times (before vs. after) = $T(B)$	$2(t-1)$		
Residual	$2t(n-1)$		
Total	$2m-1$		
(b) BACI: A single time of sampling at two locations, one control and one potentially impacted (see Fig. 6.5c)			
Before vs. after = B	1		
Locations: control vs. impact = L	1		
Interaction $B \times L$	1	Residual	$1, 4(n-1)$
Residual	$4(n-1)$		
Total	$4n-1$		
(c) BACI: Replicated before/after sampling at two locations, one control and one potentially impacted; samples are taken at t random times before and t times after the putative impact, but at the same times in each site (see Fig. 6.5d)			
Before vs. after = B	1		
Locations: control vs. impact = L	1		
Interaction $B \times L$	1	$L \times T(B)$	$1, 2(t-1)$
Times (before vs. after) = $T(B)$	$2(t-1)$	Residual	$2(t-1), 4t(n-1)$
Interaction $L \times T(B)$	$2(t-1)$	Residual	$t-1, 4t(n-1)$
$L \times T(B)$ before	$t-1$	Residual	$t-1, 4t(n-1)$
$L \times T(B)$ after	$t-1$	Residual	$T-1, 4t(n-1)$
Residual	$4t(n-1)$		
Total	$4n-1$		

Source: Reproduced from Underwood (1991), with kind permission from CSIRO Publishing

[a]In each case, analysis of variance is used to provide a standard framework for all designs. In all cases, n replicate samples are taken at each time and site of sampling

The problem of confounding (the "pseudoreplication" of Hurlbert 1984) – caused by comparing abundances in two sites, one impacted and the other a control – should be overcome by having several replicated impacted and control sites. However, unplanned impacts are seldom replicated! However, it will usually be possible to select replication control sites (Fig. 6.6b). Under this scenario, it is hoped that the temporal variation among the control sites will be more similar than the change in the temporal pattern on the impacted site (caused by the disturbance). McDonald et al. (2000) summarized some of the approaches that have been taken to analyze BACI data, and criticized the frequent introduction of pseudoreplication into these analyses. Pseudoreplication can occur because single, summarized values on control and impacted sites through time (time periods) are often used as replicates; or, the plot or transect are used as the experimental unit. Repeated measures analyses are seldom applicable because of the limited data sets that are usually available. McDonald et al. (2000) present an alternative approach based on generalized linear mixed models. In practice, however, these approaches do not provide any additional insight into the response and recovery of biological resources to perturbation.

Replicated control sites are selected that have the appropriate set of physical characteristics, mix of species, abundance of the target species, and other factors deemed important by the investigator. It will not, of course, be possible to find "identical" control sites. The goal is to find areas that are "similar" enough to be

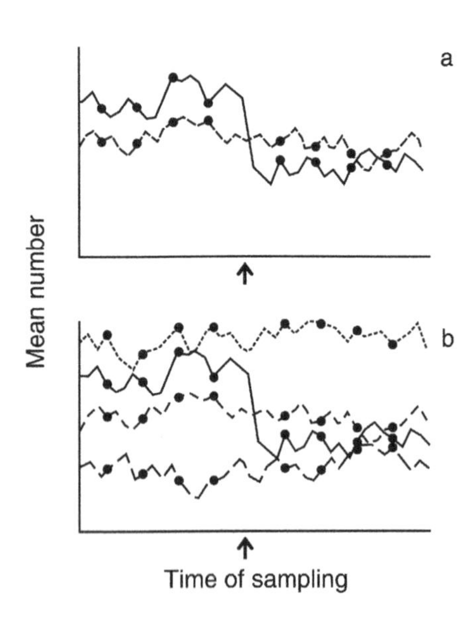

Fig. 6.6 Sampling to detect environmental impacts. (**a**) BACI design – replicated samples are taken several times in a single control (*dashed line*) and in the potentially impacted location (*solid line*) before and after a planned disturbance (at the time indicated by the *arrow*). (**b**) Sampling three control locations to provide spatial replication. Reproduced from Underwood (1994), with kind permission from Springer Science + Business Media

considered replicates. The definition of *similar* can, however, be problematic. Although researchers in nonimpact-related studies often describe their choice of sampling locations as "similar," they seldom provide the specific basis for this selection. We recommend that "similar" be defined and justified; this will be especially important if the study concludes that an impact has occurred, and this conclusion becomes the focus of debate or even litigation. For example, it would be appropriate to select controls in a hierarchical fashion of increasing spatial specificity, such as:

- Identify potential sites based on size and geographic location
- Conduct preliminary sampling of many potential control areas to determine the status of abiotic and biotic conditions of interest
- If enough potential sites are available after the above steps, randomly select a set of controls for study

It is not necessary to find sites that have nearly identical conditions or abundances of target species. The sites chosen as controls must simply represent the range of conditions like the one that might be impacted (Underwood 1994).

To aid in planning a BACI study, it would be helpful to find previous BACI studies conducted under a similar intensity of disturbance in a similar environment, and review the results for variability and effect size. This would allow the investigator to approximate the number of sampling dates needed to achieve a given level of power. Obtaining an adequate number of sampling sites and dates in the before period is crucial in any BACI assessment, since once the perturbation begins it is no longer possible to obtain before samples.

There are, however, few BACI studies that permit this type of analysis; but other data can be used to guide a BACI design. First, long-term studies that document spatial and temporal variability provide estimates of the natural variability of the changes. Second, after-only studies (described below) that assess impacts using a postimpact survey of sites can suggest the size of the effects that might occur. Third, any investigation that provides an estimate of the variability inherent in the variables of interest will provide at least preliminary indications of the sampling intensity necessary to determine an impact. Osenberg et al. (1994) detail how data from long-term studies and after-only studies can be combined to help plan BACI studies.

Osenberg et al. (1994) evaluated the sampling effort necessary to achieve reasonable estimates of environmental impacts. They found that sampling effort varied widely depending on the type of parameter being used. Following their analysis of a marine environment, we can generalize their strategy as follows (1) chemical–physical parameters include measures of the quality of air, water, and soil, (2) population-based parameters include estimates of density and species richness, and (3) individual-based parameters include the size, condition, and performance (e.g., reproductive output) of individual animals. Osenberg et al. (1994) found that most individual-based parameters required <20, and usually <10, sampling dates, whereas many of the chemical–physical and population-based parameters required >100 sampling dates to reach a power of 80% (Fig. 6.7). They concluded that relatively few of the population-based and chemical–physical parameters could provide adequate power given the time constraints of most studies. They felt that greater emphasis on individual-based parameters is needed in field assessments of environmental impacts. This is a disturbing finding given our emphasis on population-level parameters, especially density and species richness, in most impact assessments.

A serious constraint in study design is the time available to collect data before the disturbance. The before period is often short and beyond the control of the observer. Therefore, parameter selection and sampling design should consider the low number of temporal replicates that can be collected in the before period. These constraints are most likely to negatively influence the detection of impacts on population abundance and chemical–physical properties, and least likely to affect detection of effects on individual performance (Osenberg et al. 1994). We will usually want to sample population parameters, however, because such parameters are of interest to resource managers and regulators. Nevertheless, analysis of individual performance should be carefully examined in addition to population parameters. For example, while estimates of deer density are extremely difficult to estimate with adequate precision, measures of productivity and body condition are far more precise.

6.4.2 *Before–After/Control–Impact–Pairs Design*

Osenberg et al. (1994) developed the BACI design with paired sampling, or BACIP. The *BACIP* design requires paired (simultaneous or nearly so) sampling several times before and after the impact at both the control and impacted site. The measure

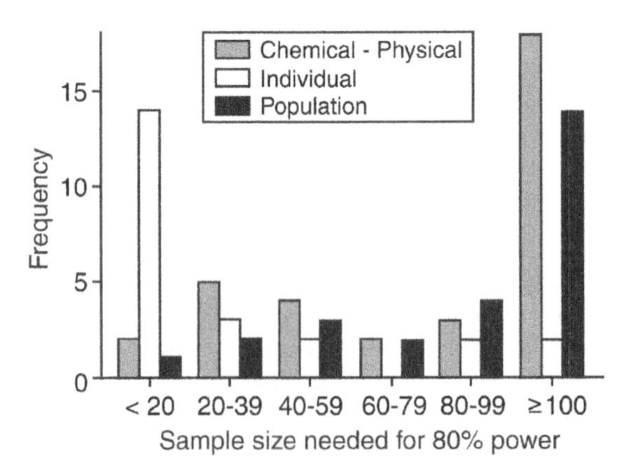

Fig. 6.7 Frequency distribution of the sample size (number of independent sampling dates) for parameters in each group that is required to achieve 80% power. Reproduced from Osenberg et al. (1994), with kind permission from Springer Science + Business Media

of interest is the difference ("delta" in statistical terms) in a parameter value between the control and impacted site as assessed on each sampling date. The average delta in the before period is an estimate of the average spatial difference between the two sites, which provides an estimate of the expected delta that should exist in the after period in the absence of an environmental impact (i.e., the null hypothesis). The difference between the average before and after deltas provides a measure of the magnitude of the impact. Confidence in this estimate is determined by the variation in the deltas among sampling dates within a period, as well as the number of sampling dates (i.e., replicates) in each of the before and after periods.

When choosing sites to use as pairs (or blocks) you hope that, because of their correlation with respect to the response variable, the natural (without treatment) differences *within* a pair is smaller than the differences *between* pairs. You thus hope that the reduction in variance in the response variable that you achieve more than compensates for the degrees of freedom you give up (by pairing). If the matched sites are not similar enough, you do not achieve this tradeoff.

6.4.3 Beyond BACI

An impact must be detectable as a different statistical pattern of statistical interaction from before and after it starts, between the impact and control sites than occurs among the control locations. Underwood described the "asymmetrical analysis of variance" as a means of improving impact assessment – his "beyond BACI" series of papers (Underwood, 1991, 1992, 1994). Use of several control locations and asymmetrical analyses also allows a solution to the problem of not being able to detect impacts in populations that have spatial and temporal interactions in their abundance.

The basic BACI design can be summarized in the form of an analysis of variance table (Table 6.2). The F-ratio in this table is the t test described by Stewart-Oaten et al. (1986) and others. Table 6.2 is the same analysis, but extended to compare abundances at more than two sites.

The repeated-sampling design developed above (Stewart-Oaten et al. 1986) does have problems regarding interpretation of results. If a difference between control and impact sites is found, one cannot conclude that the disturbance caused the impact. There is still no reason to expect that two sites will have the same time course of change in mean abundance. What we need is a set of control sites, and then to demonstrate that the pattern of change in abundance of the population being sampled from before to after is greater in the impact site relative to the control sites. Table 6.3 describes the asymmetrical design recommended by Underwood (1991) for analyzing this situation. Here, the contrast of the impacted vs. control sites and its interactions with time can be extracted from the variation among all sites and its interaction with time. An impact should now be evident in the simplest case as an interaction between the mean abundance in the impacted site and that in the control sites before compared to after the disturbance began (see Table 6.3, $B \times I$).

Table 6.2 Analyses of variance in sampling designs to detect environmental impacts

Source of variation	Degrees of freedom	*F*-ratio vs.	Degrees of freedom for *F*-ratio
(a) BACI design: data are collected in two locations (impact and control) at *t* randomly chosen times before and after a planned disturbance, *n* replicate samples are taken at each time in each location. *B* and C are fixed factors; times are a random factor, nested in either before or after			
Before vs. after = *B*	1	*T*(*B*)	
Control vs. impact = *C*	1	*T*(*B*)	
$B \times C$	1	$T(B) \times C$	$1, 2(t-1)$
Times (before or after) = *T*(*B*)	$2(t-1)$	Residual	
$T(B) \times C$	$2(t-1)$	Residual	$2(t-1), 4t(n-1)$
Residual	$4t(n-1)$		
(b) Similar design, but there is a total of l locations sampled: locations are a random factor, otherwise details are as above. There is no formal test for comparing before vs. after. This is irrelevant because an impact must cause an interaction ($B \times L$ or $T(B) \times L$); see text for details			
Before vs. after = *B*	1	No test	
Among locations = *L*	$l-1$	$T(B) \times L$	
$B \times L$	$l-1$	$T(B) \times L$	$(l-1), 2(t-1)(l-1)$
Times (before or after) = *T*(*B*)	$2(t-1)$	$T(B) \times L$	
$T(B) \times C$	$2(t-1)(l-1)$	Residual	
Residual	$2lt(n-1)$		$2(t-1)(l-1), 2lt(n-1)$

Source: Reproduced from Underwood (1994), with kind permission from Springer Science + Business Media

Table 6.3 Asymmetrical analysis of variance of model data from a single impact (*I*) and three control locations sampled at six times before and six times after a disturbance that causes no impact

Source of variation	Degrees of freedom	Mean square	*F*-ratio	*F*-ratio vs.
Before vs. after = *B*	1	331.5		
Among locations = *L*	3	25,114.4		
[a]Impact vs. controls = *I*	1	3,762.8		
[a]Among controls = *C*	2	35,790.2		
Times (*B*) = *T*(*B*)	10	542.0		
$B \times L$	3	375.0		
[a]$B \times I$	1	454.0	1.51	Residual
[a]$B \times C$	2	335.5	1.12	Residual
$T(B) \times L$	30	465.3		
[a]Times (before) × *L*	15	462.2		
[b]*T* (before) × *I*	5	515.6		
[b]*T* (before) × *C*	10	435.9		
[a]Times (after) × *L*	15	468.2		
[b]*T* (after) × *I*	5	497.3		
[b]*T* (after) × *C*	10	453.6	1.51	Residual
Residual	192	300.0		

Reproduced from Underwood (1991), with kind permission from CSIRO Publishing

[a, b] Represent repartitioned sources of variation to allow analysis of environmental impacts as specific interactions with time periods $B \times I$ or *T* (After) × *I*

Alternatively, if the impact is not sustained or is not sufficient to alter the mean-abundance in the impacted site over all times of sampling after the disturbance, it should be detected in the pattern of statistical interaction between the time of sampling and the contrast of the impacted and control sites (see Table 6.3, $T(\text{After}) \times I$)). Alternatively, a more conservative approach would be to develop error terms based on (1) $B \times L$, averaging out time or (2) $T(B) \times L$, incorporating time.

Thus, a difference is sought between the time course in the putatively impacted site and that in the controls. Such a difference would indicate an unusual event affecting mean abundance of the population in the single disturbed site, at the time the disturbance began, compared with what occurred elsewhere in the undisturbed controls. The impact will either be detected as a different pattern of interaction among the times of sampling or at the larger time scale of before to after the disturbance.

The manner in which system dynamics interact with the design of an impact study are summarized in Table 6.4. This table is largely self-explanatory. The column headed "Baseline" is defined as a study that compares pre- and postdata from the impact area only. This is analogous to Green's (1979) Main Sequence 2, where the impact is inferred from temporal variation only. Recall that reference areas in the classic (original) BACI design are not required. However, because natural factors usually vary temporally, results from baseline studies are seldom sufficient to determine

Table 6.4 Three design strategies for assessing recovery from environmental impacts on biological resources in temporally and spatially varying environments

			Multiyear	
Attributes	Baseline	Single year	No reason to reject/suspect assumptions[a]	Reason to reject/suspect assumptions[a]
When to use	Temporally invariant taxa	Spatial equilibrium achievable, short recovery period	Temporally variant taxa, long recovery period, taxa on multiple recovery periods, information on recovery process desired	
Data needs	Pre- and postimpact only	Impact and reference sites, covariates	Time series for impact and reference areas or for gradient[b]	
Comparison	Pre- vs. postimpact	Impact vs. reference, matched pairs, gradient[b]	Impact vs. reference and gradient over time[b]	
Equilibrium assumption	Steady-state	Spatial	Dynamic	Reject or suspect assumptions
Breakdown in assumptions	Temporal variation confounds with recovery	Spatial variation confounds with recovery	Temporal variation differs for impact and reference categories	NA

(continued)

Table 6.4 (continued)

Attributes	Baseline	Single year	Multiyear: No reason to reject/suspect assumptions[a]	Multiyear: Reason to reject/suspect assumptions[a]
Statistical methods[c]	*t* test: Student's paired; BACI[d]	ANCOVA, paired *t* test, gradient[b]	Level-by-time, trend-by-time, repeated measures	Gradient (with or without covariates), impact/ref, others
Conditions need for recovery	Equal pre- and postmeans	Impact and reference means equal, no impact on gradient[b]	Difference in means constant, gradients constant	Failure to reject multiple assessments of impact effect
Advantages	Reference sites not required (though useful)	Single year of data, extrapolation reasonable	Nonrandom site selection	
Disadvantages	Equilibrium assumption not reasonable, preimpact data required	Recovery snap-shot, covariables needed, matched sites for matched pairs	Multiyear data required, difficult to extrapolate from nonrandom samples	
Comments	Use with multi- or single-year studies, provides partial information on recovery process	Corroborate with contamination and toxicity (triad approach)	Use preimpact data, validate assumption	Verify with habitat changes, use alpha level > 0.05

Reproduced from Parker and Weins (2005), with kind permission from Springer Science+Business Media

NA Not applicable

[a]Reasons may include zero means. Entries that span the last two columns pertain to both situations

[b]Gradients are dose–response regressions of biological resources vs. gradients (i.e., continuous measures) of exposure

[c]Methods addressed in Wiens and Parker (1995)

[d]BACI uses prespill data at impact and reference sites and relies on the assumption of dynamic equilibrium

if recovery has occurred. "Single-year studies" compare impact and reference areas but within a single year. These designs approximate spatial equilibrium through the use of multiple sampling areas, which requires a close matching of natural conditions across sites (e.g., matched pairs design). Recovery occurs when impact and reference means are similar. "Multiyear studies" reduce the effects of temporal and spatial variation by removing (subtracting out) naturally varying temporal effects. If the impact and reference areas are in a dynamic equilibrium, recovery occurs when differences in annual means become constant (trend lines become parallel as explained above for Fig. 6.4C).

6.5 Suboptimal Designs

Designs classified as suboptimal apply to the true impact situation; namely, where you had no ability to gather preimpact (pretreatment) data or plan where the impact was going to occur. Such situations are frequent and involve events such as chemical spills and natural catastrophic events. After-only impact designs also apply, however, to planned events that resulted from management actions, such as timber harvest, road building, and restoration activities, but were done without any monitoring plan. As noted by Parker and Wiens (2005), it is critical in impact assessment to separate the recovery signal from natural variation and of verifying the ecological assumptions on which detecting recovery depends.

6.5.1 After-Only Designs

The need to determine the effects of unplanned disturbances on resources is commonplace. Such disturbances include a wide variety of causes, from accidental releases of hazardous materials to large-scale natural phenomena (e.g., forest fires, floods). Because of the unplanned nature of these events, pretreatment data are seldom directly available. Thus, the task of making a determination on the effects the disturbance had on wildlife and other resources is complicated by (1) natural stochasticity in the environment and (2) the usually unreplicated nature of the disturbance.

There are, however, many options for evaluating impact effects in these suboptimal, "after-only" designs. Below, we present many of these designs, divided by the use of single- or multiple-sampling periods postdisturbance. Much of our discussion is taken from a survey of these designs presented by Wiens and Parker (1995). The reader should be aware, however, that the statistical power associated with all of these designs is likely to be severely limited.

6.5.2 Single-Time Designs

The assumption that the sampling interval is short enough so that no changes in site conditions occurred during the sampling period is especially critical in all single-time designs. You must also assume that natural forces were acting in a similar manner across all treatment and reference sites; using multiple (replicated) sites enhances the chance that interpretable data will be gathered. In all of the following designs, analysis of covariance using nonimpact-related variables may help identify the effect such variables are having on determination of the severity of the impact.

6.5.2.1 Impact–Reference Designs

The basic *impact–reference design* mimics a classical experimental treatment and control design, where random samples are taken from sites within the disturbed area and from other, nondisturbed reference sites. A difficulty here is gathering replicated samples from within the disturbed area. If the impacted area is large relative to the animals involved (i.e., a large area relative to range size), then a case can be made for placing multiple sampling sites within the disturbance, thus gaining independent samples. For example, a fire that occurs over 10,000 ha has impacted numerous plant associations, many slopes and aspects, and other variations in environmental conditions. Likewise, waste runoff from a mine can contaminate an entire watershed.

If the impacted sites can be classified by several levels of exposure (from the example above: differing fire intensity or decreasing contamination from the mine source), the analysis is strengthened because we would expect a relationship between the severity of impact and the response by animals. The impact can be tested using a one-way ANOVA with levels of impact as treatments or a regression analysis with quantification of level of impact as an independent variable. Then, a significant relationship between mean disturbance and animal response (e.g., density, body weight, reproductive performance) is evidence of an impact.

6.5.2.2 Matched Pair Designs

Matched pair designs reduce the confounding of factors across sites. Under this design, sites within the impacted area are randomly selected and nonrandomly matched with similar reference sites. Such matching reduces the variability between pairs, thus statistically enhancing the difference between impacted and reference pairs. These differences are then compared using paired *t* tests.

You must assume under this design that the matched sites do not vary systematically in some important manner that either masks an impact or falsely indicates that an impact has occurred. As noted by Wiens and Parker (1995), this design is open to criticism because the reference sites are chosen nonrandomly.

6.5.2.3 Gradient Designs

Gradient designs analyze an impact along a continuous scale and use regression techniques to test for an association between level of impact and response by the animal. For example, data on animal condition and impact intensity can be taken at a series of points along a transect through a disturbed area, and the results regressed to look for a significant association.

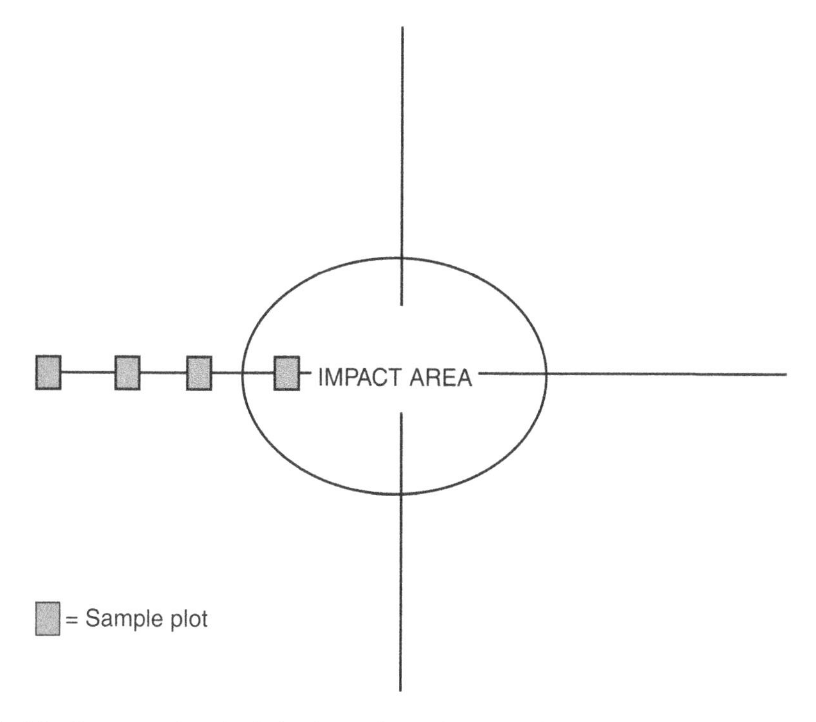

Fig. 6.8 Representation of a design layout for gradient analysis

Under this design you must assume that the complete or nearly complete range of impact, including none (which becomes a reference site embedded in the analysis), has been sampled about evenly. This ensures that regression analysis can be run properly, and increases the likelihood that other natural factors are balanced across sampling locations.

The gradient approach is especially applicable to localized impacts because it allows you to quantify that the response of elements vary with distance from the impact. Further, you might be able to identify the mechanism by which the impacted site recovers, such as through a gradual lessening of the effect of the impact along the gradient from distal to proximal ends. The ends of the gradient serve, in essence, as nonimpacted reference sites. Figure 6.8 depicts a simple but effective gradient design.

6.5.3 Multiple-Time Designs

Multiple-time designs are those where repeated sampling can occur following the disturbance. Such a sampling regime allows evaluation of temporal dynamics that could indicate both an impact and a subsequent recovery. Gathering samples repeatedly from the same locations usually requires that specialized statistical analyses (e.g., repeated measures analysis) be used to interpret recovery rate.

6.5.3.1 Time-Series Designs

In *time-series designs*, it is expected that the response of the animals to the disturbance will decrease over time; the animals are sampled at the same sites over time.

Under this design we assume that similar methods are used to gather data at all sites over the sampling time. In addition, we must assume that the ecological system is in a steady-state equilibrium: that is, the effects of natural factors are similar over the replicated sites, factors affecting the resource do not change over the period of the study in ways other than a decrease in the level of impact, and prior to the disturbance the resource was not reduced by some natural condition (and is thus returning naturally from a cause other than the impact we are interested in quantifying) (Wiens and Parker 1995).

6.5.3.2 Impact Level-By-Time Interaction Designs

Due to natural variation, the factors of interest may change in value from year to year. A difference in magnitude of these annual changes between sites is evidence that a change has occurred. If the change results from a human impact (e.g., logging), a natural catastrophe (e.g., fire, disease), or even an experimental treatment, the use of multiple sites means that reference ("control") sites will be available for comparative purposes. Following the sites for an extended period of time (likely 2–5 year) will reveal how the trajectory of recovery compares with that on reference sites; interpretation of change will not be masked by the overriding ecological process (i.e., the assumption of steady-state dynamics is relaxed, as described above). Factors can differ between sites, but temporal changes in the resource are expected to be similar to reference sites in the absence of the impact. It is thus assumed that a dynamic equilibrium exists between factors affecting the resource and the state of the resource. It also assumes that some recovery occurs during the course of the study.

The term "level" (*level-by-time interaction*) refers to the fact that specific categories (levels) of the impact are designated. For example, a chemical spill could directly kill animals or force them to leave an impacted area. Here the chemical impacts specific locations at different general levels (e.g., light, moderate, and heavy spills). As the chemical dissipates, or after it has been cleaned-up, animals might start returning to the site, or those that remained during the spill might begin to reproduce. In either case, the abundance of animals on the impacted site should recover to the pattern of abundance being shown by the reference sites during the same time period, assuming no residual effects of the chemical. If a change in abundance or other population parameter persists, then this change in pattern between impact and reference sites indicates effect. The asymmetrical design of Underwood (1994) described above is similar to this level-by-time design, except that the BACI uses a before–after comparison.

6.5.3.3 Impact Trend-By-Time Interaction Design

In the impact *trend-by-time interaction* design, continuous variables are used (rather than distinct levels) to compare trends between measures of the resource and levels of change (or impact) over time. In the absence of change, one expects that although resource measures may differ over time because of natural variation, they will show no systematic relationship to any gradient in change (or impact). This design is superior to the level-by-time design only in that the use of a gradient means that specific exposure levels need not be determined a priori.

Interactions are perhaps the most applicable design to many after-only situations in resource management, including both planned management activities and unplanned catastrophes. Here you establish multiple reference sites matched with the impacted site and gather samples over a period of time, the length of which depends on the impact and the elements under study. It is understood that the elements will vary in magnitude (of the response measurement) through time; impact is inferred when the pattern of change in the element differs significantly from that of the reference sites. Statistically, this is a factorial anova where a significant time interaction infers an impact (as graphically shown in Fig. 6.9a; no impact is inferred from Fig. 6.9b).

6.6 Supplemental Approach to Impact Assessment

Because it is unlikely that powerful "before" sampling will be available in most circumstances, we need to select alternative approaches to understanding ecological processes and the effects of planned and unplanned impacts on them. We need to determine the rates of change and the magnitude of spatial differences for various populations. An initial step could involve the selection of priority species and vegetation types. For example, we know that logging, agriculture, housing, recreation, and hunting will continue. Thus, it makes sense to monitor examples of such systems that are still in a relatively undisturbed state. A set of locations could be monitored where effects of human disturbance are minimal. These sites would collectively constitute baseline information that could then be used to contrast with perturbed areas, when the opportunity or need arose in other geographic locations.

Thus, baseline information would already exist to address a number of specific needs. This would lessen the need for specifically acquired "before" data. Such an analysis would be useful because (1) many different locations would be used, (2) it would lower the overall cost of conducting environmental assessments, and (3) it would reduce the need for location- and time-specific "before" data. Additionally, it would also improve our ability to predict the likely impact of proposed developments through the accumulation of data necessary to develop population models (Underwood 1994). Of course, it is difficult to know exactly where to place such sampling sites. This stresses the need for careful development of an appropriate sampling design, including assessment of the number and placement of locations to be monitored. Such considerations are typically ignored, even though

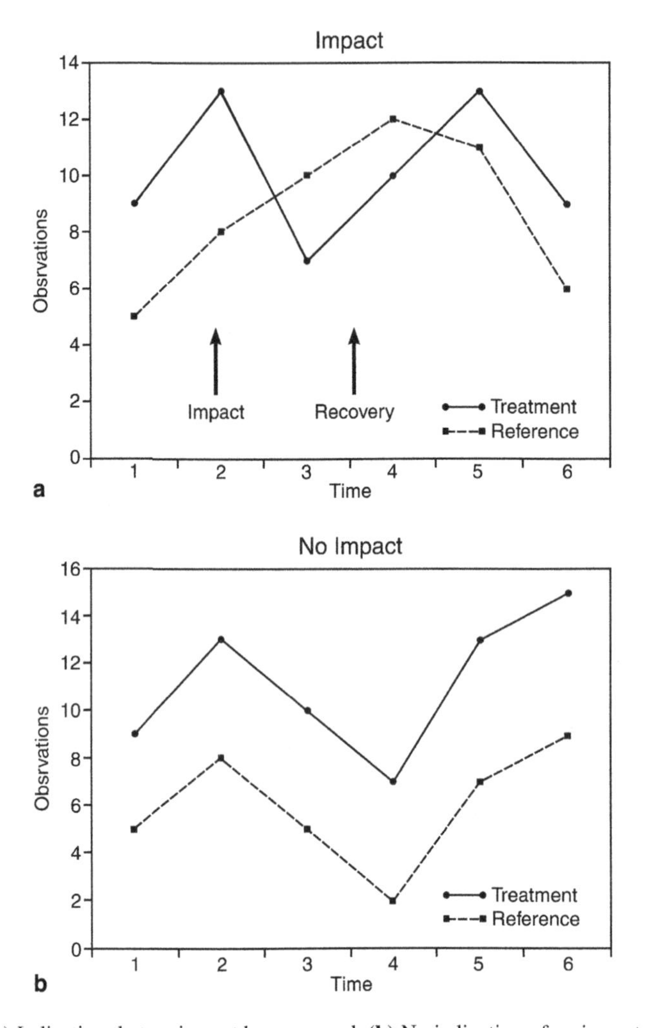

Fig. 6.9 (**a**) Indication that an impact has occurred. (**b**) No indication of an impact occurring

most statistical tests emphasize adequate randomization and replication of the sampling units (see Chap. 2). This type of design will be most effective at the broader spatial scales of analysis, such as the landscape or macrohabitat scales, where we are concerned with changes in presence–absence, or at most, general direction of trend.

This suggestion partially addresses the problem of implementing long-term research programs by reducing the time- and location-specificity of data collection. The multiple-time designs outlined above would benefit from such a program by being placed in the context of estimating of variance of response variables and system dynamics over long time frames. In addition, these designs can incorporate before–after data (Green 1979; Osenberg et al. 1994; Underwood 1994; Wiens and Parker

1995). Very rare species, especially those that are state and federally threatened and endangered, will likely require more intensive, site- and time-specific sampling because of the relatively small margin of error involved in managing these groups. In addition, there are often sample size problems with rare species because of their small numbers. Also, there may be limits placed by regulatory agencies (US Fish and Wildlife Service and the state wildlife agency) on the methodologies used to study rare species.

6.7 Epidemiological Approaches

Epidemiology is the study of the occurrence of disease, injury, or death, usually in reference to human populations. The occurrence of disease is studied in relation to factors relating to the individual and his or her environment and lifestyle with the goal of establishing the cause of the disease (Barker and Hall 1991; Ahlbom 1993). Epidemiological approaches, by focusing on determining incidence rates (of disease), lend themselves to applications in other fields.

Epidemiology studies are somewhat unique among statistical studies in the attention given to choosing the dependent or outcome variable. In such studies, the outcome variable is almost always a rate related to the frequency of disease, injury, or death. The outcome variable chosen should be that considered most likely to provide information on the hypothesis about the mechanism of disease. The outcome variable depends heavily on the mechanism hypothesized for the cause of the disease. Once a measure of the frequency of disease is chosen, a measure of effect must be chosen. This measure is used to summarize the difference between two or more populations (Mayer 1996).

As an example of application of an epidemiological approach in wildlife impact assessment, we will highlight the research program being developed to evaluate the impact of wind turbines on birds. By using such a "theme" example, the reader can see the relationships between various parts of a comprehensive research program that is designed to both assess impact and evaluate means of impact reduction. Electricity generation from wind turbines has been expanding throughout the world, including the United States. Major new plants have been constructed or planned for California, Washington, Wyoming, Minnesota, Texas, Iowa, and various locations in the Northeast. Developments range from a few dozen turbines to over 5,000. However, in certain sites, birds have been killed by flying into the rotating blades. This issue was of special concern at the Allamont Pass Wind Resource Area (WRA) because of the death of golden eagles (*Aquila chrysaetos*) and other raptors caused by wind turbines. Because of concern over bird mortalities, and because of the proposed expansion of wind plants, various government agencies initiated research programs designed to determine if wind plants were adversely affecting bird populations, and if so, to design and test methods of reducing this impact. The situation confronting these agencies was not unlike that encountered in many impact-related studies:

- An impact had already occurred (bird deaths).
- Numerous interacting environmental factors are known to impact the abundance and health of birds (e.g., food, weather, disease).
- Defining the "affected population" was problematic (i.e., where to draw study area boundaries).
- Although deaths occurred, they were not common (e.g., perhaps 30 eagles per year at Altamont).
- Scavengers could bias results by removing the evidence of death before being seen by observers.
- Turbines are usually distributed across different slopes, aspects, and microsites.
- Most large wind developments use several different types of turbines.
- Bird activity varies substantially depending upon time of day and time of year.
- Because of variations in geographic locations and size of development, extrapolation of findings among developments will be problematic.

Thus, researchers were confronted with a host of interacting factors and many major constraints when designing studies of impact assessment. We are interested in quantifying both the use of a site and the frequency of deaths associated with that use. The ratio of death and use becomes a measure of mortality; mortality is thus expressed as a "rate." Following the epidemiological approach, the outcome variable will be a rate related to the frequency of death (or injury). The outcome variable is the variable that the researcher considers most likely to shed light on the hypothesis about the mechanism of injury or death. The choice of this outcome variable (frequency of injury or death) depends heavily on the mechanism hypothesized as – the cause of injury or death. Determining the mechanism of injury or death allows the development of appropriate methods to reduce the risk to an animal of being affected by an impact.

6.7.1 Attributable Risk

Attributable risk is defined as the proportional increase in the risk of injury or death attributable to the external factor (e.g., wind turbine, pesticide, noise). It combines the relative risk (risk of natural mortality) with the likelihood that a given individual is exposed to the external factor. Attributable risk (AR) is calculated as

$$AR = (PD - PDUE) / PD,$$

where PD is the probability of death for the entire study population, and PDUE the probability of death for the population not exposed to the risk. That is, PD incorporates all causes of death or injury that the study population is experiencing, be it related to the impact of interest or another (natural or human-related) cause. The PDUE, then, is simply PD without inclusion of the impact of interest.

For example, suppose that the probability of death for a randomly chosen individual in the population is 0.01, and the probability of death in a control area for a bird flying through a theoretical rotor plane without the presence of blades is 0.0005. The AR is thus (0.01–0.0005)/0.01 = 0.95. Thus, about 95% of the risk of dying while crossing the rotor plane is attributable to the presence of blades. As noted by Mayer (1996), it is this large attributable risk that stimulates the concern about the impact of wind development on birds, regardless of the absolute number of bird deaths. Testing a preventive measure in a treatment–control experiment allows us to determine the change in risk due to the prevention.

6.7.2 Preventable Fraction

The proportion of deaths removed by a preventive step is termed the *preventable fraction* and is defined as the proportion of injuries or deaths that would be removed if all birds were able to take advantage of the *preventive intervention*. Preventable fraction (PLF) is calculated as

$$\text{PLF} = (\text{PD} - \text{PDI}) / \text{PD},$$

where PDI is the probability of injury or death given the preventive intervention. For example, studies have shown that birds are killed by wind turbines. Thus, the need arises to test various preventive measures, such as removing perches and painting blades. If population mortality in the wind farm is 0.01, and mortality for those using the area with the preventive intervention is 0.005, then the preventable fraction is (0.01–0.005)/0.01 = 0.5. Thus, about 50% of the risk would be removed if all of the perches were removed. Note that the attributable risk and preventable fraction would be the same value if the intervention removed the risk entirely.

6.7.3 Prevented Fraction

Prevented fraction is the actual reduction in mortality that occurred because of the preventive intervention. Prevented fraction (PFI) is calculated as:

$$\text{PFI} = (\text{PDAI} - \text{PD}) / \text{PDAI},$$

where PDAI is the probability of injury or death in the absence of intervention. For example, suppose that 25% of the perches are removed in a treatment–control experiment. Field studies determine that mortality is 0.01 for the population and 0.015 for those living without the prevention (e.g., perches). The prevented fraction is (0.015–0.01)/0.015 = 0.33. Thus, about 33% of the risk has been removed by removing 25% of the perches.

It is important to remember that these three measures of effect remove emphasis from the risk to individuals and place emphasis on the risk to the population. These measurements do not, however, tell us what influence a change in prevented fraction has on population abundance or trend, or other measures of success. But they are extremely useful in evaluating the potential and realized influence that preventive measures have on proportional injury or death.

6.7.4 *Measuring Risk*

The choice of the use factor, or denominator, is more important than the numerator. The choice arises from the preliminary understanding of the process of injury or death. In fact, the treatment effect is usually small relative to the variability that would arise from allowing alternative measures of risk. For example, should the denominator be bird abundance, bird flight time in the farm, bird passes through the rotor plane, or some other measure of use? Unless these measures are highly intercorrelated – which is unlikely – then the measure selected will result in quite different measures of mortality. Further, the choice of denominator is important in that it should express the mechanism causing the injury or mortality. If it does not, then we will not be able to accurately measure the effectiveness of a risk reduction treatment.

For example, suppose that bird use or abundance is the denominator, bird deaths are the numerator, and painted blades are the treatment. A treatment–control study determines that death decreases from 10 to 7 following treatment, but use actually decreases from 100 to 70 (arbitrary units). It thus appears that the treatment had no effect because both ratios are 0.1 (10/100 and 7/70). This study is seriously flawed because, in fact, no conclusion should be drawn. This is because there is no direct link between the number of birds using the area and flights near a turbine. There are numerous reasons why bird use of a wind farm could change (up or down) that are independent of the farm, for example, changes in prey availability, direction of prevailing winds, environmental contaminants, and so on. In this case, recording bird flights through the rotor plane of painted blades would have been a more correct measure of effect. In addition, the use of selected covariates can help focus the analysis on the treatment effects. Naturally, the hypothetical study noted above should be adequately replicated if implemented.

It is, of course, prohibitive from a practical standpoint to record every passage of a bird through a zone of risk (be it a rotor plane or the overall wind farm). Further, it is usually prohibitive to accurately census the population and tally all deaths. As such, we must usually rely on surrogate variables to use as indices of population size and death. A surrogate variable is one that replaces the outcome variable without significant loss in the validity or power of the study. An example would be using the number of birds observed during 10-min point counts as a measure of utilization (of a treatment or control). Utilization is an indicator of the level of at-risk behavior. Thus, adopting a measure of utilization requires the assumption that the higher the utilization the higher the risk. If feasible, assumptions should always be tested early in the study.

Once a measure of mortality is chosen, a measure of effect must be selected. This measure could be the risk ratio, defined as the ratio of mortality in one area (e.g., wind farm) to that in another area (e.g., control). Thus, if mortality in the wind farm is 0.01 and that in the control is 0.001, the risk ratio is 10; the relative risk of death is ten times greater for a randomly chosen bird in the farm vs. one in the control. Ideally, such a study should be adequately replicated, because controls are not perfect matches to their associated treated sites. An alternative is to use one of the measures of attributable risk described above. These measures have the advantage of combining relative risk with the likelihood that a given individual is exposed to the external factor. It is the proportional change in the risk of injury or death attributable to the external factor. The use of attributable risk implies that the importance of the risk is going to be weighed by the absolute size of the risk. The risk ratio ignores the absolute size of the risk.

6.7.5 Study Designs

6.7.5.1 Basic Experimental Approaches

As outlined by Mayer (1996), there are four tasks that the investigator must accomplish when designing a study of impact assessment. The logic is sequential and nested; each choice depends on the choice made before:

1. Isolate the hypothesis of mechanism that is being tested. For example, one might be testing the hypothesis that birds strike blades when attempting to perch on a turbine. The hypothesis should be simple and readily testable.
2. Choose a measure of injury–death frequency that best isolates the hypothesis being tested. The two components of this choice are to choose an injury–death count to use as a numerator and a base count (likely utilization) to use as a denominator. It is critical that a relevant measure of use be obtained (e.g., passes through the rotor plane; occurrence by flight-height categories).
3. Choose a measure of effect that uses the measure of injury–death frequency and isolates the hypothesis being tested. Here, decide whether the relative risk (risk ratio), attributable risk, or another measure of effect should be used.
4. Design a study that compares two or more groups using the measure of effect applied to the measure of injury–death frequency chosen. The goal here is to isolate the effect, control for confounding factors, and allow a test of the hypothesis. Replication is essential.

The ideal denominator in epidemiology is the unit that represents a constant risk to the animal. The unit might be miles of flight, hours spent in the farm, or years of life. If the denominator is the total population number, then we are assuming that each bird bears the same risk by being alive. In human epidemiological studies, the total population size is usually used because we cannot estimate units of time or units of use. In wildlife studies, however, actual population density is extremely difficult to estimate. If the risk is caused by being in the area, then deaths per hour in the area is probably the best epidemiological measure in wildlife studies. It is then extrapolated to the population by estimating the utilization rate of the area for the entire population. Measuring utilization is difficult, however, and must be approached carefully.

Thus, we have two major ways to calculate mortality rate:

$$\text{Number of dead birds / number of birds in population,} \tag{6.1}$$

$$\text{Number of dead birds / bird use.} \tag{6.2}$$

Equation (6.1) is ideal, but as discussed above, usually impractical. Equation (6.2) is feasible, but will vary widely depending upon the measure of bird use selected. In addition, for (6.2), the background (control) mortality rate must also be determined for comparative purposes. Thus, (6.2) should be the center of further discussion.

Consultations with personnel in the wind industry have led to the conclusion that a measure of operation time that is easily standardized among wind farms and turbine types would be preferable. It has been suggested that a measure that considers differences in blade size and operation time would be most appropriate. As such, the concept of "rotor swept area" has been developed, which is simply the circular area that a turning blade covers. Rotor swept area is then converted to an index that incorporates operation time as follows:

$$\text{Rotor swept hour} = \text{rotor swept area} \times \text{operation hours.}$$

An index of risk is then calculated by using a measure of risk:

$$\text{Rotor swept hour risk} = \text{risk measure / rotor swept hour.}$$

Here, "risk measure" could be flight passes through rotor plane or any other appropriate measure of use (as discussed above). Here again, we emphasize the need to test assumptions. For example, an assumption here is that the probability of a bird being struck is equal among all turbines, which may or may not be the case. Numerous factors, such as placement of the turbine in a string of turbines or placement of a turbine along a slope, could influence the probability of a bird collision.

6.7.5.2 Variable Selection

One primary variable will usually drive the study design; thus, the initial sample size should be aimed at that variable. It is thus assumed that at least a reasonable sample size will be gathered for the other, secondary variables. Sampling can be adjusted as data are collected (i.e., sequential analysis of sample size).

Designing treatment–control studies for inferences on measures of use is feasible. Determination of mortality (using (6.2)) is possible, but statistical power to conclude that treatment and control sites have different mortality rates will be low. For example, in a randomized pairs design, most pairs are expected to result in 0 mortalities, with tied values and no mortalities on either member of a pair. The high frequency of zero values effectively reduces the sample size for most analyses.

6.7.5.3 Case Study Approach

Case studies have high utility in evaluating mortality. Here, one collects dead or injured animals inside and outside the impacted area, and conducts blind analyses to determine the cause of death. Unfortunately, from the standpoint of study design, under most situations very few dead animals will be found outside the impacted area. However, all dead animals found in a study should be subjected to blind analyses because this information will assist with evaluation of observational data.

The case study approach suggests that epidemiological analysis can often be combined with clinical analysis to extend the inferential power of a study. Here the clinical analysis would be the necropsies of the animals. Suppose that we are successful at finding dead birds inside a wind farm. If we look at *proportional mortality* – the proportion of the birds killed by blunt trauma, sharp trauma, poisoning, hunting, and natural causes – then the proportions should differ significantly between the farm and the control area. The assumption is that the differential bias in finding dead birds within the two areas is uniform across the causes of mortality and thus the proportions should be the same even if the counts differ (i.e., relatively few dead birds found outside the farm). An inherent problem with this approach is the difficulty in finding dead birds in the control area(s).

6.7.5.4 Behavioral and Physiological Studies

Obtaining information on the sensory abilities of animals is a key step in designing potential risk reduction strategies. For example, although it makes common sense to paint blades so birds can more readily see them, there are many possible designs and colors to select from. For example, what colors can birds see, and how do birds react to different patterns? If painting blades causes a bird to panic and fly into another turbine, then painting has not achieved its intended goal. The electric power industry confronted the problem of birds flying into transmission lines through a series of studies on tower design and devices to divert birds from transmission lines; many of these designs were successful in reducing mortality (e.g., see review by Savereno et al. 1996).

Many of these questions are best investigated initially in a laboratory setting. Unfortunately, translating laboratory to field is an age-old problem in behavioral ecology, and success in the laboratory using tame and trained birds does not necessarily mean success in the field, where a myriad of other environmental factors come into play and the physical scales are different. However, initial laboratory studies at least help to narrow the scope of field trials. A sequential process of initial laboratory testing of treatments, followed by field trials, followed by additional laboratory trials as indicated, and so forth (e.g., an adaptive management design), is recommended.

6.8 Modeling Alternatives

A central part of impact assessment is development of a model that shows the survival rates required to maintain a constant population. The strategy is to determine survival rates required to sustain populations exhibiting various combinations of other parameters governing population size. To be useful in a wide range of environmental situations and useable for people with varying expertise, the model must be based on simple mathematics.

The use of models (of all types) has soared in the past 20 years. In fact, modeling is now a focus of much interest, research, and management action in wildlife and conservation biology. But as in all aspects of science, models have certain assumptions and limitations that must be understood before results of the models can be properly used. Modeling *per se* is neither "good" nor "bad"; it is the use of model outputs that determines the value of the modeling approach.

The use of population models to make management decisions is fairly common. For example, models play a role in management plans for such threatened and endangered species as the spotted owl (*Strix occidentalis*, all subspecies), desert tortoise (*Gopherus agassizi*), Kirtland's warbler (*Dendroica kirklandii*), various kangaroo rats (*Dipodomys* spp.), and so forth. Models are valuable because they analyze the effects of management proposals in ways that are usually not possible using short-term data or professional opinion. Models can be used in impact assessment to predict how a system (e.g., species, group of species, environmental measures) should have behaved under nonimpacted conditions, and also how a system might have behaved under various impact scenarios.

6.8.1 Resource Selection Models

As well summarized elsewhere (e.g., Manly et al. 2002; Morrison et al. 2006), documentation of the resources used by animals is a cornerstone – along with quantifying distribution and abundance – of animal ecology. Thus, much literature is available on how to identify, quantify, and interpret the use of resources by animals. In this section we briefly review some of the terminology associated with resource use, and provide some guidance on how studies of resources have been categorized in the literature. The specific statistical procedures and models used in resource selection studies are basically the same as those used in other studies of wildlife ecology, and have been well presented by Manly et al. (2002).

The use of resources is obviously critical to all organisms, and *resource use* is defined as the quantity of the resource that is used in a specific period of time. The amount of a resource in the environment that is accessible to an animal is termed *resource availability*; whereas the absolute amount of that resource in the environment is termed *resource abundance*. Lastly for our purposes here, *resource selection* is defined as the use of a resource relative to the availability or abundance of that resource.

Resource selection is conceptualized to occur as a hierarchical, decision making process by an animal (e.g., Manley et al. 2002, pp. 1–2; Morrison et al. 2006, pp. 155–158). Thus, when designing a study of resource selection you must consider how the animal and resources interact across spatial scales, from the broad (landscape) to the local (e.g., feeding site). In many cases studies must be designed to account for multiple scales of selection. Additionally, resource selection will vary by season, and sex and age class.

As noted above, selection can be analyzed by comparing two of the three possible sets of resource units, namely used, unused, and available. Manley et al. (2002, pp. 5–6) used these sets to identify three common sampling protocols:

A. Available units are either randomly sampled or censused and used resource units are randomly sampled
B. Available resource units are either randomly sampled or censused and a random sample of unused units is taken
C. Unused resource units and used resource units are independently sampled

Three general study designs for evaluating resource selection have been identified in the literature (see especially Thomas and Taylor 1990). Each of the three sampling protocols in the preceding paragraph (A, B, C) can be used for each of the following study designs and the specific combination of protocol and design used to gather the data determines some of the underlying assumptions required for subsequent analyses.

Design 1. The availability and use for all items are estimated for all animals (population), but organisms are not individually identified, and only the item used is identified. Availability is assumed to be equal for all individuals. Habitat studies often compare the relative number of animals or their sign of presence in each vegetation type to the proportion of that type in the study area.

Design 2. Individual animals are identified, and the use of each item is estimated for each animal. As for Design 1, availabilities are assumed equal for all individuals and are measured or estimated for the entire study area. Studies that compare the relative number of relocations of marked animals in each vegetation type to the proportion of that type in the area fall into this category.

Design 3. This design is the same as Design 2, except that the availability of the items is also estimated for each individual animal. Studies in this category often estimated the home range or territory for an individual and compared use and availabilities of items within that area.

Thomas and Taylor (1990) and Manly et al. (2002) provided a good review of studies that fit each of these categories, as well as guidelines for sample sizes necessary to conduct such analyses. Studies using Design 1 tend to be inexpensive relative to Designs 2 and 3 because animals do not need to be identified individually. Designs 2 and 3 allow for analysis of resource selection on the individual, thus estimates calculated from observations may be sued to estimate parameters for the population of animals and estimates of variability of these estimates.

Virtually all classes of statistical techniques have been used to analyze use–availability (or use–nonuse) data, depending upon the objectives of the researcher, the structure of the data, and adherence to statistical assumptions (i.e., univariate parametric or nonparametric univariate comparisons, multivariate analyses, Bayesian statistics, and various indices), and these techniques have been well reviewed (see summary in Morrison et al. 2006, pp. 166–167 and Manly et al. 2002). Compositional analysis, only recently applied to habitat analysis, should be considered for use in these studies (Aebischer et al. 1993).

6.8.2 *Synthesis*

The goal should be to present a realistic and unbiased evaluation of the model. It is preferable to present both a best and worst case scenario for model outputs, so that the range of values attainable by the model can be evaluated. For example, with a basic Leslie matrix model of population growth, knowing whether the confidence interval for the predicted (mean) value for λ (rate of population growth) includes a negative value provides insight into the reliability of the predicted direction of population growth.

The process of model development and evaluation may show that the predictions of the model are sufficiently robust to existing uncertainties about the animal's behavior and demography that high confidence can be placed in the model's predictions. Even a poor model does not mean that modeling is inappropriate for the situation under study. Rather, even a poor model (i.e., a model that does not meet study objectives) will provide insight into how a population reacts to certain environmental situations, and thus provide guidelines as to how empirical data should be collected so that the model can be improved. Modeling is usually an iterative process.

6.9 Applications to Wildlife Research

In the study of wildlife we are constantly confronted with the need to examine ecological relationships within a short timeframe – a few years – and within a small spatial area. The restriction of studies to short temporal and spatial scales usually arises because of limited funding, the specific needs of a funding agency to study a localized situation (e.g., a wildlife management area), and the fact that much research is conducted by graduate students who must complete thesis work within 2–3 years. Faculty, as well as agency, scientists are not immune from the need to develop results from research within a few years.

Despite the temporal and spatial constraints confronted by most researchers, the need to publish research results remains strong. Additionally, much of the research that is conducted under time and space constraints will be used to guide management of animals, including those hunted and those considered rare or endangered. Thus, the research that is conducted must be rigorous. In Chap. 5 we discuss many of the strategies that can be used to, in essence, make the best out of a bad set of research constraints; here we synthesize some of the steps you can take related to impact assessment studies.

Recall (Sect. 6.3.1) our definition and discussion of "recovery": a temporal process in which impacts progressively lessen through natural processes and/or active restoration efforts. Using the concept of recovery and the assumptions about the temporal and spatial variability of a natural (nonimpacted) system – steady state, spatial, or dynamic equilibrium – Parker and Wiens (2005) outlined strategies for assessing "recovery from environmental impacts." Here we use the Parker and Wiens rationale to specify a strategy for assessing the state or condition of a wildlife population (or segment thereof) over a short timeframe and in a limited area. Thus, as we noted when we opened this chapter, the broad field of impact assessment can be applied to virtually any research goal because all systems are under constant change.

Studies of recovery (impact assessment) are simply trying to separate the signal from the noise; this is the same thing most researchers are trying to do. For example, say you want to determine what food source(s) are used by deer on a wildlife management area, and you want to maximize your ability to be confident your results are not overly time and space constrained. Applying an impact assessment strategy, you can view your area of interest as the "impacted site" and multiple other, similar (nearby locations under similar environmental conditions) locations as your "reference sites." Under this impact–reference design you can compare, say, feeding rates in several categories of vegetation types or by particular species of plants on the impacted site with your reference (no impact) sites. Because deer are not randomized spatially (i.e., you selected the impact site and locations to study deer therein), random sampling alone can only reduce the confounding effects of spatial variation. ANCOVA can be used to further reduce confounding effects given you can identify factors influencing deer activity, such as distance from roads or development, availability of water, and related key factors. By including reference site(s), you will maximize you ability to identify the signal – what the deer are using on your focal area – from the noise of the environment.

Of course, the longer duration you can study the better for deciphering signal from noise. But, this strategy can be applied to studies of even one season. Table 6.4 specifies many of the strengths and weaknesses of this approach. Single-year studies provide only a brief glimpse of environmental variation, and the results of such studies must fully acknowledge such a limitation. But, applying the impact–reference strategy strengthens what you can say about the ecology under study. Using multiple study areas allows you to improve your knowledge on the amount of spatial variation that exists in the environment under study.

Improving what you can say about temporal variation, without studying for multiple years, can be achieved by increasing the number of reference sites and the spread of the sites across space. By venturing into "less similar" areas, but those that still harbor the species of interest, you begin to implement the gradient approach to impact assessment (but for a nonimpact assessment study). By moving further away from the primary area of interest, you generally begin to witness how the target animal responds to changes in biotic and abiotic factors. By studying the animals across such gradients, you are able to make assumptions about how they will react in your area of interest when confronted by such resource and environmental conditions. Morrison et al. (2006, pp. 59–60) summarized how the distribution, abundance, and activity of animals varies across environmental gradients, and how such a phenomenon can be used to identify resource requirements.

Returning to our example above, you would be asking how the deer would respond to changes in plant species composition on your area based on their current response to different plant compositions elsewhere. You are, in essence, substituting space for time and making the assumption that animals on your area of interest will respond likewise. Naturally, all results must be interpreted in the context of a good foundation of knowledge about the biology of the species (but that applies to all studies).

6.10 Summary

The field of impact assessment is expanding rapidly as new study designs and analytical techniques are applied. Because most of the impacts are not planned, suboptimal designs are usually required. As such, the need for replication and exploration of confounding factors is critical. In many cases, statistical power will remain low. In such cases, it is incumbent on the researcher to clearly acknowledge the weaknesses of the design and analyses, and fairly represent the available conclusions.

As summarized by Skalski and Robson (1992, p. 211), impact studies are among the most difficult to properly design and analyze. Impact assessments typically must include a temporal dimension to the design. Skalski and Robson (1992, p. 212–213) offered the following considerations that are unique to designing impact assessments:

- Identification of constraints imposed by the investigation with regard to randomization and replication
- Incorporation of all prior knowledge as to where, when, and how the impact is to occur (if known) into the design of the field investigation
- Expression of the impact hypothesis in statistical terms as a function of model parameters
- Use of a preliminary survey that is consistent with the objective of the consummate field design to estimate variance components for sample size calculations
- Evaluation of economic and inferential costs of conducting a constrained investigation relative to other design options
- Establishment of a field design whose spatial and temporal dimensions permit model-dependent estimates of effects of impact
- Where possible, conducting auxiliary investigations of stressors to provide ancillary data for establishing cause–effect relationship

References

Aebischer, N. J., P. A. Robertson, and R.E. Kenward. 1993. Compositional analysis of habitat use from animal radio-tracking data. Ecology 74:1313–1325.

Ahlbom, A. 1993. Biostatistics for Epidemiologists. Lewis, Boca Raton, FL.

Barker, D. J., and A. J. Hall. 1991. Practical Epidemiology, 4th Edition. Churchill Livingstone, London.

Bender, E. A., T. J. Case, and M. E. Gilpin. 1984. Perturbation experiments in community ecology: Theory and practice. Ecology 65: 1–13.

Carpenter, S. R., T. M. Frost, D. Heisey, and T. K. Kratz. 1989. Randomized intervention analysis and the interpretation of whole-ecosystem experiments. Ecology 70: 1142–1152.

DeMeo, Committee, c/o RESOLVE, Inc., Washington, DC.

Green, R. H. 1979. Sampling Design and Statistical Methods for Environmental Biologists. Wiley, New York, NY.

Hurlbert, S. J. 1984. Pseudoreplication and the design of ecological field experiments. Ecol. Monogr. 54: 187–211.

Manly, B. F. J., L. L. McDonald, D. L. Thomas, T. L. McDonald, and W. P. Erickson. 2002. Resource Selection by Animals: Statistical Design and Analysis for Field Studies, 2nd Edition. Kluwer Academic, Dordrecht, The Netherlands.

Mayer, L. S. 1996. The use of epidemiological measures to estimate the effects of adverse factors and preventive interventions. In Proceedings of National Avian-Wind Power Planning Meeting II, pp. 26–39. Avian Subcommittee of the National Wind Coordinating Committee. National Technical Information Service, Springfield, VA.

McDonald, T. L., W. P. Erickson, and L. L. McDonald. 2000. Analysis of count data from before–after control–impact studies. J. Agric. Biol. Environ. Stat. 5: 262–279.

Morrison, M. L., B. G. Marcot, and R. W. Mannan. 2006. Wildlife-habitat Relationships: Concepts and Applications, 3rd Edition. Island Press, Washington, DC.

Osenberg, C. W., R. J. Schmitt, S. J. Holbrook, K. E. Abu-Saba, and A. R. Flegal. 1994. Detection of environmental impacts: Natural variability, effect size, and power analysis. Ecol. Appl. 4: 16–30.

Parker, K. R., and J. A. Wiens. 2005. Assessing recovery following environmental accidents: Environmental variation, ecological assumptions, and strategies. Ecol. Appl. 15: 2037–2051.

Savereno, A. J., L. A. Saverno, R. Boettcher, and S. M. Haig. 1996. Avian behavior and mortality at power lines in coastal South Carolina. Wildl. Soc. Bull. 24: 636–648.

Skalski, J. R., and D. S. Robson. 1992. Techniques for wildlife Investigations: design and Analysis of Capture Data. Academic Press, San Diego, CA.

Stewart-Oaten, A., W. M. Murdoch, and K. R. Parker. 1986. Environmental impact assessment: Pseudoreplication in time? Ecology 67: 929–940.

Stewart-Oaten, A., J. R. Bence, and C. W. Osenberg. 1992. Assessing effects of unreplicated perturbations: No simple solutions. Ecology 73: 1396–1404.

Thomas, D. L., and E. Y. Taylor. 1990. Study designs and tests for comparing resource use and availability. J. Wildl. Manage. 54: 322–330.

Underwood, A. J. 1991. Beyond BACI: Experimental designs for detecting human environmental impacts on temporal variation in natural populations. Aust. J. Mar. Fresh. Res. 42: 569–587.

Underwood, A. J. 1992. Beyond BACI: The detection of environmental impacts on populations in the real, but variable, world. J. Exp. Mar. Biol. Ecol. 161: 145–178.

Underwood, A. J. 1994. On beyond BACI: Sampling designs that might reliably detect environmental disturbances. Ecol. Appl. 4: 3–15.

White, G. C., and R. A. Garrott. 1990. Analysis of wildlife radio-tracking data. Academic, San Diego, CA.

Wiens, J. A., and K. R. Parker. 1995. Analyzing the effects of accidental environmental impacts: Approaches and assumptions. Ecol. Appl. 5: 1069–1083.

Chapter 7
Inventory and Monitoring Studies

7.1 Introduction

Inventory and monitoring are probably the most frequently conducted wildlife studies. Not only are they conducted in the pursuit of new knowledge (e.g., to describe the fauna or habitats [see Sect. 1.5 for definition of habitat and related terms] of a given area, or understand trends or changes of selected parameters), but also they are cornerstones in the management of wildlife resources. In general terms, *inventories* are conducted to determine the distribution and composition of wildlife and wildlife habitats in areas where such information is lacking, and *monitoring* is typically used to understand rates of change or the effects of management practices on wildlife populations and habitats. In application to wildlife, inventory and monitoring are typically applied to species' habitats and populations. Because sampling population parameters can be costly, habitat is often monitored as a surrogate for monitoring populations directly. This is possible, however, only if a clear and direct linkage has been established between the two. By this, we mean that a close correspondence has been identified between key population parameters and one or more variables that comprise a species' habitat. Unfortunately, such clear linkages are lacking for most species.

The need for monitoring and inventory go well beyond simply a scientific pursuit. For example, requirements for monitoring are mandated by key legislation (e.g., National Forest Management Act [1976], National Environmental Policy Act [1969], Endangered Species Act [1973]), thus institutionalizing the need for conducting such studies. Even so, monitoring is embroiled in controversy. The controversy is not so much over the importance or need to conduct monitoring, but surrounds the inadequacy of many programs to implement scientifically credible monitoring programs (Morrison and Marcot 1995; White et al. 1999; Moir and Block 2001). Unfortunately, few inventory/monitoring studies are conducted at an appropriate level of rigor to precisely estimate the selected parameters. Given that inventory and monitoring are key steps in the management process and especially adaptive management (Walters 1986; Moir and Block, 2001), it is crucial to follow a credible, repeatable, and scientific process to provide reliable knowledge (cf. Romesburg 1981). The purpose of this chapter is to outline basic steps that should be followed for inventory and monitoring studies.

M.L. Morrison et al., *Wildlife Study Design.*

THEME: Mexican Spotted Owl (Strix occidentalis lucida)

Throughout this chapter, we will use a theme based on the Mexican spotted owl to illustrate inventory and monitoring concepts. The Mexican spotted owl is a less renown relative of the northern spotted owl (*S. o. caurina*). Like the northern spotted owl, the Mexican subspecies is listed as threatened under the Endangered Species Act, which prompted development of a recovery plan (USDI Fish and Wildlife Service 1995). Much of the information presented in this chapter is gleaned from that recovery plan and the deliberations underlying its development. Box 7.1 provides a brief summary of the salient points of the owl's ecology and management as they relate to points discussed in this chapter.

Box 7.1 Background on the Mexican Spotted Owl

Owl Ecology

Detailed reviews of various aspects of the owl's ecology are provided in the recovery plan (USDI Fish and Wildlife Service 1995). Our intent here is to present salient points about the owl that were key considerations in developing management recommendations. Although the Mexican spotted owl occupies a broad geographic range extending from Utah and Colorado south to the Eje Neovolcanico in Mexico, it occurs in disjunct localities that correspond to isolated mountain and canyon systems. The current distribution mimics its historical extent, with exception of its presumed extirpation from some historically occupied riparian ecosystems in Arizona and New Mexico. Of the areas occupied, the densest populations of owls are found in mixed-conifer forests, with lower numbers occupying pine-oak forests, encinal woodlands, rocky canyons, and other vegetation types. Habitat-use patterns vary throughout the range of the owl and with respect to owl activity. Much of the geographic variation in habitat use corresponds to differences in regional patterns of vegetation and prey availability. Forests used for roosting and nesting often exhibit mature or old-growth structure; specifically, they are uneven-aged, multistoried, of high canopy closure, and have large trees and snags. Little is known of foraging habitat, although it appears that large trees and decadence in the form of logs and snags are consistent components of forested foraging habitat. The quantity and distribution of potential owl habitat, specifically forests that possess relevant habitat correlates, is largely unknown. Existing data sets on forest structure are too variable in quality and in terms of coverage to permit even ballpark guesses.

With the exception of a few population demography studies, little is known of the population ecology of the Mexican spotted owl. The recovery team recognized the limitations of existing data and the inferences that could be drawn from them. Consequently, the recovery team reviewed and reanalyzed those data to estimate appropriate population parameters needed for development of the population monitoring approach that would provide more rigorous and defensible estimates of population trend.

Recovery Plan Management Recommendations

The recovery plan is cast as a three-legged stool with management recommendations as one of the three legs. Three areas of management are provided under the general recommendations: protected areas, restricted areas, and other forest and woodland types. Protected areas receive the highest level of protection. Guidelines for restricted areas are less specific and operate in conjunction with existing management guidelines. Specific guidelines are not proposed for other forest and woodland types.

Protected areas are all occupied nest or roost areas, mixed-conifer and some pine-oak forests with >40% slope where timber harvest has not occurred in the past 20 years, and all legally administered reserved lands (e. g., wilderness). Active management within protected areas should be solely to alleviate threats of catastrophic stand-replacing fires by using a combination of thinning small trees (<22 cm dbh) and prescribed fire.

Restricted areas include mixed-conifer forests, pine-oak forests, and riparian areas not included in protected areas. Management for the owl should focus on maintaining and enhancing selected restricted areas to become replacement nest and roost habitat, and abating risk of catastrophic fire in much of the restricted habitat. The amount of restricted area to be managed as replacement habitat varies with forest type and location, but ranges between 10 and 25% of the restricted area landscape. Thus, between 75 and 90% of restricted areas can be managed to address other resource objectives.

No specific guidelines are provided for other forest and woodland types – primarily ponderosa pine (*Pinus ponderosa*) and spruce-fir (*Picea* spp.-*Abies* spp.) forests, and pinyon-juniper (*Pinus* spp.-*Juniperus* spp.) and trembling aspen (*Populus tremuloides*) woodlands – outside of protected areas. However, some relevant management of these vegetation types may produce desirable results for owl recovery. Examples of extant guidelines include managing for landscape diversity, mimicking natural disturbance patterns, incorporating natural variation in stand conditions, retaining special habitat elements such as snags and large trees, and using fire appropriately.

In addition, some guidelines were proposed related to specific land use, such as grazing and recreation, and these guidelines apply to all management areas. The team recognized that effects of such activities on spotted owls are not well known, and advocated monitoring potential effects to provide a basis for more specific recommendations if warranted.

Because aspects of owl ecology, biogeography, and management practices varied geographically, the recovery team divided the range of the Mexican spotted owl into 11 recovery units: six in the United States and five in Mexico (Rinkevich et al. 1995). Recovery units were based on physiographic provinces, biotic regimes, perceived threats to owls or their habitat, administrative boundaries, and known patterns of owl distribution.

(continued)

Box 7.1 (continued)

By and large, the management recommendations allowed resource agencies considerable latitude in designing and implementing activities. The general philosophy of the team was to protect habitat where it existed, and to enhance habitat where appropriate. Whether or not the management recommendations are successful could be measured only through habitat and population monitoring, the other two legs of the stool. Without monitoring, there would be no empirical and objective basis for determining whether management guidelines led to desired outcomes and whether the owl should be delisted.

Delisting Criteria

Delisting the Mexican spotted owl will require meeting five specific criteria (USDI Fish and Wildlife Service 1995, pp. 76–77). Three of these criteria pertain to the entire US range of the owl, and two are recovery unit specific. The three range-wide delisting criteria are:

1. The populations in the Upper Gila Mountains, Basin and Range-East, and Basin and Range-West recovery units must be shown to be stable or increasing after 10 year of monitoring, using a design with power of 90% to detect a 20% decline with a Type I error rate of 0.05.
2. Scientifically valid habitat monitoring protocols are designed and implemented to assess (1) gross changes in habitat quality across the range of the Mexican spotted owl and (2) whether microhabitat modifications and trajectories within treated stands meet the intent of the Recovery Plan.
3. A long-term, US-range-wide management plan is in place to ensure appropriate management of the subspecies and adequate regulation of human activity over time.

Once these three criteria have been met, delisting may occur in any recovery unit that meets the final two criteria.

4. Threats to the Mexican spotted owl within the recovery unit are sufficiently moderated and/or regulated.
5. Habitat of a quality to sustain persistent populations is stable or increasing within the recovery unit.

Implicit to the delisting criteria is the need for reliable, defensible data to (1) assess population status, (2) habitat trends, and (3) develop long-term management guidelines. Without such information, the recovery team felt that risks to the threatened owl would be too great. As an example of the team's philosophy, we detail the population monitoring approach presented in the recovery plan, and discuss ramifications of failure to implement population monitoring (see Box 7.5).

7.2 Selection of Goals

Inventory and monitoring studies entail similar, but distinct processes. Although some steps are clearly identical, others are specific to the type of study being done (Fig. 7.1). The first step, which is universal to any study, is to clearly state the goals. For example, why conduct the study? What information is needed? How will the information be used in this or future management decisions? Clearly answering these questions will help to define a study design that addresses them adequately. That is, establishing inventory and monitoring goals is critical for defining what will be monitored (e.g., selected species or all species, population variables or habitat variables), setting target and threshold values, designing appropriate protocols for collecting data, and determining the appropriate methods for data analysis.

7.3 Basic Design Applications

Researchers and managers conduct inventory and monitoring to meet a variety of needs. These can range from basic needs such as characterizing species occurrence to monitoring effects of management activities on population status and trends of species of interest. We elaborate on basic applications of inventory and monitoring below.

7.3.1 *Inventory*

An inventory assesses the *state* or status of one or more resources. It should be designed to provide information on an environmental characteristic, such as the distribution, population, or habitat of a given species or suite of species. An inventory

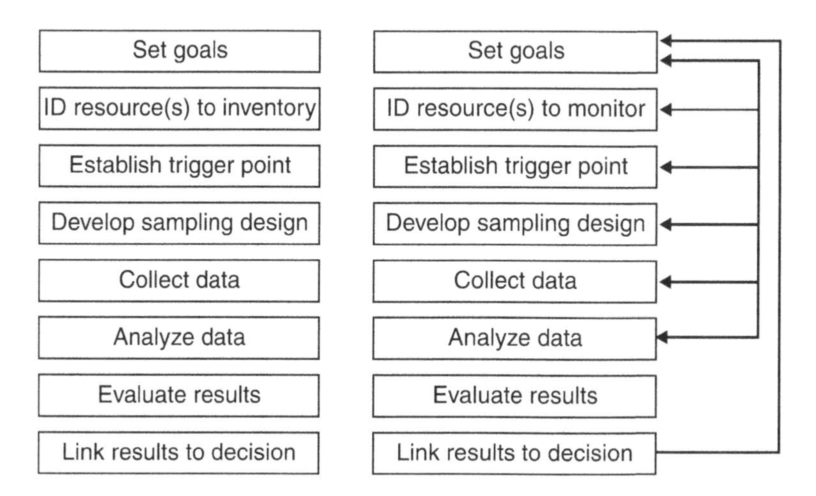

Fig. 7.1 Simplified sequence of steps involved with inventory and monitoring

provides a quantitative or qualitative description of a resource or set of resources for a given point or interval of time. Inventories are typically confined within a specific area or set of areas to determine the number and perhaps relative abundance of the species present, and they must be conducted at appropriate spatial scales and for appropriate durations depending on the resource(s) under study and the question(s) being asked. Inventories may take many years and require spatially extensive sampling to meet study goals. For example, inventories to estimate the density of rare species such as a far-ranging predator may require sampling more area than needed to estimate the density of a common species. Developing a list of breeding birds will require sampling only during the breeding season, whereas acquiring a list of all birds that use an area requires sampling year-round to record migrating and wintering birds. Further, sampling the breeding bird community will require a certain sampling effort (e.g., sampling points, duration) to provide an unbiased estimate of the species using an area. As an example, Block et al. (1994) used a bootstrap technique (Efron 1982) to estimate the number of counting stations and number of years needed to account for all species using oak woodlands during the spring breeding season (Fig. 7.2). They found that 56 counting stations sampling about 175 ha were required to record all species detected during a 2-year period, but

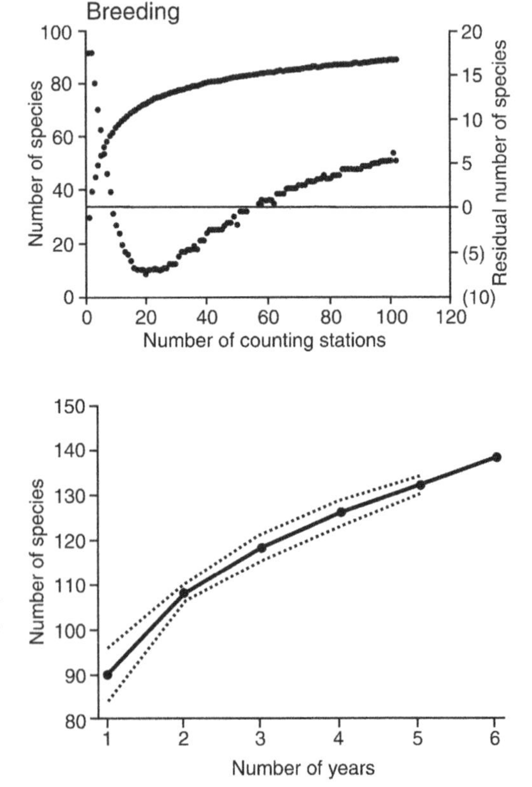

Fig. 7.2 Example of the number of (**a**) counting stations and (**b**) number of years to detect most birds using oak woodlands in California's Sierra Nevada foothills. Reproduced from Block et al. (1994), with kind permission from The Wildlife Society

that new species were being detected even after 6 years of sampling, likely because it is extremely difficult to detect all rare and incidental species.

Inventories are typically done in areas or conditions for which data are lacking, or across a range of areas or conditions to more clearly define ecological distribution of a species (e.g., define both presence and absence) (Heyer et al. 1994). A typical goal of inventories is to assess the presence or infer absence of species within an area prior to initiating a habitat-altering activity. Note that we state "infer absence." Verifying presence is straightforward; if you see or otherwise detect a species then it uses the area. However, failure to detect a species does not necessarily translate into it being absent when you sampled or that it never uses the area. This is where designing and implementing an appropriate study design is critical. The study design must be such that the probability of detecting a species or individuals using the area is high. Some important associated components are use of the proper field methodology and sampling during the appropriate period and with adequate intensity. We cannot reiterate these points enough because proper study design is the foundation of a valid wildlife study (cf. Chaps. 1 and 2).

Thus, inventories are critical tools to aid in resource planning and species conservation. Even basic information on the distribution of species and habitats in an area can then help design management to protect or enhance conditions for desired species, whether they are threatened or endangered species or those valued for consumptive or nonconsumptive reasons.

7.3.2 Monitoring

In contrast to inventory, monitoring assesses change or trend of one or more resources. The difference between change and trend is subtle in that change is evaluated by testing for differences between two points in time, whereas trend typically requires sampling for more than two occasions to evaluate the direction and consistency of change. Either way, both change and trend measure or index the dynamics as opposed to the state of a resource. Thus, monitoring requires repeated sampling of the variable(s) of interest to measure the change or trend. The variables measured and techniques used to measure them often overlap with those used for inventories. Some variables, however, may be unique to monitoring, especially those that measure rates, such as survival, and those that require repeated measures such as habitat succession. As with inventories, monitoring studies must be scaled to the variable and question being addressed. Thus, if one is assessing changes in forest structure and composition as they relate to a species' habitat, monitoring must be scaled temporally to vegetative processes. Monitoring a population to determine population trend must occur over a long enough time to be sure that the population has been subjected to an appropriate range of environmental variations. For example, populations studied during favorable weather conditions may exhibit positive trends, whereas those studied during unfavorable weather may show just the opposite. To guard against this potential bias, it is important to scale study duration long enough to include these variations.

Monitoring can include studies specifically to evaluate effects of a particular environmental treatment or impact on the resource of interest, or it could entail examining general trend without considerations of any specific activity. Impact assessment was discussed in detail in Chap. 6 so we refer you to that chapter for more detailed discussion of that particular topic. A more common monitoring study is to examine population or habitat trend, regardless of the causal factors. For example, is abundance of the Mexican spotted owl stable or increasing? Are macrohabitat and microhabitat of the owl stable or increasing? Answering these basic questions is key to the management process. If these trends are determined to be negative, one then could conduct more directed impact assessment monitoring to evaluate potential causal mechanisms.

Some broad objectives for conducting monitoring include (from Spellerberg 1991):

1. To provide guidance to wildlife management and conservation
2. To better integrate wildlife conservation and management with other land uses
3. To advance basic knowledge in addition to applied knowledge
4. To track potential problems before they become real problems

These objectives are often addressed by conducting monitoring studies (from Gray et al. 1996; Miller 1996):

1. To determine wildlife use of a particular resource (e.g., herbaceous forage) or area
2. To evaluate effects of land use on populations or habitats, measure changes in population parameters (e.g., size, density, survival, reproduction, turnover)
3. To evaluate success of predictive models
4. To assess faunal changes over time

Monitoring can be classified into four overlapping categories: implementation, effectiveness, validation, and compliance monitoring. *Implementation monitoring* is used to assess whether a directed management activity has been carried out as designed. For example, a prescribed fire is done as a habitat improvement project and the goal of the fire is to reduce fine ground fuels by 50%. Implementation monitoring would be done to evaluate whether that goal would be met. *Effectiveness monitoring* is used to evaluate whether or not the action met its stated objective. Say, for example, that the ultimate objective of the prescribed fire was to increase population numbers of the deer mouse (*Peromyscus maniculatus*). Effectiveness monitoring would involve a study to evaluate the response of the deer mouse population to the treatment. *Validation monitoring* is used to evaluate whether established management direction (e.g., National Forest Plans) provides guidance to meet its stated objectives (e.g., sustainable forest management) (Morrison and Marcot 1995). It is also used to test assumptions of models or prescriptions used to develop management decisions. This type of monitoring can be the most difficult to categorize as it often involves multiple resources and ambiguous goals. For example, forest sustainability is a laudable goal, but typically is nebulously defined. Determining exactly what to measure, and how to measure it, can be difficult indeed. On the other hand, management plans often contain specific and measurable criteria, such as the desired amount of forest in a given structural class (e.g., mature or old-growth forest), or the number of a given habitat element that should occur across the landscape. For these criteria, establishing a valid monitoring study is not nearly as challenging. *Compliance monitoring* is done when mandated by statute (see Sect. 1.3.2). An example of compliance is monitoring established within a biological opinion provided by the US Fish and Wildlife Service during interagency consultation under the Endangered Species Act. Typically, this monitoring, referred to as *take monitoring*, assesses whether an activity adversely affects the occupancy or habitat of a threatened or endangered species. If so, the action agency is charged with a "take," meaning that the activity had an adverse impact on a specified number of the species. To illustrate further the different types of monitoring, we draw upon our theme, the Mexican spotted owl (Box 7.2).

Box 7.2 Monitoring for a Threatened Species: The Mexican Spotted Owl

Different monitoring goals are illustrated in the spotted owl example. The extent to which management activities are actually applied on the ground and the degree to which those activities are in accord with recovery plan guidelines would be evaluated by implementation monitoring. For example, consider a silvicultural prescription with the ultimate objective of creating owl nesting habitat within 20 year (the criteria for nesting habitat were provided in the recovery plan). The prescription entailed decreasing tree basal area by 15% and changing the size class distribution of trees from one skewed toward smaller trees to an equal distribution of size classes. Further, the recovery plan specifies the retention of key correlates of owl habitat – trees >60 cm dbh, large snags, and large downed logs – during active management practices such as logging and prescribed burning. In this case, implementation monitoring must have two primary objectives. One is to determine if losses of key habitat elements exceeded acceptable levels, and the second is to determine if tree basal area was reduced as planned and the resultant size class distribution of trees was even. Recall that the ultimate objective of the treatment was to produce a stand in 20 year that had attributes of owl nesting habitat. Whether or not the prescription achieved this objective is the goal of effectiveness monitoring.

The owl recovery plan (USDI Fish and Wildlife Service 1995) provided five delisting criteria that must be met before the owl should be removed from the list of threatened and endangered species. One criterion was to demonstrate that the three "core populations" were stable or increasing, and another required habitat stability across the range of the subspecies. The recovery plan became official guidance for the US Fish and Wildlife Service, and then for the US Forest Service as they amended Forest Plans for all forests in the southwestern region to incorporate the recovery plan recommendations (USDA Forest Service 1996). For a little background, National Forests are mandated to develop Forest Plans by the National Forest Management Act (NFMA) of 1976, thus making

(continued)

Box 7.2 (continued)

Forest Plans a legal requirement. Guidance in Forest Plans is provided by a series of standards and guidelines, which must be followed in planning and conducting management activities. The ultimate goal of Forest Plans with respect to the Mexican spotted owl was to implement the recovery plan, and ultimately delist the owl. Whether or not implementing Forest Plans provides conditions for a viable population of owls and results in delisting is measured through validation monitoring. Two tangible measures for the owl would be to demonstrate that both owl habitat and owl populations were stable or increasing.

Compliance monitoring is done as part of the terms and conditions set forth in a biological opinion resulting from interagency consultation. For example, a form of compliance monitoring would be to monitor for "take" of owls or habitat. Take of owls could be assessed by abandonment of a territory or change in reproductive output. Take of habitat would involve reduction of key habitat components below some minimum threshold.

Monitoring can be used to measure natural or intrinsic rates of change over time or to understand effects of anthropogenic or extrinsic factors on population or habitat change or trends. By intrinsic changes, we refer to those that might occur in the absence of human impact, such as trends or changes resulting from natural processes (e.g., succession) or disturbances (fire, weather, etc.) (Franklin 1989). Anthropogenic factors are those that may alter or disrupt natural processes and disturbances and potentially affect wildlife habitats or populations. In most management situations, monitoring is conducted to understand effects of anthropogenic factors (e.g., water diversions, livestock, logging, fire suppression) on wildlife. However, recognizing trends even in the absence of anthropogenic factors is complicated by the dynamic and often chaotic behavior of ecological systems (Allen and Hoekstra 1992). Because intrinsic and extrinsic factors more often than not act synergistically to influence trend or change, the effects of either may be difficult to distinguish (Noon et al. 1999). Again, this is where application of an appropriate study design plan is critically important. A well-conceived and well-executed study may allow the investigator to partition sources of variation and narrow the list of possible factors influencing identified trends (see previous chapters).

7.4 Statistical Considerations

A premise underlying most of what we present in this volume is that study designs must permit valid treatment of the data. For inventory studies, we must be able to characterize accurately the species or habitat variables of interest. For monitoring, we must know the effort needed to show a trend over time or to document a specified effect size in a parameter from time t_1 to t_2.

In this regard, the investigator should be well aware of concepts of statistical power, effect size, and sample size, and how they interact with Type I and Type II errors (see Chaps. 2 and 3 for detailed discussion of these concepts). Typically, investigators focus on the Type I error rate or alpha. However, in the case of sensitive, threatened, endangered, or rare species, consideration of Type II error rate is equally, if not more, relevant. A Type II error would be failure to detect a difference when it indeed occurred, an error that should be kept to a minimum. With threatened, endangered, or rare species, overreaction and concluding a negative impact or negative population trend when it is not occurring (Type I error) may have no deleterious effects on the species because additional protections would be invoked to guard against any negative management actions. In contrast, failing to conclude a significant decline in abundance when it is occurring (Type II error) may allow management to proceed without change even though some practices are deleterious to the species. The potential risk to the species could be substantial.

7.4.1 Effect Size and Power

Effect size and power go hand in hand when designing a monitoring study. Simply stated, *effect size* is a measure of the difference between two groups. This difference can be quantified a number of ways using various indices that measure the magnitude of a treatment effect. Steidl et al. (1997) regarded effect size as the absolute difference between two populations in a select parameter. Typically, investigators establish effect a priori and should be the minimum level that makes biological difference. For example, a population decline of 10% for a species of concern might be biologically relevant, so you would need a study with adequate sensitivity to show that decline when it occurs.

Three common measures of effects size are Cohen's *d*, Hedges' *g*, and Cohen's f^2 (Cohen 1988, 1992; Hedges and Olkin 1985). Cohen's *d* measures the effect size between two means, where *d* is defined as the difference between two means divided by the pooled standard deviation of those means. To interpret this index, Cohen (1992) suggested that $d = 0.2$ indicates a small, 0.5 a medium, and 0.8 a large effect size. Hedges' $\hat{g}$ incorporates sample size by both computing a denominator which looks at the sample sizes of the respective standard deviations and also makes an adjustment to the overall effect size based on this sample size. Cohen's f^2 is analagous to an *F* test for multiple correlation or multiple regression. With this index, f^2 of 0.02 is considered a small effect size, 0.15 is medium, and 0.35 is large (Cohen 1988).

Simply stated, *statisical power* is the probability that you will correctly reject a null hypothesis (Steidl et al. 1997). Recall from Chap. 2 that failure to reject correctly the null hypothesis is termed Type II error. As power increases, Type II error decreases. Power analysis can be done before (prospective) or after (retrospective) data are collected. Preferably, a researcher conducts prospective power analysis to determine sample sizes needed to have adequate power to detect the effect size of interest. The strength of this approach is that you can evaluate the interactions among power, effect size, and sample size to evaluate what is attainable. Stedl et al. (1997) provided an example of such analysis for two common birds species – hairy woodpecker (*Picoides villosus*) and chestnut-backed chickadee (*Poecile refuscens*) – found in Oregon forests (Fig. 7.3). They generated four curves for each species corresponding population increases of 50, 100, 150, and 200% across 3–9 replicate treatment pairs (treated and untreated). They applied the general rule that power >0.80 was acceptable. That was not achieved until there were at least eight replicates for the woodpecker and, even then, there was adequate power to detect only a 150% increase in the population (effect size). Had the population increased only 50%, a study with eight replicates would have been insufficient. By comparison, the more common chickadee required fewer replicates (seven) to detect a smaller increase (100%) in its population. Unfortunately, populations rarely show this level of response to habitat change caused by management unless, of course, the change

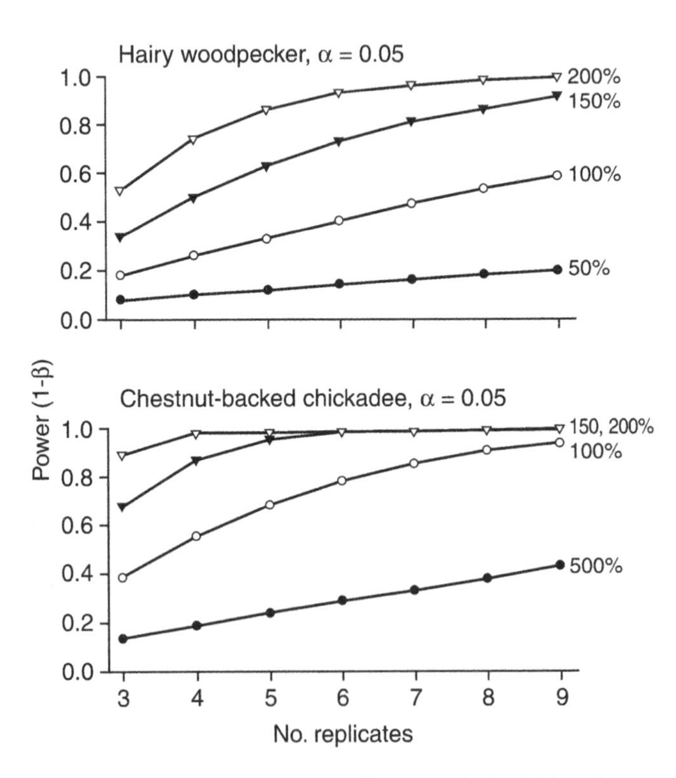

Fig. 7.3 Power analysis for hairy woodpecker and chestnut-backed chickadee to evaluate number of replicates needed to detect population increases of 50, 100, 150, and 200%. Reproduced from Steidl et al. (1997), with kind permission from The Wildlife Society

is severe. Thus, we are interested in more subtle population changes, which may go undetected given this experimental design.

It has become common practice to conduct retrospective power analysis in situations where results of a test are nonsignificant. Basically, such tests are used more as a diagnostic tool to evaluate what effects size might have been detected given a certain power, or vice versa, what power might be achieved given certain effect size or sample size. Steidl et al. (1997) caution about taking the results of retrospective power analyses too far. Effectively, their primary use is to evaluate hypothetical scenarios that may help to inform similar studies conducted sometime in the future. In some cases, they might also be used to test hypothesized effects sizes thought to be biologically relevant or to calculate confidence intervals around the observed effect size (Hayes and Steidel 1997; Thomas 1997).

7.4.2 Balancing Response Variables with Study Goals

Resources can be measured directly or indirectly. For example, if the study is to address the effects of a management activity on population trend of a species, then a direct approach would involve measuring the appropriate population attribute, such as abundance or density. However, populations of many species or other ecosystem attributes are difficult to sample because of their rarity or secretiveness that precludes obtaining enough samples even with a huge effort. In these cases, investigators often resort to indirect measures. These can include indices, indicator species, and stressors. Indirect measures should only be used if a clear and interpretable relationship has been established between the resource being studied and the surrogate measure (Landres et al. 1988).

Direct measures are variables that link clearly and directly to the question of interest. If they exist and are feasible to obtain, direct measures are preferred over indirect measures. Concerning inventories of species presence or faunal composition for community surveys, direct measures are used to assess presence or infer absence of the species of interest. Direct measures for populations can be measures of abundance, density, or of other population parameters of interest (e.g., survival, reproduction). Inventories or monitoring of habitats often focus on variables established as strong correlates of use by a species, or strong correlates to some measure of fitness.

Indirect measures are widely used for inventory and monitoring studies. *Indicator species* are used to index or represent specific environmental conditions or the population status of another ecologically similar species. They can be divided into two major categories: ecological indicators and management indicators. This concept was initially proposed by Clements (1920) to explain plant distributions based on specific environmental conditions, primarily soil and precipitation. Vertebrates are also tied to specific environmental conditions as this is the basis for describing species' habitats (Block and Brennan 1993). Many wildlife species, however, are vagile, and can adjust to variations in environmental conditions simply by moving or migrating. Thus, relationships between environmental conditions and most wildlife species may not be quite as strong as they are for many plants, and their predictive value of environmental conditions may be limited (Morrison 1986). If indicators are used, they should meet rigorous standards (Landres et al. 1988). These include (1) clearly stating what the indicator indicates about the environment or resource, (2) selection of indicators should be objective and quantitative, (3) all monitoring programs using indicators should be reviewed (a standard that should apply to all monitoring, not just indicators), and (4) indicators must be used at the appropriate spatial and temporal scales. Thus, use of indicators should not be simply a matter of convenience, but must be based on strong empirical evidence that supports their usage.

Stressors are another group of surrogate variables that can be measured in lieu of measuring a resource directly. *Stressors* are natural and anthropogenic events that affect resource distribution or abundance. Examples of stressors are loss of late seral forest due to fire; alterations of hydrologic regimes by water diversions; reduction, loss, or fragmentation of habitat; increased sediment loads following storms; or overharvesting of game or commercial species (Noon et al. 1999). Thus, rather than inventorying or monitoring a population or habitat directly, inferences are made based on some metric applied to the stressor. As with indicator species, the validity of stressors and their relationships to the variables of interest must be firmly established prior to their use.

As mentioned earlier, habitat is often monitored as a surrogate for monitoring an animal population directly. Costs of monitoring a population sufficiently to have acceptable statistical power to detect a trend can be rather high (Verner 1983). The estimated annual costs for conducting population monitoring for the Mexican spotted owl, for example, was about $1.2 to 1.5 million (USDI Fish and Wildlife Service 1995). When projected for 10–15 years, the costs could exceed $20 million for just this one subspecies! Consequently, macrohabitat or microhabitat (sensu Block and Brennan 1993) is often monitored to index population trend for a species. Unfortunately, information that documents clear and strong relationships between habitat components and population trend is lacking for most species. Thus, caution is advised when extrapolating habitat trends to populations.

If an indicator or stressor is monitored, then justification based on previous work must be provided to demonstrate that the variable studied is a good measure of ecosystem status or health. If the literature is unclear and cannot support the selection of a surrogate for study, then you should conduct a pilot study to test whether or not the variable you select measures what you intend it to, or abandon use of a surrogate and monitor the variable of interest directly. We recognize, however, that the use of surrogates such as indicators or stressors in monitoring, specifically their applicability and validity, is the subject of debate (Morrison 1986; Landres et al. 1988).

Another group of indirect measures or indices is *community metrics*. These indices provide little information about individual species, but provide quantitative values that are related to numbers, degree of association, diversity, and evenness of species (see Sect. 1.5.2). They can be applied to animals and their habitats.

Whereas *species richness* is a fairly straightforward concept in that it is simply a count of the number of species present, numerous algorithms are available for estimating degrees of association, diversity, and evenness (Hayek 1994; Pielou 1977). Measures of association include similarity coefficients, matching coefficients, and more traditional association coefficients (Hohn 1976; Hayek 1994). Similarity (e.g., Sorensen (1948) or Jaccard (1901)) and matching coefficients (e.g., Sokal and Michener (1958)) are not test statistics and are not based on a presumed sampling distribution. At best, they can be used in qualitative comparisons between different areas or comparisons of the same place but at different times. Traditional association coefficients include chi-square and contingency statistics, and can be evaluated against a probability distribution. Hayek (1994) reviewed various measures of species diversity and concluded that the concept is "variously and chaotically defined in the literature." Generally, measures include estimates of species richness and evenness. Evenness refers to the distribution of individuals among species. Differences among diversity algorithms often relate to how they weight diversity and evenness in calculation of their index value. A plethora of algorithms has been proposed; the two most often used are Shannon–Weiner and Simpson's indices. Often, it is difficult or impossible to ascribe a biological interpretation to diversity indices because nobody really knows what they measure. Thus, we recommend caution in using these indices as valid measures for inventory and monitoring studies.

7.5 Distinguishing Inventory from Monitoring

The answer to the question of what makes inventorying and monitoring different is basic. The difference between the two is largely a function of time; inventory measures the status of a resource at a point in time, whereas monitoring assesses change or trend over time in resource abundance or condition. Inventory and monitoring follow different processes to meet their goals, especially the series of feedback loops inherent to monitoring (see Fig. 7.1). Both require that you set goals, identify what to measure, and, in the case of management, state a value that when exceeded will result in a management decision. However, because inventory is to assess resource state whereas monitoring is to assess resource dynamics, they will often require different study designs. For example, the sampling design for a study to inventory Arizona to determine the distribution of spotted owls would be much different from a study to monitor population trend. Each study would be designed to estimate different parameters and would entail application of different statistical procedures, thus requiring different approaches to collect the relevant data. One basic principle common to both inventory and monitoring is that both should be scientifically valid. Thus, concepts discussed in Chaps. 1 and 2 regarding adequate sample sizes, randomization, replication, and general study rigor are critically important to any inventory or monitoring study. Failure to incorporate these considerations will result in misleading information, and potentially inappropriate conclusions and deleterious management decisions. To provide an example of how the goals of inventory and monitoring differ, consider the inventory and monitoring goals presented below in the Mexican spotted owl recovery plan (Box 7.3).

As we can see from this example, goals of inventory and monitoring can be quite different. Inventories are often done with the goal of assessing the status of a species

Box 7.3 Inventory and Monitoring Goals for the Mexican Spotted Owl

Inventories are used in two basic ways for the Mexican spotted owl. One is part of project planning and the other is to increase basic knowledge about owl distribution. The Mexican Spotted Owl Recovery Plan requires that all areas with any chance of occupancy by owls be inventoried prior to initiating any habitat-altering activity. The reason why is to determine if owls are using the area and if so, to modify the management activity if necessary to minimize impact to the bird. Thus the goal is straightforward: to determine occupancy (or infer nonoccupancy) of owls to help guide the types and severity of habitat-modifying management that might impact the owl. The second goal of inventory is to understand the distribution of the owl better. Most inventories for owls have been conducted in areas where management (typically timber harvest and prescribed fire) is planned as part of the process described above. These areas represent only a subset of the lands that the owl inhabits. Thus, to increase knowledge of owl distribution and population size, the plan calls for inventories in "holes in the distribution" or in potential habitats where no records of owls exist.

The recovery plan also requires both population and habitat monitoring. The reasons for monitoring are multifaceted. First, the owl was listed as threatened based on loss of habitat and the concern that habitat would continue to be lost given current management practices. Although not explicitly stated in listing documents, it was assumed that there was a population decline concomitant with habitat decline. Thus, a very basic reason to monitor is to evaluate whether or not these trends are indeed occurring and if they are correlated. A second objective for monitoring is to evaluate whether or not implementation of management recommendations in the recovery plan were accomplishing their intended goal, namely recovering the subspecies. This would entail (1) implementation monitoring to determine if management activities were done as designed and (2) effectiveness monitoring to evaluate whether following management recommendations is sustaining owl populations and habitats. This would be tested by examining both habitat and population trends to ensure that owl populations persist into the future. A third objective of monitoring, validation, would provide measurable, quantitative benchmarks that when met would allow the bird to be removed from the list of threatened species (i.e., delisted).

on an area planned for management activities. By status, we mean presence/absence, abundance, density, or distribution. With threatened or endangered species such as the owl, inventories are often used to permit modify, or curtail habitat-altering activities. Other goals of inventory might be to document species presence within a management unit to aid in resource planning or conservation (Hunter 1991; Scott et al. 1993), or evaluate habitat suitability of an area for a given species to determine if it has the potential for occupancy (Verner et al. 1986), or the goal might be simply for increasing scientific knowledge by inventorying new areas and documenting species that were previously undescribed. Certainly faunal inventories by early naturalists such as Wallace, Darwin, Audubon, Xantu, and others provided key baseline information for addressing many interesting and complicated ecological questions.

7.6 Selection of a Design

Monitoring and inventory projects require an adequate sampling design to ensure unbiased and precise measures of the resource(s) of interest. To do so requires a priori knowledge of the resource under study, including its behavior, distribution, biology, and abundance patterns (Thompson et al. 1998). It is also necessary to understand the statistical properties of the population from which a sample is to be taken. Once these basic properties are known, the investigator must determine the appropriate sampling methodology to meet inventory or monitoring objectives, given available funds and personnel.

A sampling design for an inventory or monitoring study consists of four interrelated components (see Morrison et al. 1998 for detailed discussion). An investigator must first decide *what* it is that he or she wants to measure, *where* to sample (the sampling universe), *when* to study (timing and length of time), and, finally, *how* to collect data. We discuss these components below.

7.6.1 Identifying the Resources to Be Measured

Selecting the variables to measure should be supported by previous knowledge or established information. Hopefully, the investigator possesses a certain expertise in the species or system being inventoried or monitored and can draw on that knowledge to select variables or specific resources to study. Often this is not the case and the investigator will need to do some background work, such as a literature review, consulting with established experts, or using results of similar studies to establish the basis for measuring a given variable(s).

When monitoring populations, it is important to determine the parameters most sensitive to change and focus on those. Typically, investigators focus on population abundance or density of breeding individuals. This might be misleading, however, if there exists a large number of nonterritorial animals (e.g., nonbreeding individuals) not easily sampled using traditional methods (e.g., auditory surveys). In this case, it is possible that you can have high mortality of territorial animals that are immediately replaced by surplus, floating individuals. The population of territorial animals may appear stable while the over all population is declining. Information on the age of initial territorial occupancy or the age class distribution might be needed to more fully understand the status of the population. Again, the point here is that you must understand the biology and population dynamics of the species being monitored to make better decisions on exactly what to monitor.

7.6.2 Selection of Sampling Areas

Once the study objective is established, the scale of resolution chosen by ecologists is perhaps the most important decision in inventory and monitoring because it predetermines procedures, observations, and results (Green 1979; Hurlbert 1984). A major step in designing an inventory or monitoring study is to establish clearly the target population and the sampling frame. Defining the target population essentially defines the area to be sampled. For example, if an area was to be inventoried to determine the presence of Mexican spotted owls on a national forest, sampling areas should include general areas that the owl uses (mature conifer forests and slickrock canyons) but not include areas that the owl presumably would not use (grasslands, desert scrub) based on previous studies. This first step establishes the sampling universe from which samples can be drawn and the extent to which inferences can be extrapolated. Thus, the results of these owl surveys apply only to the particular national forest and not to all national forests within the geographic range of the owl.

Although this seems rather straightforward, the mobility of wildlife can muddle the inferences drawn from the established area. Consider, for example, the case of the golden eagle example presented in Chap. 6. A somewhat arbitrary decision was made to define the "population" potentially affected by wind turbine mortality as the birds found within a fixed radius of 30 km of the wind farm. The basis for this decision included information on habitat use patterns, range sizes, movement patterns, and logistics of sampling a large area. The primary assumption is that birds within this radius have the greatest potential of encountering wind turbines and are the birds most likely to be affected. Not measured, however, were cascading effects that may impact eagles beyond the 30 km radius, because eagles found within this radius were not a distinct population. Thus, factors that influenced birds within this arbitrary boundary may have also affected those outside of the boundary. The point here is that even though considerable thought went into the decision of defining the sampling universe for this study, the results of the monitoring efforts may be open to question because mortality of birds within the 30 km may be affecting the larger population, including birds found beyond the 30 km radius.

7.6.3 Study Duration

A key aspect in the design of any study is to identify when to collect data. There are two parts to this aspect: the timing of data collection and the length of time over which data should be taken. The choice of timing and length of study is influenced by the biology of the organism, the objectives of the study, intrinsic and extrinsic factors that influence the parameter(s) to be estimated, and resources available to conduct the study. Overarching these considerations is the need to sample adequately for precise estimates of the parameter of interest.

Timing refers to when to collect data and it depends on numerous considerations. Obviously, studies of breeding animals should be conducted during the breeding season, studies of migrating animals during the migration period, and so on. Within a season, timing can be critically important because detectability of individuals can change for different activities or during different phenological phases. Male passerine birds, for example, are generally more conspicuous during the early part of the breeding when they are displaying as part of courtship and territorial defense activities. Detection probabilities for many species will be greater during this period than at other times. Another consideration is that the vary population under study can change within a season. For example, age class structures and numbers of individuals change during the course of the breeding season as juveniles fledge from nests and become a more entrenched part of the population. Population estimates for a species, therefore, may differ substantially depending on when data are collected. Once the decision is made as to when to collect data, it is crucial that data are collected during the same time in the phenology of the species during subsequent years to control for some of the within season variation.

Objectives of a study also dictate when data should be collected. If the study is an inventory to determine the presence of species breeding in an area, sampling should occur throughout the breeding season to account for asynchrony in breeding cycles and heterogeneity in detectabilities among species. Sampling spread over the course of the season would give a greater chance of recording most of the species using the area. If a monitoring study is being conducted to evaluate population trend of a species based on a demographic model, sampling should be done at the appropriate time to ensure unbiased estimates of the relevant population parameters. Demographic models typically require fecundity and survival data to estimate the finite rate of population increase. Sampling for each of these parameters may be necessary during distinct times to ensure unbiased estimates for the respective measures (USDI Fish and Wildlife Service 1995).

Length of the study refers to how long a study must be done to estimate the parameter of interest. It depends on a number of factors including study objectives, field methodology, ecosystem processes, biology of the species, budget, and feasibility. A primary consideration for monitoring and inventory studies should be temporal qualities of the ecological process or state being measured (e.g., population cycles, successional patterns). Temporal qualities include frequency, magnitude, and regularity, which are influenced by both biotic and abiotic factors acting both stochastically and deterministically (Franklin 1989). Further, animals are subjected to various environmental influences during their lifetimes. A study should engage in data collection over a sufficiently long period to allow the population(s) under study to be subjected to a reasonable range of environmental conditions. Consider two hypothetical wildlife populations that exhibit cyclic behaviors, one that cycles on average ten times per 20 years, and the other exhibiting a complete cycle just once every 20 year (Fig. 7.4). The population cycles are the results of various intrinsic and extrinsic factors that influence population growth and decline. A monitoring program established to sample both populations over a 10-year period may be adequate to understand population trends in the species with frequent cycles, but may be misleading for the species with the long population cycle. Likely, a longer timeframe would be needed to monitor the population of species with the lower frequency cycles.

However, considering only the frequency of population cycles may be inadequate as the amplitude or magnitude of population shifts may also influence the length of a study to sort out effects within year variation from between year variation. Consider two populations that exhibit ten cycles in 20 years, but now the magnitude of the change for one is twice that of the other (see Fig. 7.4). Sampling the population exhibiting greater variation would require a longer period to detect a population trend or effect size should one indeed occur.

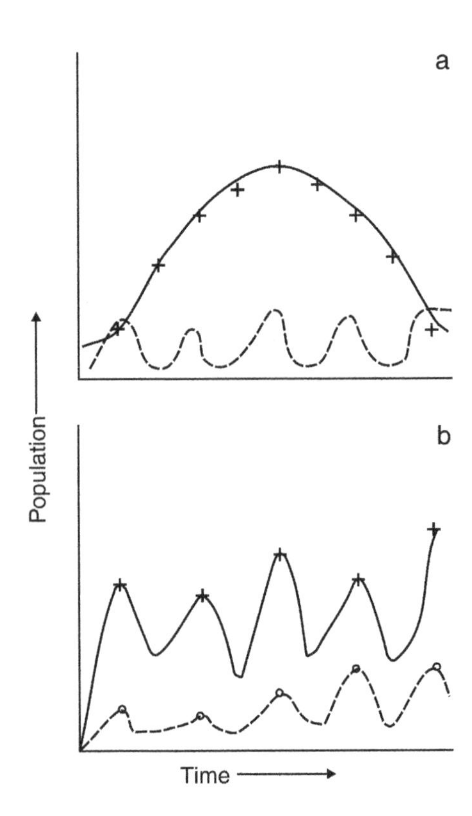

Fig. 7.4 Theoretical population cycles comparing (**a**) species with high (*dashed line*) and low (*solid line*) frequency cycles, and (**b**) species of low-amplitude (*dashed line*) and high-amplitude (*solid line*) population cycles

Typically, biologists do not have a wealth of information to draw upon prior to deciding the duration of study. In these situations, they must draw upon the best available information, and perhaps structure the study to adjust the length, as data are collected. In the case of Mexican spotted owls, for example, a wealth of information was available from both the literature and unpublished reports for developing population monitoring. Based on this information, a period of 10–15 years to delist the owl and 5 year postdelisting was chosen as the period for population monitoring (USDI Fish and Wildlife Service 1995). The basis for 10 years until delisting was that this would be ample time for 70% of the adult population to turn over, and that 10 years would allow the owl population to be subjected to variations in environmental factors that might influence its population. The additional 5 year of postdelisting monitoring would provide additional time to reaffirm the trend measured after 10 years. If the population trend is negative after 10 years of monitoring, the birds would not be delisted and monitoring should continue. The point here is that the length of study must have a biological basis. Failure to conduct a study for an adequate length of time might lead to erroneous conclusions of trends or effects.

In reality, however, costs, personnel, logistical constraints, and shifting priorities add a great deal of difficulty to first committing to and then continuing monitoring over the long term (Morrison and Marcot 1995; Moir and Block 2001; White et al. 1999). Consequently, innovative approaches are required to attempt to achieve unbiased results from suboptimal monitoring designs. The compromise typically made is to find alternatives to long-term studies. These approaches are discussed in Sect. 7.7.

7.6.4 Monitoring Occupancy vs. Abundance

Gathering abundance and demographic data can be costly, entail extensive field sampling, and require highly skilled personnel. Cost is higher largely because of the number of the samples needed for precise point estimates of the relevant parameters. Often, these costs are beyond the budget of many funding agencies, thus requiring more cost-effective approaches. Even if cost is not the primary constraint, the feasibility of obtaining enough samples to estimate abundance for rare species may be limiting.

New advances for estimating detection probabilities and using this information to adjust occupancy rates are largely responsible for the renewed interest in occupancy monitoring. In addition, one can model covariates as they relate to occupancy rates. These models can serve as descriptive tools to explain variation in occupancy rates. Although occupancy estimation is not new and is the basis for numerous indices, it has gone through a recent resurgence as a viable monitoring approach, especially for rare and elusive species (MacKenzie et al. 2004). Generally, occupancy monitoring is cost-efficient, can employ various indirect signs of occupancy, and does not always require as highly skilled personnel. Occupancy can be useful for a number of different studies including those investigating metapopulation structures, changes in geographic distribution, patch use, and species diversity patterns. However, occupancy does not convey the same information as abundance or density estimates. Hopefully, occupancy will index abundance but those relationships are likely species, time, and location specific.

Ganey et al. (2004) evaluated the feasibility of implementing the mark–recapture design for monitoring Mexican spotted owls presented in the recovery plan for this subspecies. Their evaluation included logistical aspects of implementing the study and statistical considerations of the sampling effort needed to show population decline. They concluded that the expense and personnel needs to conduct mark-recapture monitoring were daunting. More troublesome, however, was that random variation in the population was so great that it was difficult to ascribe a 20% decline in the population to anything more than chance. Given high costs and logistical hurdles of implementing this approach, the Mexican Spotted Owl Recovery Team revised their approach to population monitoring by focusing on occupancy.

7.6.5 Sampling Strategies

We focus extensively on sampling design and applications in Chaps. 4 and 5, so we will not repeat them here. Clearly, the design and execution of monitoring and inventory studies depends on the same basic considerations as other studies.

In some cases, the sampling universe is small enough to permit a complete enumeration (e.g., census) of the entire area. More typically, the entire sampling universe cannot be surveyed, thus you need to establish sample plots. Primary considerations with establishing plots are (1) their size and shape, (2) the number needed, and (3) how to place them within the sampling universe (See Chap 2). Size and shape of plots depend on numerous factors, such as the method used to collect data, biological edge effects, distribution of the species under study, biology of the species, and logistics of collecting the data. Thompson et al. (1998, pp. 44–48) summarize the primary considerations and tradeoffs in choosing a plot design. For example, long and narrow plots may allow for more precise estimates, but square plots will have less edge effect. They concluded that no single design is optimal for all situations, and they suggested trying several in a pilot study. Plot size depends largely on the biology and distribution of the species under study. Larger plot sizes are needed for species with larger home ranges and for species with clumped distributions. For example, larger plots would be needed to survey the spotted owl (home range size about 800 ha) than would be needed for the dark-eyed junco (*Junco hyemalis*) (home range about 1 ha). Further, larger plots are needed for species with clumped distributions, such as quail, than might be needed for species with more even distributions, such as the plain titmouse (Fig. 7.5). Note that the species in Fig. 7.5b will not be sampled adequately using the same plot size as used for the species in Fig. 7.5a. Larger-sized plots will be needed to sample the species with the clumped distribution (Fig. 7.5b).

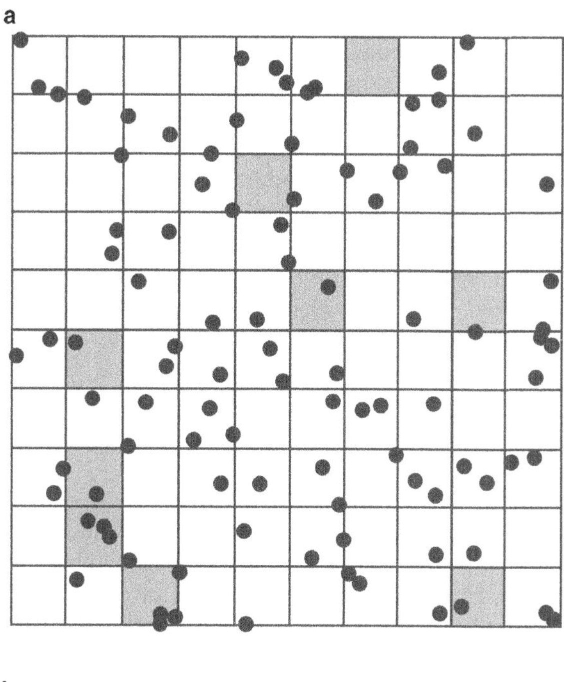

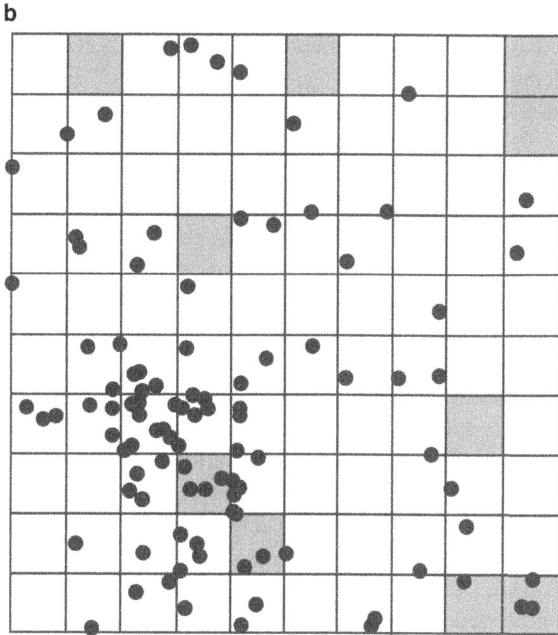

Fig. 7.5 Simple random samples of ten plots (gray plots) from sampling frames containing (**a**) a random distribution of individuals and (**b**) a clumped distribution of individuals. Reproduced from Thompson et al. (1998), with kind permission from Elsevier

The number of sample plots and placement of plots within the study area depend on a number of sampling considerations, including sampling variances and species distributions and abundances. Sample size should be defined by the number of plots to provide precise estimates of the parameter of interest. Allocation of sample plots should try to minimize sampling variances and can be done a number of ways. Survey sampling textbooks are a good source of discussion of the theoretical and practical considerations. Basic sampling designs include simple random, systematic random, stratified random, – cluster sampling, two-stage cluster sampling, and ratio estimators (Thompson 2002; Cochran 1977). Chapters 4 and 5 presented some of these basic sampling designs with examples of how they are typically applied.

7.6.6 Use of Indices

Historically, wildlife biologists have made heavy use of indices as surrogates for measuring populations. These can include raw counts, auditory counts, track surveys, pellets counts, browse sign, capture per unit of effort, and hunter success. Indices are often used to address inventory and monitoring questions (see Sect. 7.1). Implicit to indices is that they provide an unbiased estimate of the relative abundance of the species under study. This assumption, however, rests heavily on the assumption that capture probabilities are homogeneous across time, places, and observers (Anderson 2001).

Although indices are widely used, they are not widely accepted (Anderson 2001; Engeman 2003). Primary criticisms are that they fail to account for heterogeneous detection probabilities (Anderson 2001), employ convenience samples which are not probabilistic samples (Anderson 2001, 2003) typically lack measures of precision (Rosenstock et al. 2002), and when provided they have large confidence intervals (Sharp et al. 2001).

However, few investigators have enough resources to feed the data hungry analyses that permit raw counts to be adjusted by detection probabilities (Engeman 2003), thereby relegating investigators to using indices. McKelvey and Pearson (2001) noted that 98% of the small mammal studies published in a 5-year period had too few data for valid mark–recapture estimation. Verner and Ritter (1985) found that simple counts of birds were highly correlated with adjusted counts, but simple counts were possible for all species whereas adjusted counts were possible only for common species with enough detection.

Index methods are efficient and their use will likely continue (Engeman 2003). Engeman (2003) notes that the issue with indices is not so much the method as it is with selecting and executing an appropriate study design and conducting data analysis to meet the study objective. Methods exist to calibrate indices by using ratio estimation techniques (see Chap 5; Eberhardt and Simmons 1987), double sampling techniques (Bart et al. 2004), or detection probabilities (White 2005). These calibration or correction tools may reduce bias associated with indices and render indices more acceptable as inventory and monitoring tools.

7.7 Alternatives to Long-Term Studies

Four phenomena necessitate long-term studies (1) slow processes, such as forest succession or some vertebrate population cycles, (2) rare events, such as fire, floods, diseases, (3) subtle processes where short-term variation exceeds the long-term trend, and (4) complex phenomena, such as intricate ecological relationships (Strayer et al. 1986). Unfortunately, needs for timely answers, costs, changing priorities, and logistical considerations may preclude long-term studies. In such cases, alternative approaches are sought to address inventory or monitoring objectives. Various alternatives to long-term sampling have been proposed, such as retrospective sampling (Davis 1989), substitution of space for time (Pickett 1989), the use of systems with fast dynamics as analogies for those with slow dynamics (Strayer et al. 1986), modeling (Shugart 1989), and genetic approaches (Schwartz et al. 2007).

7.7.1 Retrospective Studies

Retrospective studies have been used to address many of the same questions as long-term studies. A key use of retrospective studies is to provide baseline data for comparison with modern observations. Further, they can characterize slow processes and disturbance regimes, and how they may have influenced selected ecosystem attributes (Swetnam and Bettancourt 1998). Perhaps the greatest value of retrospective studies is for characterizing changes to vegetation and wildlife habitats over time. Dendrochronological studies provide information on frequencies and severities of historical disturbance events (Swetnam 1990) (Fig. 7.6). This information can be used to reconstruct ranges of variation in vegetation structure and composition at various spatial scales. These studies can also be used to infer short- and long-term effects of various management practices on habitats, as well as effects of disruptions of disturbance regimes on habitats.

Other potential tools for retrospective studies include databases from long-term ecological research sites, forest inventory databases, pollen studies, and sediment cores. They are also used in epidemiological and epizootiological studies. With any of these studies, one must be aware of the underlying assumptions and limitations of the methodology. For example, dendrochronological methods often fail to account for the small trees because they are consumed by fire and not sampled. This limitation may result in a biased estimate of forest structure and misleading inferences about historical conditions. If the investigator understands this idiosyncrasy, then he or she can consider this during evaluation.

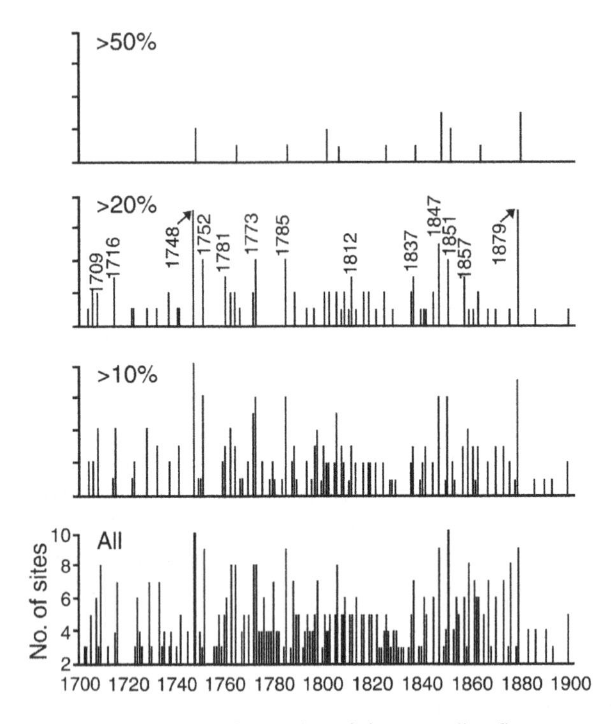

Fig. 7.6 Fire-area index computed as the number of sites recording fires per year for the period 1700–1900. Fires recorded by any tree within the sites are shown on the bottom plot, whereas fires recorded by 10, 20, or 50% of the trees are shown above (from Swetnam 1990)

7.7.2 Substitutions of Space for Time

Substituting space for time is achieved by finding samples that represent the range of variation for the variable(s) of interest in order to infer long-term trends (Pickett 1989, Morrison 1992). The assumption is that local areas are subjected to different environments and different disturbance histories that result in different conditions across the landscape. Thus, rather than following few samples over a protracted period to understand effects of slow processes, random events, or systems with high variances, more areas are sampled hoping that they represent conditions that might exist during different phases of these processes. For example, if you wanted to understand the long-term effects of forest clear-cutting on wildlife, a logical approach would be to locate a series of sites representing a *chronosequence* of conditions rather than waiting for a recent clear-cut to go through succession. By chronosequence, we mean areas that were clear-cut at various times in the past (e.g., 5, 10, 20, 30, 50, 75, and 100 years ago). By sampling enough areas representative of vegetation structure and composition at different times following clear-cuts you could draw inferences as to possible short- and long-term effects on wildlife. To provide valid results using this approach requires that many sites with somewhat similar histories and characteristics be used (Morrison 1992). If substantial sources of variation between sampling units cannot be accounted for, then substituting space for time will fail (Pickett 1989). Even if these sources can be accounted for, space-for-time substitutions may fail to take into account mesoscale events (Swetnam and Bettancourt 1998) that affect large regions and tend to mitigate or swamp local environmental conditions. Pickett (1989) cautioned that studies that rely on spatial rather than temporal sampling are best suited for providing qualitative trends or generating hypotheses rather than for providing rigorous quantitative results. Even so, spatially dispersed studies are preferred for inventory studies.

Clearly, an empirical basis is needed to support the use of space-for-time substitutions in monitoring studies. By this, we mean that you should conduct a baseline study to evaluate whether such an approach would provide unbiased estimates of the variable(s) under study. This baseline study would require comparisons of an existing long-term data set collected as part of another study with a data set collected from multiple locations over a time. If no significant differences are observed in estimates of the variables of interest, then space-for-time substitutions may be justified. If a difference is observed, then one can explore methods to calibrate results of one approach with the other. If the differences cannot be rectified by calibration, you should reconsider the use of space-for-time substitutions in your study design.

7.7.3 Substitutions of Fast for Slow Dynamics

Applying the results of a simple system with rapid generation times or accelerated rates of succession can provide insights into how systems with inherently slower processes might behave (Morrison 1992). For example, applying results of laboratory studies on rodents might provide some insight on population dynamics of larger wild mammals. Obviously, extending results of captive animals to wild populations has obvious drawbacks, as does applying results from *r*-selected species such as rodents to larger *K*-selected species such as carnivores. At best, such substitutions might provide a basis for development of hypotheses or theoretical constructs that can be subjected to empirical tests. These tests should be designed to show the correspondence between the surrogate measure (e.g., that with fast dynamics) and the variable that exhibits slow dynamics. If the relationship is strong, then it might be acceptable to use behavior of the surrogate measure as an index for the variable of interest.

7.7.4 Modeling

Use of models has gained wide application in studies of wildlife habitats (Verner et al. 1986) and populations (McCullough and Barrett 1992). Models can be conceptual or empirical (Shugart 1989). Conceptual models are generally used to structure a scientific endeavor. As an example, one might ask, "How is the population of spotted owls influenced by various environmental factors?" A conceptual model might consist of an envirogram that depicts how owls are linked to various ecological components and processes (Verner et al. 1992). This conceptual model can provide the basis for conducting specific studies to understand the effects of one or more factors on owl population trends (Fig. 7.7). One can argue, in fact, that all scientific studies are based on conceptual models of various levels of sophistication regardless of whether the researcher is explicitly aware of this fact. The example

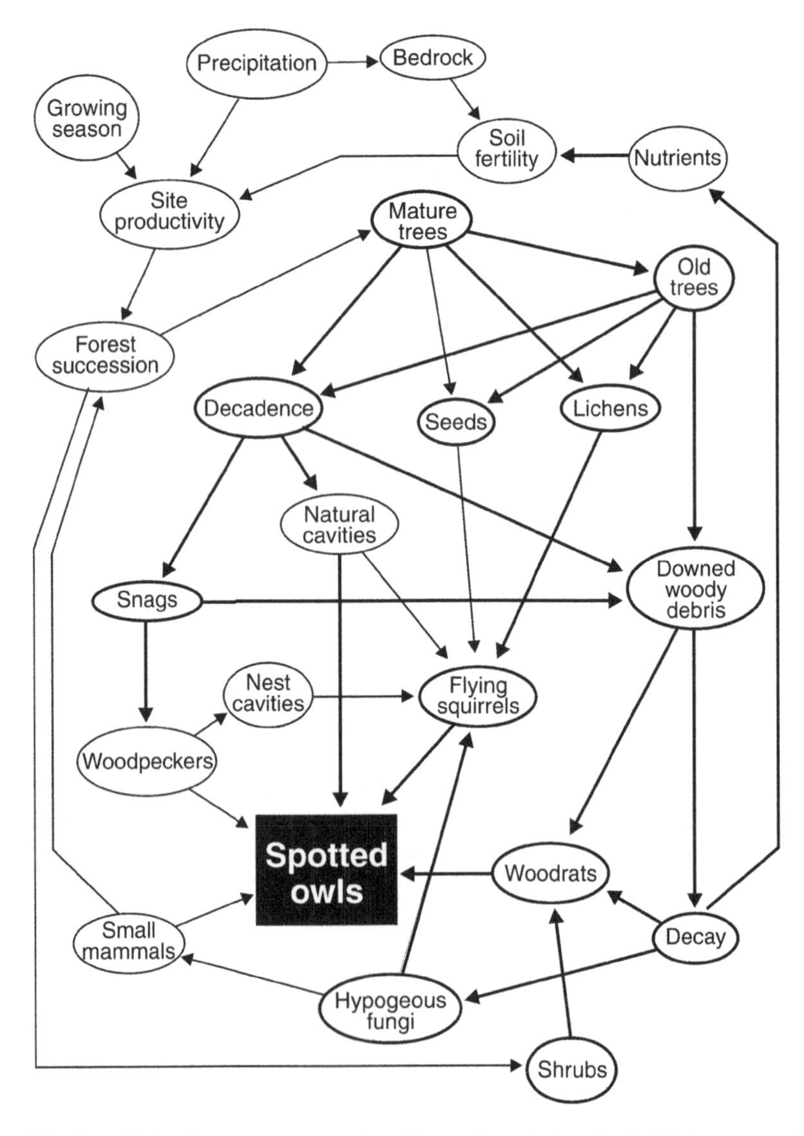

Fig. 7.7 Simplified schematic representation of some important ecological linkages associated with California spotted owls (from Verner et al. 1992)

provided in Fig. 7.7 is perhaps more detailed than most conceptual models, but it does show how a system can be characterized as interactions among many tractable and researchable components.

Quantitative forecasts from predictive models are used to provide wildlife managers with realizations of ecological processes. When structuring any modeling exercise to address population dynamics questions, an initial decision must be made concerning the proposed model's purpose (McCallum 2000). Empirical models are quantitative predictions of how natural systems behave. Models for examining population dynamics exist on a continuum from empirical models used to make predictions to abstract models that attempt to provide general insights (Holling 1966; May 1974; McCallum 2000). Predictive models require a larger number of parameters than abstract models, increasing their predictive ability for the system of interest, but reducing the generality of the model and thus its ability to expand results to other systems.

Ecological modeling in wildlife studies encompasses a broad range of topics, but most often relates to two topics, demographic (parameter estimation) and population modeling. Demographic modeling is directed toward developing a model which best explains the behavior and characteristics of empirical data, and then using that model to predict how that or similar systems will behave in the future (Burnham and Anderson 2002). The use and sophistication of demographic modeling has increased along with increases in personal computing power (White and Nichols 1992) and development of statistical programs specifically for ecological data (Sec 2.7.2).

Population modeling is directed towards development of predictive models, based on the aforementioned demographic parameters, which we use to forecast the response of wildlife populations to perturbations. Population models come in many forms: population viability analysis, matrix population models, individual based models, and so on (Caswell 2001; Boyce 1992; DeAngelis and Kross 1992) each structured with the intent of describing and predicting population dynamics over time and space (Lande et al. 2003). To be realistic, population models must include simultaneous interactions between deterministic and stochastic processes (Lande et al. 2003), which lends uncertainty to predictions of population trajectories. Because the fundamental unit in animal ecology is the individual (Dunham and Beaupre 1998), many population models incorporate individual variability (e.g., stochasticity in estimates of demographic parameters).

7.7.5 Genetics

Genetic techniques represent a new and burgeoning field providing novel approaches to monitoring. Schwartz et al. (2007) provide an insightful overview of these techniques. They separated genetic monitoring into two categories (1) markers used for traditional population monitoring and (2) those used to monitor population genetics.

Most genetic materials are obtained through noninvasive samples – hair, scat, feathers, and the like – thus, obviating the need to capture or even observe the species under study. Individual animals are identified using genetic markers, thus permitting estimates of abundance and vital rates. For rare species, abundance indices are possible, which are adjusted subsequently for small population size or detection probability (White 2005). For more abundant species, capture–recapture analyses can be applied (see Chap. 4). These samples can also be used to estimate survival and turnover rates. Survival rates are often difficult to estimate using traditional mark–capture techniques, especially when detection or capture rates vary with time. For example, male northern goshawks are detected more easily using traditional techniques during years when they breed than in years when they do not (Reynolds and Joy 2006). Survival estimates based on years when the goshawks do not breed may be underestimates given lower capture probabilities. This bias might be reduced using molted feathers and genetic markers to estimate survival.

Genetics can also be used to identify species, the presence of hybrids, and the prevalence of disease or invasive species. For example, genetics has been used to identify the historical geographical range of fisher (*Martes pennanti*) (Aubry et al. 2004; Schwartz 2007), the presence of Canada lynx (*Lynx canandensis*) (McKelvey et al. 2006), hybridization between bobcats (*Lynx rufus*) and lynx (Schwartz et al. 2004), and hybridization between northern spotted owls (*Strix occidentalis caurina*) and barred owls (*Strix varia*) (Haig et al. 2004).

Genetics can also be used to estimate effective population size and changes in allele frequencies. This information is critical to understanding patterns of gene flow and effects of habitat fragmentation on populations. The insight provided by these approaches and others has tremendous implications for present and future management of these species. Ultimately, the success of that management can only be assessed with continued monitoring in the mode of adaptive management.

7.8 Adaptive Management

The concept of *adaptive management* rests largely on monitoring the effects of implementing land management activities on key resources, and then using monitoring results as a basis for modifying those activities when warranted (Walters 1986; Moir and Block 2001). It is an iterative process whereby management practices are initiated and effects are monitored and evaluated at regular intervals. Effectively, land management activities are implemented incrementally and desired outcomes are evaluated at each step. If outcomes are consistent with or exceed predictions, the project continues as designed. If outcomes deviate negatively from predictions, then management can proceed in one of three directions: continue, terminate, or change.

This general scenario can be characterized by a seven-step process that includes a series of feedback loops that depend largely on monitoring (Moir and Block 2001) (Fig. 7.8). The primary feedback loops in Fig. 7.8 are between steps 5–6–7, 2–7, and 7–1. The 5–6–7 feedback loop is the shortest and perhaps the fastest. It implies that management prescriptions are essentially working and need only slight, if any, adjustments. Perhaps the primary obstacle in this loop is the lag time

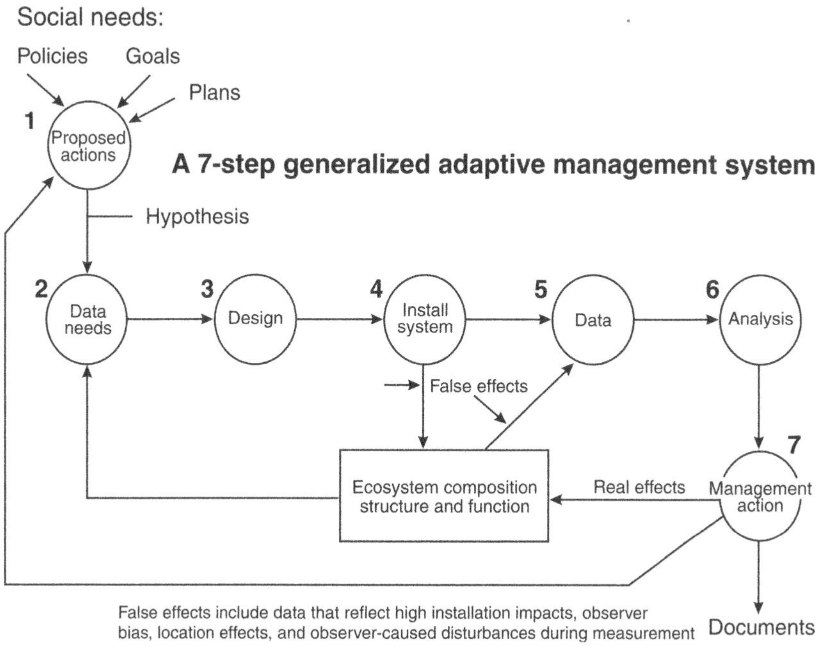

Fig. 7.8 A seven-step generalized adaptive management system illustrating the series of steps and feedback loops. Reproduced from Moir and Block (2001), with kind permission from Oxford University Press

between project implementation and completion of monitoring. Because this timeframe can be prolonged, numerous factors may complicate the ability or willingness of the organization to complete monitoring (Morrison and Marcot 1995). Consequently, the loop is often severed and feedback is never provided. The second feedback loop, 2–7, indicates that monitoring missed the mark. By this, we mean the monitoring study was poorly designed, the wrong variables were measured, or monitoring was poorly executed. Regardless of exactly what went wrong, monitoring failed to provide reliable information to permit informed conclusions on the efficacies of past management, or in making decisions for future management direction. The 7–1 feedback loop is the one typically associated with adaptive management; it is when a decision must be made regarding the course of future management and monitoring activities. If monitoring was done correctly, then informed decisions can be made for future management direction. If monitoring was not conducted or was done poorly, then another opportunity was lost to provide a scientific basis for resource management. Unfortunately, the latter is more the rule than the exception (White et al. 1999; Moir and Block 2001).

If adaptive management is to be the paradigm followed in the future as espoused by most contemporary resource management agencies, it is only possible by conducting credible monitoring. Inventory and monitoring provide critical information on resource status and trends needed to make informed management decisions. Failure to incorporate these studies will doom adaptive management to failure.

7.8.1 Thresholds and Trigger Points

In designing inventory or monitoring studies for management applications, you must establish some benchmark that signals a need for subsequent actions. Benchmarks can signify success as well as failure. Thus, actions taken in response to reaching a benchmark may range from a cessation of activities to engaging in the next step in a management plan. Regardless of exactly how the benchmark is used, it provides a measurable criterion for management actions.

Benchmarks also play a role in study design, particularly in determining sampling intensity. In monitoring, effect size establishes the amount of change that you want to detect if it indeed occurs. Thus, effect size is closely interrelated with statistical power, sample size, and Type I error, as all three of these will define the minimal size of an effect that can be detected. Figure 7.9 shows the tradeoff between power and effect size to detect a population trend. Note that as effect size increases, statistical power increases. This essentially means that it is easier to statistically show effect when the change is big than it is to statistically show effect when change is small. The tradeoff is one that must be carefully evaluated and decided upon at the onset of designing a study.

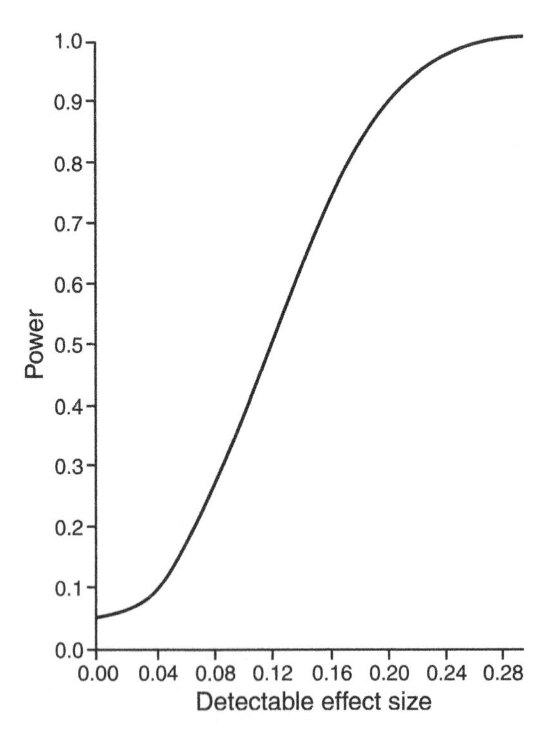

Fig. 7.9 Hypothetical curve of the statistical power needed to detect a population trend in a population (from USDI Fish and Wildlife Service 1995)

Thresholds and trigger points represent predetermined levels that when exceeded will lead to an action or response. The action or response could be termination or modification of a particular activity. For example, consider a prescribed fire project to be conducted in spotted owl habitat. The plan calls for treating 5,000 ha, spread across six separate fires. The fire plan calls for no special protection to trees, but predicts that no trees >60 cm dbh will be killed because of the fire. Two statistical tests are developed, one to test the spatial extent of loss and the other to test for the absolute magnitude of the percentage of large trees lost. A monitoring plan is developed following a standard protocol (see Box 7.4). Following postfire monitoring, it was determined that too many large trees were lost exceeding prefire predictions, resulting in feedback into the system. In this case, the loss of any trees signified a threshold that was exceeded. If actions were then developed and initiated to mitigate the loss of trees, then the threshold becomes a trigger point. In this case, future prescription may require removing litter and flammable debris from the base of large trees to minimize the probability of tree mortality.

Box 7.4 Mexican Spotted Owl Microhabitat Implementation Monitoring

The Recovery Plan for the Mexican Spotted Owl (USDI Fish and Wildlife Service 1995) allows treatments in forested landscapes. However, the extent and type of treatments are limited in mixed-conifer, part of the pine-oak, and riparian vegetation types. The Plan also calls for monitoring of macrohabitat and microhabitat as part of the recovery process. Delisting criterion 2 in the Recovery Plan specifically requires habitat monitoring to demonstrate that habitat across the range is stable or increasing (USDI Fish and Wildlife Service 1995, p. 77).

This protocol partially addresses the microhabitat monitoring requirement (referred to as implementation monitoring) by assessing the retention of Key Habitat Components (described below) in protected and restricted forest types following habitat-altering activities.

The purpose of monitoring is to index the change of key components in owl habitat in treated areas. Losses are predicted at two scales. One is the total percentage change to the component across the entire project area (project-level monitoring). Analysis of total percentage change will provide information on the magnitude of change across the project. The second scale is the percentage loss of the component on a plot-level basis. Analysis on plot-level basis will provide spatial information on treatment effects.

This protocol applies to silviculture, thinning, management-ignited fire, and other activities directed at modifying forests and woodlands (excluding prescribed natural fire) in protected and restricted areas as defined in the Recovery Plan (USDI Fish and Wildlife Service 1995, pp. 84–95).

(continued)

Box 7.4 (continued)

What, Where, and When Monitoring Should Occur

The approach described below focuses only on implementation monitoring and not on effectiveness monitoring as required in the Recovery Plan. Implementation monitoring addresses (1) whether treatments within threshold areas were successful in maintaining habitat attributes at or above the levels shown in Table 7.1 and (2) posttreatment changes in Key Habitat Components (defined below) as the direct or indirect result of management activities were roughly equivalent to predicted changes.

It is also important to know how well treatments in restricted areas (including target and threshold) and protected areas retain Key Habitat Components. Key Habitat Components of Mexican spotted owl habitat include large trees, snags, logs, and hardwoods, and must be retained in adequate quantities and distributions (USDI Fish and Wildlife Service 1995, pp. 94–95). The objectives of monitoring treatments in these areas is to evaluate whether actual losses in the Key Habitat Components exceed the losses predicted during project planning, to quantify the loss of these components, and then adjust future prescriptions as appropriate.

Table 7.1 Target/threshold conditions for restricted area mixed-conifer and pine-oak forests. Table III.B.1. from USDI (1995)

	% Stand density				
Recovery Units Forest Type	trees 30-45 cm dbh	trees 30-45 cm dbh	trees >60n dbh	Basal area (m/ha)	Tree>45 cm (number/ha)
Basin and Range - East RU					
Mixed-conifer	10	10	10	32	49
Mixed-conifer	10	10	10	39	49
All RUs, except Basin and Range - East RU					
Mixed-conifer	10	10	10	32	49
Mixed-conifer	10	10	10	39	49
Colorado Plateau, Upper Gila Mountains, Basin and Range - West RUs					
Pine-oak[a]	15	15	15	32	49

[a]For pine-oak, 20ft^2/acre of oak must be provided as a threshold/target condition
Source: table III.B.1. from USDI Fish and Wildlife Service (1995)

Variables Assessed in Microhabitat Monitoring

The variables assessed in these protocols are those identified in the Recovery Plan to be habitat correlates of Mexican spotted owls and their prey. These variables include Key Habitat Components and Fine Filter Factors that apply to all protected and restricted areas, and variables derived from Table 7.1 that apply only to target/threshold areas. Generally, all variables listed below are directly from the Recovery Plan or are based on our interpretation of the Recovery Plan.

Key Habitat Components

The variables listed in the Recovery Plan (USDI Fish and Wildlife Service 1995, pp. 94, 107) are of importance to the habitat of Mexican spotted owls and their prey. These variables include:

- Number of trees >60 cm diameter at breast height (dbh) for conifers, or diameter at root collar (drc) for hardwoods
- Number of trees 48–60 cm dbh/drc
- Number of logs >30 cm at 1.5 m from the large end and 1.3-m long
- Number of live hardwood stems >12 cm drc
- Number of snags >30 cm dbh and >1.3 tall
- Total basal area of trees >12 cm dbh/drc

Table 7.1 Variables

Additional variables must be measured in threshold areas to evaluate whether threshold values (see Table 7.1) were maintained following treatment. The variables must also be measured in target areas to evaluate how close post-treatment conditions are relative to values in Table 7.1. The variables needed in addition to the Key Habitat Components include:

- Number of live trees 30–45 cm dbh/drc
- Number of live trees 12–29.9 cm dbh/drc
- Number of live trees 2.5–11.9 cm dbh/drc

These measurements will also allow for calculations of total tree basal area, the distribution of stand density across diameter classes, and the density of large trees (i.e., those >45 cm dbh/drc).

Procedures for Monitoring Key Habitat Components

Planning/Project Design

The purpose of monitoring Key Habitat Components is to index their change in treated areas. Thus, treatment plans must state treatment objectives and quantify projected changes to each Key Habitat Component (such as the expected percentage loss of each component) as result of the treatment. Losses are considered in two ways. One is the total percentage loss of the component across the project area. The other loss is the percentage loss of the component on a plot-level basis. If the loss of Key Habitat Components during implementation exceeds those predicted during the analysis, then prescriptions should be adjusted to mitigate excessive losses in future projects.

Two criteria are considered when evaluating project implementation. One is the spatial extent of the loss of each Key Habitat Component. Thus, the number of plots in which this change occurs provides an index of how much area was affected. The other is assessing the magnitude of the loss of each component across the project area. Both should be considered simultaneously because the plots where a component (e.g., large trees, large logs) was lost

(continued)

Box 7.4 (continued)

may have been where it mostly occurred. Thus, even though only a few plots may be affected, the actual loss of a component may have been large. Considering losses both spatially and in magnitude is important when evaluating project implementation and for planning future treatments.

Analysis and Rationale for Number of Plots Needed

The minimum number of plots needed to be sampled will probably differ between the plot- and project-level analyses because different statistical analyses will be used. Plot-level analyses are based on a one-sided chi-square test, whereas project-level analyses are based on a paired-sample *t* test.

Plot-Level Analysis

A one-sided chi-square test (Marascuilo and McSweeny 1977, pp. 196–198) is the basis for the plot-level analysis. This is applied as a one-tailed test with the level of significance at 0.05. The test assesses if implementation of the prescription resulted in excessive loss (i.e., more than specified in the treatment plans; thus a one-tailed *t* test) of each Key Habitat Component on a plot-level basis.

Two proportions are considered in this analysis: null hypothesis proportion of plots (hereafter null proportion) and the observed proportion. The null proportion is the predicted proportion of plots where the loss of a Key Habitat Component will exceed a predetermined threshold value. The observed proportion of plots is that where loss of a Key Habitat Component was exceeded from posttreatment measurements.

Necessary sample size is based on the null proportion and the statistical power of detecting an increase over the null proportion (Table 7.2). Only null proportions between 0 and 10% were considered because monitoring is conducted on key components of spotted owl habitat; thus, a "light trigger" is needed to measure excessive losses of these components. Statistical power was set at $P = 0.9$ for detecting small increases for the same reason.

Table 7.2 Minimum sample sizes for plot-level analysis based on the null hypothesis of the proportion of plots affected by treatment

Type I error	Statistical power	Null proportion	Sample size
0.05	0.90	0	25
0.05	0.90	0.10	50

Table 7.2 specifies necessary minimum sample sizes for two null proportions. Application of these sample sizes will depend on the particular Key Habitat Component and the number of acres treated. See below for more specific guidelines.

This analysis involves a two-step process to evaluate whether the treatment was implemented correctly. The first is to compare the observed proportion of plots where losses exceeded predictions under the null hypothesis. If the observed proportion is less than the null proportion, then the project was successful from a spatial standpoint. If the observed proportion is greater than the null proportion, the analysis should proceed to the second step.

In the second step, $P = 0.95$ one-sided confidence limits on the observed proportion are compared to the null proportion. Figure 7.10 contains confidence limits plotted for a range of observed proportion with sample specified at $n = 25$ and $n = 50$, respectively. In Figure 7.10a, if an observed proportion is 0.05 or larger, the lower confidence limit exceeds the null proportion of 0 and the project should be judged as unsuccessful. Also, based on the upper confidence limit, the "true" proportion of plots exceeding predicted losses might be 20% or more, an unacceptable level. In other words, we estimated the proportion to be 0.05 based on a sample, but the real effect could be much higher, 20% or more.

In Fig. 7.10b, if the observed proportion is 0.18 or larger, the lower confidence limit exceeds the null proportion of 0.10 and the project should be judged unsuccessful. The upper confidence limit on the "true" proportion is 0.30, also an unacceptable level.

The lower and upper confidence bounds can be calculated empirically by (Fleiss 1981, pp. 14–15)

$$\text{Lower limit} = ((2np + c^2 - 1) - c(c^2 - (2 + 1/n) + (4\,p(nq + 1)))^{1/2}))/(2n+2c^2),$$

$$\text{Upper limit} = ((2np + c^2 - 1) + c(c^2 - (2 + 1/n) + (4\,p(nq + 1)))^{1/2}))/(2n+2c^2),$$

where n = sample size, p = observed proportion, $q = 1 - p$, c = value from the normal distribution corresponding to $1 - (\alpha/2)$.

For example, authors of an environmental assessment done for the Podunk Ranger District estimate that 20% of the snags (magnitude loss) will be lost within a 603-acre project area because of prescribed fire. In this case, the null proportion would be 10 and the necessary sample 50 (see Table 7.2). Posttreatment monitoring indicated that >20% of the snags were lost on 11 of the 50 plots (22%). Since the observed proportion (0.22) was greater than 0.18, the lower confidence limit exceeds 0.10 and the project should be judged as unsuccessful. It is also worth noting that the upper confidence limit at this point is 0.34, a very high level.

Project-Level Analysis

Losses to habitat components do not occur evenly over a project area; for example, some areas in a management-ignited fire might crown out while other areas may not burn at all. Because of this, a proportion of plots should have losses that exceed what was predicted over the treatment area. Although excessive losses may occur in patches within a treatment area, it does not

(continued)

Box 7.4 (continued)

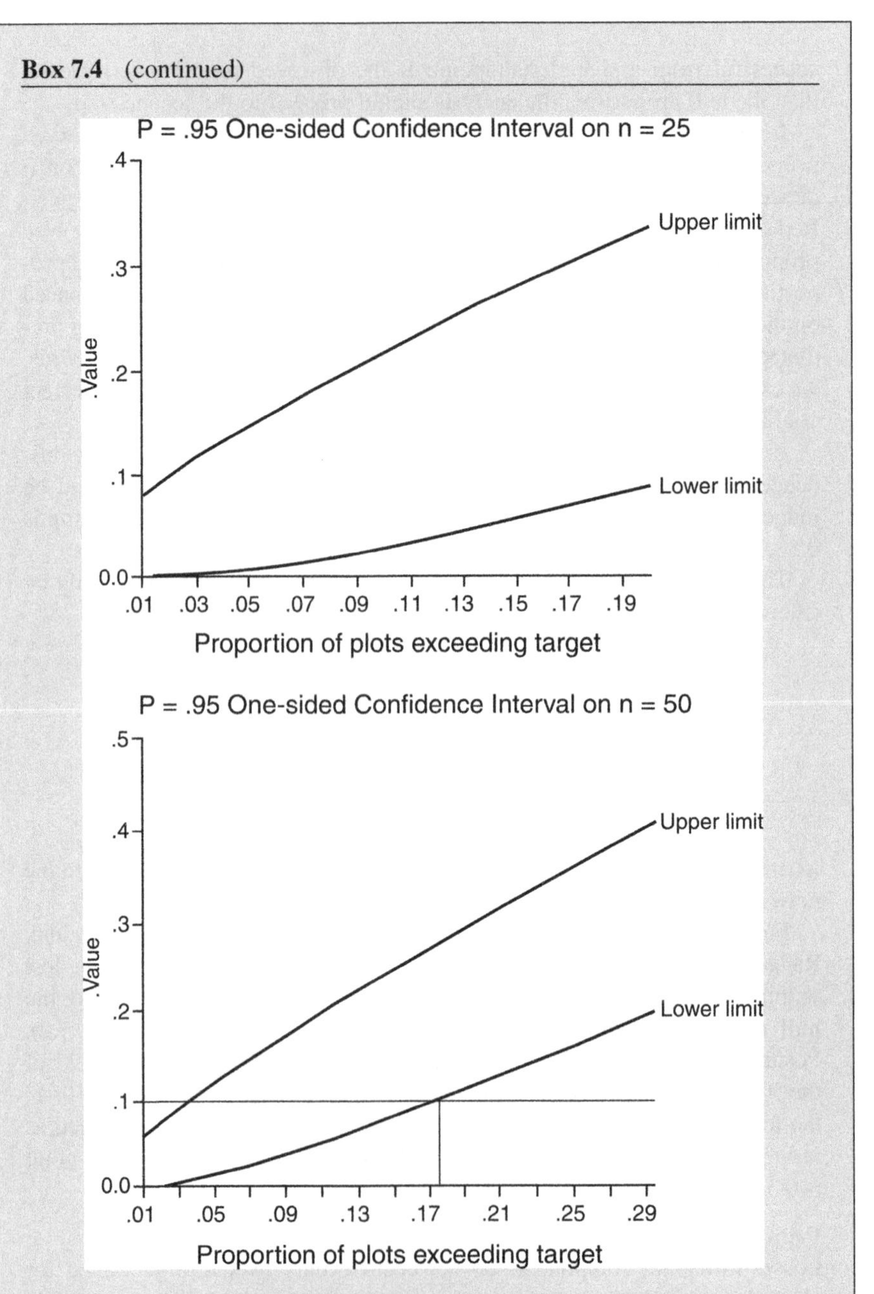

Fig. 7.10 Confidence limits plotted for a range of observed proportions with sample size specified at (**a**) $n = 25$ and (**b**) $n = 50$

mean that the treatment was implemented unsuccessfully. However, if the areas where losses occur are where most of a particular component is found, then most of that component may be lost and one may conclude statistically that the treatment was successful.

The basis for this analysis is a paired *t* test (Sokal and Rohlf 1969, p. 332). One group is the set of plots measured prior to the treatment; the other group is the same plots measured posttreatment. Sample size can be calculated empirically using the methodology presented by Sokal and Rohlf (1969, p. 247).

This approach requires that we know the standard deviation, state the difference that we want to detect in the loss of each Key Habitat Component, and the statistical power or probability that the difference will be significant. The sample size equation takes the form (Sokal and Rohlf 1969, p. 247)

$$n > 2\,(\sigma / \delta)^2\,\{t_{\alpha[v]} + t_{2(1-P)[v]}\}^2,$$

where, n is the sample size, σ is the standard deviation of the differences, δ is the smallest difference desired to be detected, α is the significance level, v is the degrees of freedom of the sample standard deviation, and $t_{\alpha[\upsilon]}$ and $t_{2(1-P)[v]}$ are the values from a two-tailed table with n degrees of freedom and corresponding to probabilities of α and $2(1 - P)$, respectively.

This test should be done as a one-tailed test with $\alpha = 0.05$. If a two-tailed table is used, then use critical values for $\alpha = 0.10$. This test also requires knowledge of the standard deviation for each variable. This can be obtained with a pilot study or from comparable existing data. A finite population correction factor should be used in calculating the standard deviation, which effectively reduces the standard deviation and needed sample sizes. Calculation of the standard deviation with a finite population correction factor takes the generalized form:

$$\sigma_c = \sigma_u\,(1 - \phi),$$

where σ_c is the finite population corrected standard deviation, σ_u is the uncorrected deviation, and ϕ is the ratio of the proportion of the area sampled out of the total area.

Box 7.5 Population Monitoring for the Mexican Spotted Owl

The Mexican Spotted Owl Recovery Plan sketched out a procedure for estimating population trend of the subspecies with the three most populated recovery units (USDI Fish and Wildlife Service 1995). Populations in the other, more sparsely populated recovery units were not monitored because of logistical difficulties in designing and implementing an unbiased sampling approach. Implicit to monitoring birds in only three recovery units was the assumption that they represented the core populations, and trends in these populations would apply to the overall trend in owl populations throughout its geographic range. We briefly describe below some of the primary components of the population monitoring sampling design.

The target population to be sampled consists of territorial owls in the three recovery units mentioned above. Thus, all potential owl habitat in these three recovery units must be included in the sampling frame. Sampling units will consist of randomly spaced quadrats 50–75 km^2 in size. The intent is to evaluate populations subjected to all factors, natural and anthropogenic, that may influence trends. Quadrats should be stratified by vegetation type and owl density within each recovery unit. A certain percentage of quadrats would be replaced each year to guard against selective management practices being practiced inside quadrats that were not being done outside of quadrats. For example, an agency may avoid cutting timber within quadrats to minimize any potential effects on owls. If timber harvest does influence owl population trends, exclusion of this practice from the sampling units may result in a biased estimate of population trends.

Within quadrats, sampling will consist of the following aspects. Survey stations will be placed to ensure adequate coverage of the quadrat. Each survey station will be sampled at night four times during the breeding season to locate territorial adult owls. Multiple visits to each survey station will allow for estimation of detection probabilities, which can be used to adjust raw counts. Adjusted counts are then transformed to density estimates for each quadrat and then aggregated for estimates within strata. Auxiliary diurnal sampling will be done to visually locate birds detected during nighttime surveys, assess reproductive status, color-band adult and juvenile owls, and provide a check on the accuracy of nighttime surveys. For example, nighttime surveys might only detect a male owl, when the territory is occupied by a pair. Furthermore, color-banding individuals may allow alternative ways to estimate population trends based on analyses of mark–recapture data.

A monitoring program of this magnitude and complexity has rarely, if ever, been conducted on a wildlife species. Thus, a pilot study was needed to evaluate sampling intensity or the number of quadrats needed for precise estimates of the relevant population parameters. Once completed, the responsible management agencies must decide whether they have the resources and commitment to implement the program and carry it through fruition (White et al. 1999).

Ganey et al. (2004) reported the results of a pilot study conducted in 1999. The study occurred with the Upper Gila Mountains Recovery Unit on 25 40–76 km^2 quadrats. Quadrats were stratified into high and low density, and field sampling followed established mark–recapture protocols for this subspecies. They concluded that the approach was possible but infeasible given costs and logistics of conducting field samples. They also found that temporal variation inherent to Mexican spotted owl populations was so large, the power to detect a population trend was relatively low. They proposed occupancy monitoring as a cost-effective alternative to mark–recapture, a proposal under serious consideration by the Mexican Spotted Owl Recovery Team.

7.9 Field Applications

7.9.1 Short-term and Small Area Applications

Many inventory and monitoring studies are short term occurring within a restricted area or both. Indeed, studies done to inventory an area for species of concern prior to implementing a habitat-altering activity do not have the luxury of a long-term study. To accommodate this situation, specific studies should be done or existing data sets should be analyzed to establish the minimum amount of time for the study to provide reliable information. For rare or elusive species, such studies would focus on the amount of time and number of sampling points needed to detect a species if it is present. For studies whose goal is to develop list of the species present, pilot studies or existing data could be used to develop species accumulation curves that can help to define the amount of effort needed to account for most species present (Fig. 7.2).

Often studies are restricted in the amount of area available for study. This may occur in an island situation either in the traditional sense or when a patch of vegetation is surrounded by a completely different vegetation type (e.g., riparian habitats in the southwest). Small areas also occur when the area of interest is restricted. An example is when development is planned for a small parcel of land and the objective is to evaluate the species potentially affected within that parcel. In these situations, you are not so much faced with a sampling problem as you are with a sample size problem. Given the small area, you should strive to detect every individual and conduct a complete census. Even so, you may have too few data to permit rigorous treatment of the data for many of the species encountered. Various tools such as rarefaction and bootstrapping can be used to compensate for small samples encountered in small areas studies.

7.9.2 Long-term and Regional Applications

Ideally, monitoring studies should occur over long periods. The objective of such studies is to document trends that can help to inform predictions of future trajectories. Many of these monitoring programs also occur over wide geographic areas such the Breeding Bird Survey, Christmas Bird Counts, and Forest Inventory and Assessment. The logistics of implementing such large-scale, long-term monitoring programs is daunting, and potentially compromises integrity of the data. Perhaps the major hurdle of long-term, regional studies is to make sure that protocols are followed consistently over time. For example, the Breeding Bird Survey is a long-term monitoring program that occurs throughout United States (Sauer et al. 2005). The survey entails conducting point counts along established road transect. Unfortunately, coverage of these transects varies from year to year, which reduces the effectiveness of the monitoring program and necessitates innovative analyses to fills in the gaps. Transects are sampled by a large number of people, with varying levels of expertise, skill, and ability. A similar situation occurs with habitat monitoring and the US Forest Service's Forest Inventory and Assessment program. This program includes vegetation plots on a 5,000-m grid with plots on lands regardless of ownership (i.e., not just on Forest Service land). For various reasons the Forest Service has altered the sampling design by changing the number of plots surveyed, revising measurement protocols, and the frequency at which they sample points. These changes effectively compromise the ability to examine long-term trends because of the difficulty of sorting out variation ascribed to changes in sampling protocols from variation resulting from vegetation change.

7.10 Summary

Inventory and monitoring are key aspects of wildlife biology and management; they can be done in pursuit of basic knowledge or as part of the management process. Inventory is used to assess the state or status of one or more resources, whereas monitoring is typically done to assess change or trend. Monitoring can be classified into four overlapping categories:

- Implementation monitoring is used to assess whether or not a directed management action was carried out as designed.
- Effectiveness monitoring is used to evaluate whether a management action met its desired objective.
- Validation monitoring is used to evaluate whether an established management plan is working.
- Compliance monitoring is used to see if management is occurring according to established law.

Selecting the appropriate variable to inventory or monitor is a key aspect of the study design; direct measures, such as population numbers, are preferred over indirect measures, such as indicator species. The length of monitoring studies depends largely on the process or variable being studied. The appropriate length often exceeds available resources, necessitating alternative approaches such as retrospective studies, modeling, genetic tools, substituting space for time, and substituting fast for slow dynamics.

Time, cost, and logistics often influence the feasibility of what can be done. Use of indices can be an effective way to address study objectives provided data are collected following an appropriate study design and data are analyzed correctly. Indices can be improved and calibrated using ratio-estimation and double-counting techniques.

Monitoring effects of management actions requires a clear and direct linkage between study results and management activities, often expressed as a feedback loop. Feedback is essential for assessing the efficacy of monitoring and for validating or changing management practices. Failure to complete the feedback process negates the intent and value of monitoring.

References

Allen, T. F. H., and T. W. Hoekstra. 1992. Towards a Unified Ecology. Columbia University Press, New York, NY.

Anderson, D. R. 2001. The need to get the basics right in wildlife field studies. Wildl. Soc. Bull. 29: 1294–1297.

Anderson, D. R. 2003. Response to Engeman: Index values rarely constitute reliable information. Wildl. Soc. Bull. 31: 288–291.

Aubry, K., S. Wisley, C. Raley, and S. Buskirk. 2004. Zoogeography, pacing patterns and dispersal in fishers: insights gained from combining field and genetic data, in D. J. Harrison, A. K. Fuller, and G. Proulx, Eds. Martins and Fishers (Martes) in Human-Altered Environments: An International Perspective, pp. 201–220. Springer Academic, New York, NY.

Bart, J., S. Droge, P. Geissler, B. Peterjohn, and C. J. Ralph. 2004. Density estimation in wildlife surveys. Wildl. Soc. Bull. 32: 1242–1247.

Block, W. M., and L. A. Brennan. 1993. The habitat concept in ornithology: Theory and applications, in D. M. Power, Ed. Current Ornithology, vol. 11, pp. 35–91. Plenum, New York, NY.

Block, W. M., M. L. Morrison, J. Verner, and P. N. Manley. 1994. Assessing wildlife–habitat-relationships models: A case study with California oak woodlands. Wildl. Soc. Bull. 22: 549–561.

Boyce, M. S. 1992. Population viability analysis. Annu. Rev. Ecol. Syst. 23: 481–506.

Burnham, K. P., and D. R. Anderson. 2002. Model Selection and Multimodel Inference: A Practical Information-Theoretic Approach, 2nd Edition. Springer-Verlag, New York, NY.

Caswell, H. 2001. Matrix Population Models: Construction, Analysis, and Interpretation, 2nd Edition. Sinauer Associates, Inc., Sunderland, MA.

Clements, E. E. 1920. Plant Indicators. Carnegie Institute, Washington, DC.

Cochran, W. G. 1977. Sampling Techniques, 3rd Edition. Wiley, New York, NY.

Cohen, J. 1988. Statistical Power Analysis for the Behavioral Sciences, 2nd Edition. Erlbaum, Hillsdale, NJ.

Cohen, J. 1992. A power primer. Psychol. Bull. 112: 155–159.

Davis, M. B. 1989. Retrospective studies, in G. E. Likens, Ed. Long-Term Studies in Ecology: Approaches and Alternatives, pp. 71–89. Springer-Verlag, New York, NY.

DeAngelis, D. L., and L. J. Gross. 1992. Individual-Based Models and Approaches in Ecology. Chapman and Hall, London.

Dunham, A. E., and S. J. Beaupre. 1998. Ecological experiments: Scale, phenomenology, mechanism, and the illusion of generality, in J. Bernardo and W. J. Resetarits Jr., Eds. Experimental Ecology: Issues and Perspectives, pp. 27–49. Oxford University Press, Oxford.

Eberhardt, L. L., and M. A. Simmons. 1987. Calibrating population indices by double sampling. J. Wildl. Manage. 51: 665–675.

Efron, B. 1982. The Jackknife, The Bootstrap, and Other Resampling Plans. Society for Industrial and Applied Mathematics, Philadelphia, PA, 92 pp.

Engeman, R. M. 2003. More on the need to get the basics right: Population indices. Wildl. Soc. Bull. 31: 286–287.

Fleiss, J. L. 1981. Statistical Methods for Rates and Proportions, 2nd Edition. Wiley, New York, NY.

Franklin, J. F. 1989. Importance and justification of long-term studies in ecology, in G. E. Likens, Ed. Long-Term Studies in Ecology: Approaches and Alternatives, pp. 3–19. Springer-Verlag, New York, NY.

Ganey, J. L., G. C. White, D. C. Bowden, and A. B. Franklin. 2004. Evaluating methods for monitoring populations of Mexican spotted owls: A case study, in W. L. Thompson, Ed. Sampling Rare and Elusive Species: Concepts, Designs, and Techniques for Estimating Population Parameters, pp. 337–385. Island Press, Washington, DC.

Gray, P. A., D. Cameron, and I. Kirkham. 1996. Wildlife habitat evaluation in forested ecosystems: Some examples from Canada and the United States, in R. M. DeGraaf and R. I. Miller, Eds. Conservation of Faunal Diversity in Forested Landscapes, pp. 407–536. Chapman and Hall, London.

Green, R. H. 1979. Sampling Design and Statistical Methods for Environmental Biologists. Wiley, New York, NY.

Haig, S. M., T. D. Mullins, E. D. Forsman, P. W. Trail, and L. Wennerberg. 2004. Genetic identification of spotted owls, barred owls, and their hybrids: Legal implications of hybrid identity. Conserv. Biol. 18: 1347–1357.

Hayek, L. C. 1994. Analysis of amphibian biodiversity data, in W. R. Heyer, M. A. Donnelly, R. W. McDiarmid, L. C. Hayek, and M. S. Foster, Eds. Measuring and Monitoring Biological Diversity: Standard Methods for Amphibians, pp. 207–269. Smithsonian Institution Press, Washington, DC.

Hayes, J. P., and R. J. Steidel. 1997. Statistical power analysis and amphibian population trends. Conserv. Biol. 11: 273–275.

Hedges, L. V., and Olkin, I. (1985). Statistical Methods for Meta-Analysis. Academic, San Diego, CA.

Heyer, W. R., M. A. Donnelly, R. W. McDiarmid, L. C. Hayek, and M. S. Foster. 1994. Measuring and Monitoring Biological Diversity: Standard Methods for Amphibians. Smithsonian Institution Press, Washington, DC.

Hohn, M. E. 1976. Binary coefficients: A theoretical and empirical study. J. Int. Assoc. Math. Geol. 8: 137–150.

Holling, C. S. 1966. The functional response of invertebrate predators to prey density. Mem. Entomol. Soc. Can. 48: 1–86.

Hunter Jr., M. L. 1991. Coping with ignorance: The coarse filter strategy for maintaining biodiversity, in K. A. Kohm, Ed. Balancing on the Brink of Extinction: The Endangered Species Act and Lessons for the Future, pp. 266–281. Island Press, Washington, DC.

Hurlbert, S. H. 1984. Pseudoreplication and the design of ecological field experiments. Ecol. Monogr. 54: 187–211.

Jaccard, P. 1901. The distribution of the flora in the alpine zone. New Phytol. 11: 37–50.

Lande, R., S. Engen, and B. -E. Sæther. 2003. Stochastic Population Dynamics in Ecology and Conservation. Oxford University Press, Oxford.

Landres, P. B., J. Verner, and J. W. Thomas, 1988. Ecological uses of vertebrate indicator species: A critique. Conserv. Biol. 2: 316–328.

MacKenzie, D. I., J. A. Royle, J. A. Brown, and J. D. Nichols. 2004. Occupancy estimation and modeling for rare and elusive populations, in W.L. Thompson, Ed. Sampling Rare or Elusive Species. pp. 142–172. Island Press, Covelo, CA.

Marascuilo, L. A., and M. McSweeny. 1977. Non-Parametric and Distribution-Free Methods for the Social Sciences. Brooks/Cole, Monterey, CA.

May, R. M. 1974. Stability and Complexity in Model Ecosystems, 2nd Edition. Princeton University Press, Princeton.
McCallum, H. 2000. Population Parameters: Estimation for Ecological Models. Blackwell, Malden, MA.
McCullough, D. R., and R. H. Barrett. 1992. Wildlife 2001: Populations. Elsevier, London.
McKelvey, K. S., and D. E. Pearson. 2001. Population estimation with sparse data: The role of estimators versus indices revisited. Can. J. Zool. 79: 1754–1765.
McKelvey, K. S., J. von Kienast, K. B. Aubry, G. M. Koehler, B. T. Maletzke, J. R. Squires, E. L. Lindquist, S. Loch, M. K. Schwartz. 2006. DNA Analysis of hair and scat collected along snow tracks to document the presence of Canada lynx. Wildl. Soc. Bull. 34(2): 451–455.
Miller, R. I. 1996. Modern approaches to monitoring changes in forests using maps, in R. M. DeGraaf and R. I. Miller, Eds. Conservation of Faunal Diversity in Forested Landscapes, pp. 595–614. Chapman and Hall, London.
Moir, W. H., and W. M. Block. 2001. Adaptive management on public lands in the United States: Commitment or rhetoric? Environ. Manage. 28: 141–148.
Morrison, M. L. 1986. Birds as indicators of environmental change. Curr. Ornithol. 3: 429–451.
Morrison, M. L. 1992. The design and importance of long-term ecological studies: Analysis of vertebrates in the Inyo-White Mountains, California, in R. C. Szaro, K. E. Severson, and D. R. Patton, Tech. Coords. Management of Amphibians, Reptiles, and Small Mammals in North America, pp. 267–275. USDA Forest Service, Gen. Tech. Rpt. RM-166. Rocky Mountain Forest and Range Experiment Station, Fort Collins, CO.
Morrison, M. L., and B. G. Marcot. 1995. An evaluation of resource inventory and monitoring programs used in national forest planning. Environ. Manage. 19: 147–156.
Morrison, M. L., B. G. Marcot, and R. W. Mannan. 1998. Wildlife–Habitat Relationships: Concepts and Application, 2nd Edition. University of Wisconsin Press, Madison, WI.
Noon, B. R., T. A. Spies, and M. G. Raphael. 1999. Conceptual basis for designing an effectiveness monitoring program, in B. S. Mulder, B. R. Noon, T. A. Spies, M. G. Raphael, C. J. Palmer, A. R. Olsen, G. H. Reeves, and H. H. Welsh, Tech. Coords. The Strategy and Design of the Effectiveness Monitoring Program in the Northwest Forest Plan, pp. 21–48. USDA Forest Service, Gen. Tech. Rpt. PNW-GTR-437. Pacific Northwest Research Station, Portland, OR.
Pickett, S. T. A. 1989. Space-For-Time Substitutions as an Alternative to Long-Term Studies, in G. E. Likens, Ed. Long-Term Studies in Ecology: Approaches and Alternatives, pp. 110–135. Springer-Verlag, New York, NY.
Pielou, E. C. 1977. Mathematical Ecology, 2nd Edition. Wiley, New York, NY.
Reynolds, R. T., and S. M. Joy. 2006. Demography of Northern Goshawks in Northern ARIZONA, 1991–1996. Stud. Avian Biol. 31: 63–74.
Rinkevich, S. E., J. L. Ganey, W. H. Moir, F. P. Howe, F. Clemente, and J. F. Martinez-Montoya. 1995. Recovery units, in Recovey Plan for the Mexican Spotted Owl (*Strix occidentalis lucida*), vol. I, pp. 36–51. USDI Fish and Wildlife Service, Southwestern Region, Albuquerque, NM.
Romesburg, H. C. 1981. Wildlife science: Gaining reliable knowledge. J. Wild. Manage. 45: 293–313.
Rosenstock, S. S., D. R. Anderson, K. M. Giesen, T. Leukering, and M. E. Carter. 2002. Landbird counting techniques: Current practices and alternatives. Auk 119: 46–53.
Sauer, J. R., J. E. Hines, and J. Fallon. 2005. The North American Breeding Bird Survey, Results and Analysis 1966–2005. Version 6.2.2006. USGS Patuxent Wildlife Research Center, Laurel, MD
Schwartz, M. K. 2007. Ancient DNA confirms native rocky mountain fisher *Martes pennanti* avoided early 20th century extinction. J. Mam. 87: 921–92.
Schwartz, M. K., K. L. Pilgrim, K. S. McKelvey, E. L. Lindquist, J. J. Claar, S. Loch, and L. F. Ruggiero. 2004. Hybridization between Canada lynx and bobcats: Genetic results and management implications. Conserv. Genet. 5: 349–355.
Schwartz, M. K., G. Luikart, and R. S. Waples. 2007. Genetic monitoring as a promising tool for conservation and management. Trends Ecol. Evol. 22: 25–33.
Scott, J. M., F. Davis, B. Csuti, R. Noss, B. Butterfield, C. Grives, H. Anderson, S. Caicco, F. D'Erchia, T. Edwards Jr., J. Ulliman, and R. G. Wright. 1993. Gap analysis: A geographical approach to protection of biodiversity. Wildl. Monogr. 123.

Sharp, A., M. Norton, A. Marks, and K. Holmes. 2001. An evaluation of two indices of red fox (Vulpes vulpes) abundance in an arid environment. Wildl. Res. 28: 419–424.

Shugart, H. H. 1989. The role of ecological models in long-term ecological studies, in G. E. Likens, Ed. Long-Term Studies in Ecology: Approaches and Alternatives, pp. 90–109. Springer-Verlag, New York, NY.

Sokal R. R., and C. D. Michener. 1958. A statistical method for evaluating systematic relationships. Univ. Kansas Sci. Bull. 38: 1409–1438.

Sokal, R. R., and F. J. Rohlf. 1969. Biometry: The Principles and Practice of Statistics in Biological Research. Freeman, San Francisco, CA.

Sorensen, T. 1948. A method for establishing groups of equal amplitude in plant sociology based on similarity of species content, and its applications to analyses of the vegetation on Danish commons. Det Kongelige Danske Viden-skkabernes Selskab, Biloogiske Skrifter 5: 1–34.

Spellerberg, I. F. 1991. Monitoring Ecological Change. Cambridge University Press, New York, NY.

Steidl, R. J., J. P. Hayes, and E. Schauber. 1997. Statistical power analaysis in wildlife research. J. Wildl. Manage. 61: 270–279.

Strayer, D., J. S. Glitzenstein, C. G. Jones, J. Kolasa, G. Likens, M. J. McDonnell, G. G. Parker, and S. T. A. Pickett. 1986. Long-Term Ecological Studies: An Illustrated Account of Their Design, Operation, and Importance to Ecology. Occas Pap 2. Institute for Ecosystem Studies, Millbrook, New York, NY.

Swetnam, T. W. 1990. Fire history and climate in the southwestern United States, in J. S. Krammes, Tech. Coord. Effects of Fire Management of Southwestern Natural Resources, pp. 6–17. USDA Forest Service, Gen. Tech. Rpt. RM-191. Rocky Mountain Forest and Range Experiment Station, Fort Collins, CO.

Swetnam, T. W., and J. L. Bettancourt. 1998. Mesoscale disturbance and ecological response to decadal climatic variability in the American Southwest. J. Climate 11: 3128–3147.

Thomas, L. 1997. Retorspective power analysis. Conserv. Biol. 11: 276–280.

Thompson, S. K. 2002. Sampling, 2nd Edition. Wiley, New York, NY.

Thompson, W. L., G. C. White, and C. Gowan, 1998. Monitoring Vertebrate Populations. Academic, San Diego, CA.

USDA Forest Service. 1996. Record of Decision for Amendment of Forest Plans: Arizona and New Mexico. USDA Forest Service, Southwestern Region, Albuquerque, NM.

USDI Fish and Wildlife Service. 1995. Recovery Plan for the Mexican Spotted Owl (*Strix occidentalis lucida*), vol. I. USDI Fish and Wildlife Service, Albuquerque, NM.

Verner, J. 1983. An integrated system for monitoring wildlife on the Sierra Nevada Forest. Trans. North Am. Wildl. Nat. Resour. Conf. 48: 355–366.

Verner, J., and L. V. Ritter. 1985. A comparison of transects and point counts in oak-pine woodlands of California. Condor 87: 47–68.

Verner, J., M. L. Morrison, and C. J. Ralph. 1986. Wildlife 2000: Modeling Habitat Relationships of Terrestrial Vertebrates. University of Wisconsin Press, Madison, WI.

Verner, J., R. J. Gutiérrez, and G. I. Gould Jr. 1992. The California spotted owl: General biology and ecological relations, in J. Verner, K. S. McKelvey, B. R. Noon, R. J. Gutiérrez, G. I. Gould Jr., and T. W. Beck, Tech. Coords. The California Spotted Owl: A Technical Assessment of Its Current Status, pp. 55–77. USDA Forest Service, Gen. Tech. Rpt. PSW-GTR-133. Pacific Southwest Research Station, Berkeley, CA.

Walters, C. J. 1986. Adaptive Management of Renewable Resources. Macmillan, New York, NY.

White, G. C. 2005. Correcting wildlife counts using detection probabilities. Wildl. Res. 32: 211–216.

White, G. C., and J. D. Nichols, 1992. Introduction to the methods section, in D. R. McCullough and R. H. Barrett, Eds. Wildlife 2001: Populations, pp. 13–16. Elsevier, London.

White, G. C., W. M. Block, J. L. Ganey, W. H. Moir, J. P. Ward Jr., A. B. Franklin, S. L. Spangle, S. E. Rinkevich, R. Vahle, F. P. Howe, and J. L. Dick Jr. 1999. Science versus reality in delisting criteria for a threatened species: The Mexican spotted owl experience. Trans. North Am. Wildl. Nat. Resour. Conf. 64: 292–306.

Chapter 8
Design Applications

8.1 Introduction

Our goal in this chapter is to provide guidance that will enhance the decision-making process throughout a given study. The often-stated axiom about "the best laid plans..." certainly applies to study design. The initial plan that any researcher, no matter how experienced, develops will undergo numerous changes throughout the duration of a study. As one's experience with designing and implementing studies grows, he or she anticipates and resolves more and more major difficulties before they have an opportunity to impact the smooth progression of the study. The "feeling" one gains for what will and what will not work in the laboratory or especially the field is difficult to impart; there is truly no substitute for experience. However, by following a systematic path of steps when designing and implementing a given study, even the beginning student can avoid many of the most severe pitfalls associated with research.

The guidelines we present below will help develop the thought process required to foresee potential problems that may affect a project. We present these guidelines in a checklist format to ensure an organized progression of steps. Perhaps a good analogy is that of preparing an oral presentation or writing a manuscript; without a detailed outline, it is too easy to leave out many important details. This is at least partially because the presenter or writer is quite familiar with their subject (we hope), and tends to take many issues and details for granted. The speaker, however, can always go back and fill in the details, and the writer can insert text; these are corrective measures typically impossible in a research project.

Throughout this chapter, we provide examples from our own experiences. We will focus, however, on a "theme" example, which provides continuity and allows the reader to see how one can accomplish a multifaceted study. The goal of the theme study is development of strategies to reduce nest parasitism by the brown-headed cowbird (*Molothrus ater*) on host species along the lower Colorado River (which forms the border of California and Arizona in the desert southwest of the United States). We provide a brief summary of the natural history of the cowbird in Box 8.1.

The theme example is multifaceted and includes impact assessment (influence of cowbirds on hosts), treatment effects (response of cowbirds to control measures),

M.L. Morrison et al., *Wildlife Study Design*.

Box 8.1

The brown-headed cowbird is a small blackbird, is native to this region, and has been increasing in both abundance and geographic range during the late 1800s and throughout the 1900s. This expansion is apparently due to the ability of the cowbird to occupy farm and rangelands and other disturbed locations. The cowbird does not build a nest. Rather, it lays its eggs in the nest of other species, usually small, open-cup nesting songbirds such as warblers, vireos, and sparrows. The host adults then raise the nestling cowbirds. Research has shown that parasitized nests produce fewer host young than nonparasitized nests. Although this is a natural occurrence, the population abundance of certain host species can be adversely impacted by parasitism. These effects are exacerbated when the host species is already rare and confined to environmental conditions favored by cowbirds. In the Southwest, host species that require riparian vegetation are the most severely impacted by cowbirds. This is because most (>90% along the lower Colorado River) of the riparian vegetation has been removed for river channelization and other management activities, thus concentrating both cowbirds and hosts in a small area.

problems in logistics (extremely hot location, dense vegetation), time constraints (at least one host is nearing extinction), and many other real-life situations. We added other relevant examples to emphasize salient points.

Numerous publications have developed step-by-step guides to study development (Cochran 1983; Cook and Stubbendieck 1986; Levy and Lemeshow 1999; Martin and Bateson 1993; Lehner 1996; Garton et al. 2005). We reviewed these publications and incorporated what we think are the best features of each into our own template (Fig. 8.1). Project development should follow the order presented in Fig. 8.1; below we follow this same progression. Not surprisingly, the steps we recommend for planning, implementing, and completing a given research project (Fig. 8.1) are components of the steps typically used during the process implementing a research program in natural science (Table 1.2), which includes multiple individual research projects. In previous chapters, we discussed in detail nearly all materials presented below. While presenting our examples, we refer you to previous sections of the book and reference key primary publications to facilitate review.

8.2 Sequence of Study Design, Analysis, and Publication

8.2.1 Step 1 – Questions

A study is designed around questions posed by the investigator (see Sect. 1.3). Although this statement may sound trivial, crafting questions is actually a difficult and critically important process. If a full list of questions is not produced before the

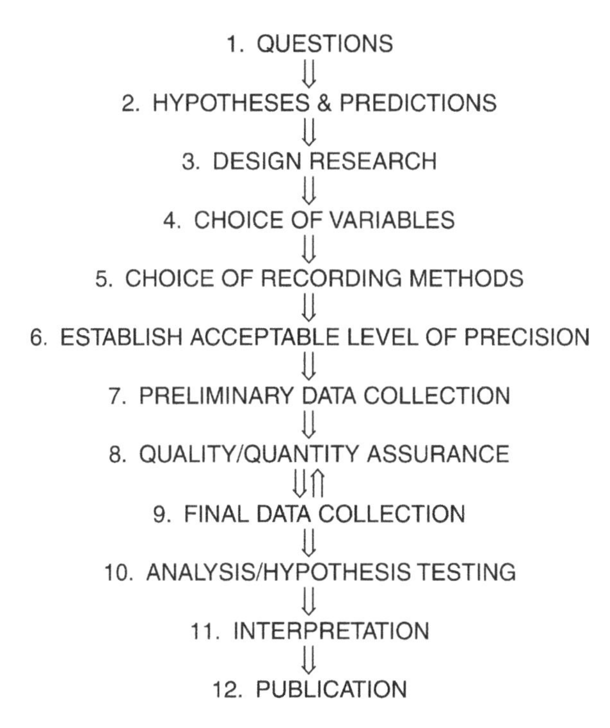

Fig. 8.1 Progression of steps recommended for planning, implementing, and completing a research project. Reproduced from Morrison et al. (2001), with kind permission from Springer Science + Business Media

study plan is developed, it is often problematic to either insert new questions into an ongoing study or answer new questions not considered until the conclusion of data collection. Thus, the first step should be listing all relevant questions that should be asked during the study. Then, these questions should be prioritized based on the importance of the answer to the study. Optimally, we then design the study to answer the aforementioned questions, in a statistically rigorous fashion, one question at a time. The investigator *must* resist the overpowering temptation to try to address multiple questions with limited resources simply because "they are of interest." Such a strategy will, at a minimum, doom a study to mediocrity. We would much rather see one question answered thoroughly than see a paper reporting the results of a series of partially or weakly addressed questions.

Guiding the selection of questions will be the availability of study locations, time, personnel, and funding. All studies will have limitations for all of these parameters, which place constraints on the number and types of questions that can be addressed.

Questions are developed using literature, expert opinion, your own experiences, intuition, and guesswork. A thorough literature review is an essential cornerstone of all studies. There is no need to reinvent the wheel; we should attempt to advance knowledge rather than simply repeating a study in yet another geographic location. Likewise, we should critically evaluate published work and not repeat biased or substandard research. We must be familiar with the past before we can advance into the future. The insights offered by people experienced in the field of interest always should be solicited, because these individuals have ideas and intuition often unavailable in their publications. You should synthesize all of these sources and develop your own ideas, blended with intuition, to devise your questions.

8.2.1.1 Theme Example

We developed the following questions by reviewing the literature, discussing research topics with cowbird researchers, considering the needs of resource agencies, and brainstorming among field personnel; no prioritization is implied by the order of questions.

1. What is the effectiveness of a single trap; i.e., the area of influence a trap exerts on parasitism and host productivity?
2. What are the movements of males/females by age within and between riparian patches?
3. What is the relative effectiveness of removing cowbird adults vs. removing their eggs and young?
4. When is the nesting cycle to remove eggs/young?
5. When is it most effective to trap adult cowbirds?
6. Is it more effective to remove adult females rather than both male and female cowbirds?
7. What is the trapping intensity of cowbirds within a year necessary to influence host productivity?
8. What is the best measure of project success (e.g., increased host reproductive success or fledgling success)?
9. What are the population abundance of cowbirds and host species?
10. Will winter cowbird trapping exert effective control of cowbirds?
11. Where do the cowbirds reside during winter?
12. Do predators significantly influence host-breeding success, thus effectively negating any positive effects of cowbird control?

Exercise: The preceding 12 questions are not arranged in any priority. As an exercise, recall the previous discussion describing the theme example. Then, on a separate sheet list what you would consider the top four questions in the priority that you would pursue them. Although there is no "correct" answer, in Box 8.2 we provide our list along with a brief explanation.

8.2.2 Step 2 – Hypotheses and Predictions

In Chaps. 1 and 2, we discussed the philosophical underpinnings of hypotheses testing, model selection, and related approaches to gaining knowledge. In his now classic paper, Romesburg (1981) argued that wildlife science was not advancing

Box 8.2

Here is our prioritization of the study questions for the theme example given in text under "Questions."

#1 – h. What is an adequate measure of project success (e.g., reduced parasitism rates, increased reproductive success; fledgling success)?

This should be the obvious choice given the project goal of determining ways of lowering parasitism and increasing breeding success of cowbird hosts. The complicating issue here, however, is that simply lowering parasitism may not increase reproductive success. This is because other factors, most notably predation by a host of species, could negate any beneficial effects of adult or egg and young removal. Thus, while documenting a reduction in parasitism is essential, it alone is not sufficient to declare the study a success.

#2 – c. Effectiveness of removal of adults vs. removal of eggs and young

It is extremely time consuming to locate host nests. Thus, it would be easier to focus management on removal of adult cowbirds. It was deemed necessary to examine both methods because of the worry that removal of adults alone would be inadequate, and the necessity of rapidly determining an effective management strategy.

#3 – a. Effectiveness of a single trap; i.e., the area of influence a trap exerts on parasitism and host productivity

The failure to detect a significant influence of adult cowbird removal on host reproduction could be criticized as resulting from inadequate trapping effort. Therefore, it was judged to be critical to ensure that an adequate trapping effort be implemented. The decision was made to "overtrap" to avoid this type of failure; future studies could refine trapping effort if trapping was shown to be an effective procedure.

#4 – i. Population abundance of cowbirds and host species

The initial reason for concern regarding the influence of cowbird parasitism on hosts was the increase in cowbird abundance and the decrease in many host species abundance. Therefore, determining trends in cowbird and host abundance will serve, over time, to determine the degree of concern that should be placed on this management issue.

Not all questions are mutually exclusive, nor is it possible to produce a perfect ordering. However, it is critical that a short list of prioritized items be developed, and that this prioritization be based on the ultimate reason for conducting the study.

our ability to make reliable predictions because, in part, of the failure to follow rigorously the H–D method (see Sect. 1.4). However, testing hypotheses does not necessarily lessen the numerous ad hoc explanations that accompany studies that fail to establish any formal and testable structure. A decent biologist can always explain a result. Our point here is that simply stating a hypothesis in no way guarantees increased knowledge or revolutionary science (Kuhn 1962). Instead, what is required is careful and thoughtful evaluation of the predictive value of any proposed hypothesis. In designing a study, the researcher must carefully determine what is needed to advance management of the situation of interest. Questions are driven by goals, and hypotheses that follow from the questions are driven by the predictive power needed to answer the questions.

8.2.2.1 Theme Example

In our theme example, we developed specific research hypotheses directed at determining whether cowbird trapping could result in a biologically meaningful increase in host productivity.

The primary hypotheses to be tested are:

H_0: No difference in parasitism rates between treated and controls plots
H_1: Parasitism rates differ significantly between treated and control plots

Given rejection of the above two-tailed null hypothesis, the following one-tailed hypotheses were tested:

H_0: Parasitism rates on treated plots are ≤30%
H_1: Parasitism rates on treated plots are >30%

Previous research indicated those parasitism rates exceeding 25–30% caused breeding failure in hosts (Finch 1983; Laymon 1987; Robinson et al. 1993). Therefore, we determined that parasitism <30% would be an appropriate measure of success for this aspect of the study.

Next, the same set of hypotheses can be stated for examining nesting success (defined here as the number of young fledged/nest):

H_0: No difference in nesting success between treated and controls plots
H_1: Nesting success differs significantly between treated and control plots

Given rejection of the above two-tailed null hypothesis, the following one-tailed hypotheses were tested:

H_0: Nesting success on treated plots was ≥40%
H_1: Nesting success on treated plots was <40%

Establishing an appropriate measure of nest success is difficult. The literature indicates that success of >40% is a general measure of population health of many passerine nests (Martin 1992). You will find that choosing an appropriate magnitude of difference that has biological relevance is one of the most difficult aspects of study design. But without setting a biological magnitude, you are allowing your study be driven by statistics (i.e., the test statistic and associated P-value), and then a posteriori trying to establish biological explanations. That is, you are letting the statistics lead the study; rather, you want your study to be led by biology as supported by statistics. Again, refer to Chaps. 1 and 2 for a thorough discussion of these design issues.

8.2.3 Step 3 – Design

Design of the project entails the remaining sections of this chapter. Indeed, even the final section on publishing should be considered in designing your study (see Sect. 1.3.1). For example, a useful exercise when designing your study is to ask, "How will I explain this method when I write this up?" If you do not have absolute confidence in your answer, then a revision of methods is called for. Put more bluntly, "If you cannot explain it, how can you expect a reader to understand it?" A well-written study plan/proposal can become, in essence, the methods section of a manuscript.

Imbedded in our previous discussion of question and hypotheses development are issues related to delineation of the study population, spatial and temporal extent of study, sex and age considerations, needed generalizability of results, and so forth. Each of these separate issues must now be placed into a coherent study design. This requires that we make decisions regarding allocation of available resources (personnel, funding, logistics, and time). At this step, it is likely that any previous failure to reduce the number of questions being asked will be highlighted.

Delineation of the study population is a critical aspect of any study (see Sects. 1.3–1.5). Such delineation allows one to make definitive statements regarding how widely study results can be extrapolated. Unfortunately, few research papers make such a determination. A simple example: The willow flycatcher (*Empidonax traillii*) is a rare inhabitant of riparian and shrub vegetation throughout much of the western United States (Ehrlich et al. 1988). It is currently divided into three subspecies, one of which (*E. t. extimus*) is classified as threatened/endangered and is restricted to riparian vegetation in the arid Southwest. The other two subspecies (*E. t. brewsteri* and *E. t. adastus*) occur in higher elevation, mountain shrub (especially willow, *Salix*) and riparian vegetation; they are declining in numbers but not federally listed (Harris et al. 1987). Thus, it is unlikely that research findings for *E. t. brewsteri* and *E. t. adastus* would be applicable to *E. t. extimus*. Further, it is unlikely that results for studies of *E. t. extimus* from the lower Colorado River – a desert environment – would be applicable to populations of this subspecies occurring farther north in less arid regions. Subspecies are often divided into ecotypes that cannot be separated morphologically. Indeed, the physiological adaptations of ecotypes are little studied; differences among ecotypes (and thus populations) are probably a leading reason why research results can seldom be successfully applied to other geographic locations.

The distribution of sampling locations (e.g., plots, transects) is a critical aspect of all studies (see Chap. 4). First, the absolute area available to place the plots is usually limited by extent of the vegetation of interest, legal restrictions preventing access to land, and differences in environmental conditions. If the goal of a study is to determine effects of a treatment in riparian vegetation on a wildlife refuge, then the study is constrained by the extent of that vegetation on a defined area (as is the case in our theme example).

Second, the behavior of the animal(s) under study will constrain the placement of sampling locations. Most studies would be severely biased by movement of animals between what are designated as independent sampling locations. For example, avian ecologists often assess the abundance of breeding passerines by placing counting points 200–300 m apart. This criterion is based on the argument that most passerines have breeding territories of <100-m radius, and that it is unlikely that individuals will move between counting points during the short period of time (usually 5–10 min) that an observer is present. These arguments are suspect, however, given that breeding birds move beyond territory boundaries, and adjacent points are likely being influenced by similar factors (e.g., disturbances, relatively microclimatic conditions, predators, competitors). The placement of small mammal trapping plots, the locations used to capture individuals for radio tagging, the selection of animals for blood sampling, and the gathering of pellets for food analysis are but a few of the other sampling decisions that, are difficult to conduct with independence (see Sect. 2.3).

The time and money available to conduct a study constrains study design. It is critical that you do not try to conduct a study that overextends available resources. You do not want to place yourself in the position of using "time and money constraints" as an excuse for not achieving reliable research results (e.g., small sample sizes).

8.2.3.1 Theme Example

The optimal design for this study would have been based on at least 1 year of pretreatment data collection, followed by random assignment of treatments (adult and/or egg–young removal). However, funding agencies required a more rapid assessment of treatment effects because of pressure from regulatory agencies. Thus, the decision was made to forego use of a BACI-type design, and instead use another impact assessment approach (see Chap. 6). Specifically, pairs of plots were selected based on similarity in environmental conditions and proximity to one another. Treatments were then assigned randomly to one member of each pair. Paired plots were placed in close proximity because of the lack of appropriate (riparian) vegetation. Because of limited riparian vegetation, it was not possible to have control plots for adult cowbird removal that were within a reasonably close proximity to treated plots. This is because previous research indicated that a single cowbird trap could impact parasitism rates up to at least 0.8 km from the trap. The decision was made to treat all plots for adult removal, but only one of the pairs would be treated for egg and young removal. Plots receiving no treatments were placed upstream far outside the influence of the cowbird traps. This design carries the assumption that the upstream reference plots will be an adequate indicator of naturally occurring parasitism rates in the area. Research on parasitism rates conducted several years prior to the new study provided additional support. Thus, the design implemented for this study is nonoptimal, but typical of the constraints placed on wildlife research. Textbook examples of experimental design are often impossible to implement in the real world of field biology (although this is not an excuse for a sloppy design; see Sect. 8.3). Thus, a certain amount of ingenuity is required to design an experiment that can still test the relevant hypotheses. Of course, some hypotheses cannot be tested in the field regardless of the design (see Chap. 1).

8.2.4 Step 4 – Variable Selection

Variables must be selected that are expected to respond to the treatment being tested, or be closely linked to the relationship being investigated. Even purely descriptive, or hypothesis generating, studies should focus sampling efforts on a restricted set of measurements. A thorough literature review is an essential part of any study, including development of a list of the variables measured by previous workers (see Sect. 1.3). However, one must avoid the temptation of developing a long "shopping list" of variables to measure: This only results in lowered precision because of smaller sample sizes. Rather, a list of previous measurements should be used to develop a short list of variables found previously to be of predictive value. Measuring numerous variables often means that some will be redundant and measure the same quality (e.g., tree height and basal area). Also, lists of variables from past studies may provide an opportunity to learn from past efforts; that is, there may be better variables to measure than those measured before.

For example, since multivariate statistical tools became readily available and easy to use, there has been a proliferation of studies that collected data on a massive number of variables. It is not unusual to see studies that gathered data on 20–30 variables or more, or read analyses of all possible subsets of ten or more variables (which results in millions of comparisons); this tendency is especially apparent in studies of wildlife–habitat relationships (see Morrison et al. 2006 for review). Preliminary data collection (see Sect. 8.2.7) can aid in developing a short list of variables. Most studies, including those using multivariate methods, identify only a few variables that have the majority of predictive power. There is ample evidence in the literature to justify concentrating on a minimum number of predictive or response variables.

Each variable measured increases the time spent on each independent sample, time that could be better spent on additional independent samples. Further, researchers often tend to use rapid measurement techniques when confronted with a long list of variables. For example, many workers visually estimate vegetation variables (e.g., shrub cover, tree height) even though research indicates that visual methods are inferior to more direct measurements (e.g., using a line intercept for shrubs and a clinometer for tree height; see Block et al. 1987). Thus, there is ample reason to concentrate on a few, carefully selected variables.

This is also the time to identify potential covariates (see Chap. 3). Analysis of covariance is an indirect, or statistical, means of controlling variability due to experimental error that increases precision and removes potential sources of bias. As a reminder, statistical control is achieved by measuring one or more concomitant variates in addition to the variate of primary interest (i.e., the response variable). Measurements on the covariates are made for adjusting the measurements on the primary variate. For example, in conducting an experiment on food preference, previous experience will likely influence results. The variate in this experiment may be a measure of the time taken to make a decision; the covariate may be a measure associated with the degree of experience at the start of the trials. Thus, covariates must be designed into the study, and should not be an afterthought. Covariates may be used along with direct experimental design control. Care must be taken in the use of covariates or misleading results can result (see Chap. 3 (see also Winer et al. 1991, Chap. 10). Step 5 (Sect. 8.2.5) incorporates further discussion of variable selection.

8.2.4.1 Theme Example

One of our research interests is the movement of males and females by age within and between riparian patches. The process of selecting variates to measure begins with prioritization of the information needed to address project goals. Where and when the birds are located and their activity (e.g., foraging, nest searching) is important because this provides information useful in locating traps and searching for host nests that have been parasitized. The sex composition of the birds (such as in flocks) is important because it provides information on (1) how the sex ratio may be changing with trapping and (2) where females may be concentrating their activities. The age composition also is important because it provides a measure of experimental success (i.e., how many cowbirds are being produced). Finally, the type of vegetation used by cowbirds provides data on foraging and nest-searching preferences.

An important aspect of the cowbird study is the effectiveness of the removal of adults in controlling parasitism. Confounding this analysis could be the location of a host nest within a study plot. This is because research has shown that cowbirds seem to prefer edge locations (e.g., Laymon 1987; Robinson et al. 1993; Morrison et al. 1999), with a decreasing parasitism rate as you move farther into the interior of a woodland. Thus, the variate may be a measure of host nesting success or parasitism rate, with the covariate being the distance of the nest from the plot edge.

8.2.5 Step 5 – Recording Methods

There are usually many methods available for recording data on a specific variable. For example, birds can be counted using fixed area plots, variable distance measures, and transects; vegetation can be measured using points, plots of various sizes and shapes, and line transects; and so forth (see Chapter 4). However, many of these decisions will be guided by the objectives of the study and, thus, the specific types of data and precision associated with the data needed. For example, if study objectives require that nesting success is determined, then an intensive, site-specific counting method (such as spot mapping) might be most appropriate. This same method, however, would not necessarily be appropriate for a study of bird abundance along an elevational gradient. Thus, there is no "best" method; rather, there are methods that are most appropriate for the objectives of any study.

It is also important that all proposed methods be thoroughly reviewed for potential biases and degree of precision attainable (see Sect. 2.6.4). Unfortunately, there is often little guidance regarding how to handle bias in publications reporting implementation of a method. It is usually necessary to go back to the original publication that reported on development of the method, or to search for the few papers that have analyzed bias. Within a single category of methods, such as "vegetation sampling" or "bird counting," there are numerous submethods that carry different levels of bias. For example, measuring tree height carries a much higher level of bias than does measuring tree dbh (Block et al. 1987). Similarly, the variable circular plot counting method for birds requires estimation of the distance from observer to bird and often determination of species identification by sound alone. Precision of the distance estimate declines with distance to the bird, and can be highly variable among observers (Kepler and Scott 1981; Alldredge et al. 2007).

Designing data forms is not a trivial matter and includes four distinct steps (Levy and Lemeshow 1999):

1. Specify the information to be collected
2. Select the data collection strategy
3. Order the recording of data
4. Structure the recording

We developed item 1 in Sect. 8.2.4, and item 2 earlier in this section. Below we develop the remaining two items.

8.2.5.1 Order the Recording

Once the information to be recorded is determined and the collection strategy established, the next step is to group the items in some meaningful and efficient manner, and order the items within the groups. Developing an efficient recording order greatly simplifies data collection, thus reducing observer frustration and increasing the quantity and quality of the data. For our theme example, some items refer to an individual bird, some to the flock the birds may be in, some to the foraging substrate, some to the foraging site, and so on. Items within each group are then ordered in a manner that will maximize data collection. For example, items related to the foraging individual include:

- Age
- Sex
- Plant species the bird is in or under
- Substrate the bird is directing foraging upon
- The behavior of the bird
- Rate of foraging

8.2.5.2 Structuring the Recording

This is the "action plan" for data collection. This plan includes three specific parts:

1. Sampling protocol
2. Data form
3. Variable key

In Box 8.3 we present an example of these three steps that was developed for a study of the distribution, abundance, and habitat affiliations of the red-legged frog (*Rana aurora draytonii*).

Box 8.3

The material given below outlines an example of the three primary products necessary to organize properly a data-collecting strategy (1) sampling protocol, (2) variable key, and (3) data form.

1. Sampling Protocol

This protocol is designed to quantify the habitat use of the red-legged frog (RLF) at three hierarchical scales. It is focused on aquatic environments, although the species is known to use upland sites during certain periods of the year. The hierarchy used is based on Welsh and Lind (1995). Terrestrial sightings are accommodated in this protocol in a general fashion, although more specific information would need to be added at the microhabitat scale than is provided herein.

The landscape scale describes the general geographic relationship of each reach sampled. Additional information that could be recorded in separate notes includes distance from artificial water containments; the distance to, and type of, human developments; and so forth. The appropriate measure of the animal associated with this broad scale would be presence or absence of the species. The macrohabitat scale describes individual segments or plots stratified within each reach by general features of the environment. The appropriate measure of the animal associated with this midscale is abundance by life stage. The microhabitat scale describes the specific location of an egg mass, tadpole, or adult.

All data should be recorded on the accompanying data form using the indicated codes. Any changes to the codes, or addition to the codes, must be indicated on a master code sheet and updated as soon as possible.

The following table summarizes the hierarchical arrangement of sampling used to determine the distribution, abundance, and habitat affinities of red-legged frogs, although the general format is applicable to a variety of species (see Welsh and Lind 1996). Variables should be placed under one classification only.

(continued)

I. Landscape Scale (measured for the reach)
- A. Geographic relationships
  1. UTM coordinates
  2. Elevation (m)
  3. Slope (%)
  4. Aspect (degrees)

II. Macrohabitat Scale (measured for the segment or plot)
- A. Water quality (average values in segment or plot)
  1. Depth (cm)
  2. Flow (cm s^{-1})
  3. Temp. (°C)
  4. Salinity (ppt)
- B. Site description
  1. Size
     - a. Width
     - b. Length (m), if applicable
  2. Predominant condition
     - a. Stream
       - (1) Pool
       - (2) Glide
       - (3) Riffle
     - b. Seep
     - c. Marsh
     - d. Upland
       - (1) Grassland
       - (2) Shrubland
       - (3) Woodland
  3. Sediment
     - a. Coarse
     - b. Medium
     - c. Fine
- C. Predominant aquatic vegetation
  1. Type 1 (open water with no vegetation)
  2. Type 2 (algae, flotsam, ditchgrass, low herbs and grass)
  3. Type 3 (tall, vertical reed-like plants [cattail, rush, sedge])
  4. Type 4 (live and dead tangles of woody roots and branches [willow, cotton wood, blackberry])
- D. Stream covering
  1. % stream covered by overhanging foliage

(continued)

Box 8.3 (continued)

E. Red-legged frogs
   1. Egg mass present
   2. Tadpole present
   3. Adult present

F. Predators and competitors
   1. Predatory fish present
   2. Bullfrogs present (adults or tadpoles)

G. Other herpetofauna

III. Microhabitat Scale (measured for the animal location or subsegment/subplot)

A. Number of red-legged frogs
   1. Egg mass
   2. Tadpole
   3. Mature:
      a. Small (<60 mm snout–urostyle length)
      b. Large (≥60 mm)

B. Animal location:
   1. Land (>50 cm from water edge)
   2. Bank (within 50 cm of water)
   3. Shore (within 50 cm of land)
   4. Water (>50 cm from land)
      a. Depth (cm)

C. Specific animal location
   1. Site description
      a. In litter
      b. Under bank
      c. Under woody debris
   2. Distance from water (m)

D. Predominant aquatic vegetation (if applicable)
   1. Cover (%) by species
   2. Vigor of II.C.1. (live, declining, dead)
   3. Height (cm)

E. Predominant terrestrial vegetation (if applicable)
   1. Cover (%) by species
   2. Vigor of II.C.1. (live, declining, dead)
   3. Height (cm)

F. Aquatic substrates
   1. Cover (%) coarse sediment (rocks, boulders)
   2. Cover (%) medium sediment (gravel, pebble)
   3. Cover (%) fine sediment (sand, silt)

2. Variable Key

The following is a measurement protocol and key for the data form given below. Each variable on the data form is described on this key. In field sampling, changes made to the sampling procedures would be entered on a master key so that data forms can be revised and procedures documented (which is especially important in studies using multiple observers).

I. Landscape Scale

A. Take measurements from a central, characteristic location in the reach.

II. Macrohabitat Scale

Each reach will be divided into sections or plots based on some readily identifiable landscape feature, or some feature that is easy to relocate (e.g., bridge, road, house).

A. Take the mean of measurements needed to characterize the segment (1–3 measurements depending on size of segment)
B. 1. a. Average width of predominant condition given in 2 below
   b. Length for nonstream condition
   2. Predominant condition: record single predominant condition in the segment
   3. Sediment: coarse (rocks, boulders); medium (gravel, pebble); fine (sand, silt)
C. Record the indicated "type" of predominant aquatic vegetation
D. Record the percentage of the stream within the segment that is obscured (shaded) by live foliage of any plant species
E. The goal is to note if any life stage of RLF is present; approximate numbers can be recorded if time allows (code 1 = 1 or a few egg masses or individuals; code 2 = 5–10; code 3 = >10).
F. The goal is to note the presence of any predator or competitor; approximate numbers can be recorded as for E above

III. Microhabital Scale

Individual animals that are selected for detailed analysis, or all animals within subsegments or subplots, can be analyzed for site-specific conditions.

A. The goal is to accurately count all RLF by life stage if working within a subsegment or subplot; or record the specific life stage if sampling microhabitat for an individual.
B. Record the location of the life stage by the categories provided (1–4); also record water depth (4.a.) if applicable.
C. Within the location recorded under B above, record the specific site as indicated (l.a. to c.); also include the distance from water if applicable (2.).
D. 1. Record the estimated percentage cover of the four predominant aquatic plant species within the subsegment/subplot, or within a 1-m radius of the individual RLF; also record the vigor (D.2.) and height (D.3.) of each plant species.

(continued)

Box 8.3 (continued)

E. As for D above, record for terrestrial vegetation. Both D and E can be used if the subsegment or subplot for the individual includes both aquatic and terrestrial vegetation (e.g., if animal is on a bank).
F. Record the approximate percentage cover for each of the indicated substrates.

3. Data Form

RED-LEGGED FROG SURVEY
DATE ___ ___ ___ ___ ___ ___ TIME ___ ___ ___ ___ ___
DRAINAGE___ ___ LOG ___ ___ ___ OBS ___ ___; ___ ___ ___

I. LANDSCAPE:
UTM ___ ___ ___. ___ ___ ___; ___ ___ ___. ___ ___ ___
ELE ___ ___ ___ ___ SLOPE ___ ___ ASPECT ___ ___ ___
II. MACROHABITAT: SITE ___ ___ ___ ___
A. WATER QUALITY: DEPTH ___ ___ ___
FLOW ___ ___ TEMP ___ ___
B. SITE: SIZE: WIDTH ___ ___ ___ LENGTH ___ ___ ___ CONDITION ___ ___ SEDIMENT ___ ___
C. VEG TYPE ___ D. COVER ___ ___ ___ E. RLF ___ ___ ___ F. PRED ___ ___ ___
G. HERPS: ___ ___; ___ ___; ___ ___; ___ ___
III. MICROHABITAT: SITE ___ ___ ___
A. RLF: EGG ___ ___ ___ TAD ___ ___ ___ AD: SM ___ ___ ___ LG ___ ___ ___
B. LOC ___ ___ If LOC = 'W', record DEPTH ___ ___ ___
C. SPECIFIC ___ ___ ___ DIST ___ ___ ___
D. AQ VEG: COVER ___ ___ ___ VIGOR ___ HT ___ ___ ___
E. TERR VEG: COVER ___ ___ ___ VIGOR ___ HT ___ ___ ___
F. SUB: CO ___ ___ ME ___ ___ FI ___ ___
NOTES:

8.2.5.3 Theme Example

Suppose we have indeed decided to record details on foraging birds, their movements, and related topics. Referring back to the section on Step 1 (Sect. 8.2.1) – Questions, we might ask include:

1. Estimate the use of foraging areas
2. Determine if use varies within and between seasons and years, and age and sex
3. Determine the energy expenditures in various crop types

From these three objectives, we might extract the following inventory list of information needs:

1. Characteristics of the bird: age, sex, foraging location, foraging rates
2. Characteristics of the foraging location: the species, size, and health of the plants used for foraging
3. Characteristics of the foraging region: plant species composition, plant density by size and health categories; slope, aspect, distance to water
4. Characteristics of the environment: weather, both regional and foraging site; wind, solar radiation

Note that each of the inventory items is needed to meet the objectives. For example, weather conditions and foraging rates are needed, in part, to calculate energy expenditures.

Data Collection Strategy: There are numerous means of recording foraging behavior. The method chosen must allow analyses that meet objectives. For example, following a single individual for long periods vs. recording many individual behaviors simultaneously; collecting data during inclement weather vs. fair weather; recording data on breeding vs. nonbreeding animals. These choices determine the types of statistical analyses that are appropriate.

8.2.6 Step 6 – Precision and Error

The critical importance of determination of adequate sample sizes and power analyses was developed in Chaps. 2.6.6 and 2.6.7, respectively (see also Thompson et al. 1998, Chap. 6). A priori power analyses are often difficult because of the lack of sufficient data upon which to base estimates of variance. Nevertheless, power calculations remove much of the arbitrariness from study design (power calculations are performed again at various times during the study, as discussed in Sect. 8.2.7). In addition, power calculations give a good indication if you are trying to collect data on too many variables, thus allowing you to refocus sampling efforts on the priority variables (see Steidl et al. [1997] for a good example of application of power analysis to wildlife research).

8.2.6.1 Theme Example

To determine the number of paired plots (treatment vs. control) needed to rigorously test the null hypothesis of no treatment effect, data on nesting biology collected during several previous years of study were available. Although the data were not collected from plots, they were collected within the same general areas. Thus, the data were collected in similar environmental conditions as would be encountered during the new study. If these data had not been available, then information from the literature on studies in similar conditions and with similar species would have been utilized.

For example, the data available on nesting success of Bell's vireos, a species highly susceptible to cowbird parasitism, indicated that nesting success of parasitized nests was 0.8 ± 1.3 (SD) fledglings/nest. Using this value, effect sizes of 0.5, 1.0, and 1.5 fledglings/nest were calculated with power of 80% and $\alpha = 0.1$. Assuming equal variances between plots, and using calculations for one-sided tests (because primary interest was in increased nest success with cowbird control), sample sizes (number of plots) were calculated as >>20 (power = ~50% for $n = 20$), 15, and 7 for the three effect sizes. Thus, we concluded that time and personnel availability allowed us to identify an effect equal to 1.5 additional fledgling per nest following treatment at 80% power.

An additional important and critical step in development of the study plan is independent peer review (see Sect. 1.3.1). Included should be review by experts in the field and experts in technical aspects of the study, particularly study design and statistical analyses. Review prior to actually collecting the data can help avoid, although not eliminate, many problems and wasted effort. In this example, experts from management agencies helped confirm that the questions being asked were relevant to their needs, several statisticians were consulted on the procedures being used, and several individuals studying bird ecology and cowbirds reviewed the design.

8.2.7 Step 7 – Preliminary Analyses

All studies should begin with a preliminary phase during which observers are trained to become competent in all sampling procedures. The development of rigid sampling protocols, as developed above (see Sect. 8.2.5) improves the chances that observers will record data in a similar manner. Training of observers should include:

1. *Testing of visual and aural acuity*. Much wildlife research involves the ability to see and hear well. For example, birds can produce calls that are near the limits of human hearing ability. Slight ear damage, however, can go unnoticed, but result in an inability to hear high-frequency calls. Hearing tests for personnel who will be counting birds using sight and sound should be conducted (e.g., Ramsey and Scott 1981).
2. *Standardization of recording methods*. Unfortunately, there is seldom a correct value with which we can compare samples. Most of our data represent indices of some "true" but unknown value (e.g., indices representing animal density or vegetation cover). Further, field sampling usually requires that the observer interpret a behavior or makes estimate of animal counts or plant cover. Thus,

there is opportunity for variation among observers, which in turn introduces variance into the data set. Training observers to ensure that they collect data in a standardized and consistent manner is thus essential in reducing variance. Studies such as those conducted by Block et al. (1987) and papers in Ralph and Scott (1981, pp. 326–391) are examples of the value of observer training.

Initial field sampling should include tests of data collection procedures; often called *pretesting period*. Such pretesting allows for redesign of data forms and sampling protocols. Pretesting sampling should cover as much of the range of conditions that will be encountered during the study. Some, but seldom all, of the data collected during pretesting might be suitable for inclusion with the final data set.

Levy and Lemeshow (1999) differentiated between the pretest and the pilot survey or pilot study, with the latter being described as a full-scale dress rehearsal. The *pilot study* includes data collection, data processing, and data analyses, and thus allows thorough evaluation of all aspects of the study including initial sample size and power analyses. Thus, a pilot study is often done with a much larger sample than a pretest. Such studies are especially useful when initiating longer term studies.

8.2.7.1 Theme Example

An essential part of the theme study was determining the abundance of cowbirds and potential host species. Because the majority of birds encountered during a formal bird count are heard but not seen, observers must possess advanced identification skills and excellent hearing capabilities. Selecting experienced personnel eases training of observers. Although a talented but inexperienced individual can usually learn to identify by song the majority of birds in an area within 1–2 mo, it takes many years of intensive study to be able to differentiate species by call notes, including especially rare or transient species. Thus, observers should not be asked to accomplish more than their skill level allows.

Even experienced observers must have ample time to learn the local avifauna; this usually involves 3–4 weeks of review in the field and the use of song recordings. Regardless of the level of experience, all observers must standardize the recording of data. In the theme study, observers went through the following steps:

1. During the first month of employment, new employees used tapes to learn songs and call notes of local birds. This included testing each other through the use of tapes. Additionally, they worked together in the field, making positive visual identifications of all birds heard and learning flight and other behaviors.
2. While learning songs and calls, observers practiced distance estimation in the field. Observers worked together, taking turns pointing out objects, with each observer privately recording their estimate of the distance; the distance was then measured. This allows observers to achieve both accurate and precise measurements.

3. Practice bird counts are an essential part of proper training. When any activity is restricted in time, such as a 5-min point count, inexperienced observers become confused, panic, and fail to record reliable data. Having the ability to identify birds (i.e., being a good bird watcher) does not mean a person can conduct an adequate count. Thus, realistic practice sessions are necessary to determine the capability of even good bird-watchers. In the theme study, observers, accompanied by the experienced project leader, conducted "blind" point counts during which each person independently recorded what they saw and heard, as well as an estimate of the distance to the bird. These practice counts were then immediately compared with discrepancies among observers discussed and resolved.

8.2.8 Step 8 – Quality Assurance

The purpose of *quality assurance* (also called quality assurance/quality control, or QA/QC) is to ensure that the execution of the plan is in accordance with the study design. It is important to the successful completion of the study that a formal program of QA/QC is instituted on both the data collection and data processing components (see Levy and Lemeshow 1999).

A principal method of assuring quality control is through resampling a subset of each data set. For example, a different team of observers can resample vegetation plots. Unfortunately, in field biology there are often no absolutely correct answers, especially when visual estimates are involved (e.g., visual estimations of canopy closure). As quantified by Block et al. (1987), sampling errors and observer bias can be minimized through by using relatively rigorous and repeatable measurement techniques whenever possible. For example, measuring dbh with a tape or using a clinometer to estimate tree height rather than through visual estimation. Despite these difficulties, quantifying the degree of interobserver error can only enhance a research project.

In addition to the use of well-trained observers, several other steps can be taken to assure data quality:

1. When conducting data recording in pairs, observers should repeat values back to one another.
2. Another observer at the close of each recording session should proof all data forms. Although this proofing cannot determine if most data are "correct," this procedure can identify obvious recording errors (e.g., 60-m tall oak trees do not exist), eliminate illegible entries (e.g., is that a "0" or a "6"?), and reduce blank entries. Because some of these errors must be corrected by recall of an observer, this proofing must take place as soon after recording as possible.
3. Place each field technician in charge of some aspect of the study. This increases the sense of responsibility and allows the project manager to identify relatively weak or strong personnel (i.e., mistakes are not "anonymous").

8.2.8.1 Theme Example

The quality of data was enhanced and controlled throughout the study duration by incorporating activities such as:

- *Banding*. Repeating of band numbers and colors, sex, age, and all other measurements between observer and recorder. This serves as a field check of data recording, and keeps field technicians alert.
- *Band resightings*. Color bands are difficult to see in the field, and many colors are difficult to differentiate because of bird movements and shadows (e.g., dark blue vs. purple bands). Observers can practice by placing bands on twigs at various distances and heights and under different lighting conditions.
- *Bird counts*. Regular testing of species identification by sight and sound, distance estimation, and numbers counted. Technicians do not always improve their identification abilities as the study proceeds. Numerous factors, such as fatigue, forgetfulness, deteriorating attitudes, etc., can jeopardize data quality.
- *Foraging behavior*. Regular testing of assignment of behaviors to categories. As described above for bird counts, many factors can negatively impact data quality. Additionally, technicians need to communicate continually any changes in interpretation of behaviors, additions and deletions to the variable list, and the like. This becomes especially important in studies using distinct field teams operating in different geographic locations.
- Data proofing: as described above, it is important that one proofs data after every sampling session.

8.2.9 Step 9 – Data Collection

There should be a constant feedback between data collection and QA/QC. Probably one of the principal weaknesses of most studies is a failure to apply QA/QC on a continuing basis. The prevalent but seldom acknowledged problem of *observer drift* can affect all studies regardless of the precautions taken during observer selection and training. The QA/QC should be ongoing; including analysis of sample sizes (are too many or too few data being collected?). Electronic data loggers can be useful in many applications. People can collect data in the field and the data loggers can be programmed to accept only valid codes. Data can then be directly downloaded into a computerized database for proofing and storage. The database then can be queried and analyses made in the statistical program of choice. Previously, data loggers often were beyond the financial reach of many projects. This is no longer true as prices have dropped precipitously and technology advanced considerably.

A weakness of most studies is failure to enter, proof, and analyze data on a continuing basis. Most researchers will quickly note that they have scarcely enough time to collect the data let alone enter and analyze even a portion of it. This is, of course, a fallacious response because a properly designed study would allocate adequate resources to ensure that oversampling or undersampling is not occurring, and that high-quality data are being gathered. Data analysis in multiyear studies should not be postponed until several years' data are obtained. Allocating resources to such analyses will often necessitate a reduction in the scope of the study. Yet, we argue that it is best to ensure that high-quality data be collected for priority objectives; secondary objectives should be abandoned. This statement relates back to our discussion of the choice of questions (Sect. 8.2.1) and the design of the research (Sect. 8.2.3).

8.2.9.1 Applications: Managing Study Plots and Data

When initiated, most studies have a relatively short (a few months to a few years) time frame. As such, maintaining the sites usually only requires occasional replacing of flagging or other markers. It is difficult to anticipate, however, potential future uses of data and the location from where they were gathered (i.e., they might become monitoring studies). In southeastern Arizona, for example, few anticipated the dramatic spreading in geographic area by the exotic lovegrasses (*Eragrostis* spp.) when they were introduced as cattle forage. However, several studies on the impacts of grazing on wildlife included sampling of grass cover and species composition, thereby establishing baseline information on lovegrasses "by accident." By permanently marking such plots, researchers can now return, relocate the plots, and continue the originally unplanned monitoring of these species. Unexpected fires, chemical spills, urbanization, and any other planned or unplanned impact will undoubtedly impact all lands areas sometime in the future. Although it is unlikely that an adequate study design will be available serendipitously, it is likely that some useful comparisons can be made using "old" data (e.g., to determine sample size for a planned experiment). All that is required to ensure that all studies can be used in the future is thoughtful marking and referencing of study plots. All management agencies, universities, and private foundations should establish a central record-keeping protocol. Unfortunately, even dedicated research areas often fail to do so; no agency or university we are aware of has established such a protocol for *all* research efforts.

It is difficult to imagine the amount of data that have been collected over time. Only a tiny fraction of these data resides in any permanent databases. As such, except for the distillation presented in research papers, these data are essentially lost to scientists and managers of the future. Here again, it is indeed rare that any organization requires that data collected by their scientists be deposited and maintained in any centralized database. In fact, few maintain any type of catalog that at least references the location of the data, contents of the data records, and other pertinent information.

Perhaps one of the biggest scientific achievements of our time would be the centralization, or at least central cataloging, of data previously collected (that which is not lost). An achievement unlikely to occur. Each research management organization can, however, initiate a systematic plan for managing both study areas and data in the future.

8.2.9.2 Theme Example

Our example here is brief because the process initiated on the study was primarily continuation, on a regular basis, of the routine established under Step 8, QA/QC (Sect. 8.2.8). Regular testing of observer data gathering was conducted on a bimonthly basis. Additionally, all study points and plots were located both on topographic maps of the study area, and by the use of Global Positioning System (GPS) technology. These data were included in the final report of the study, thus ensuring that the sponsoring agencies had a formal record of each sampling location that could be cross-referenced with the original field data.

8.2.10 Step 10 – Analyses

Testing of hypotheses and evaluating conceptual models are, of course, central features of most wildlife studies (see Sect. 1.3). However, several steps should be taken before formally testing hypotheses or evaluating models. These include:

1. *Calculating descriptive statistics*. As a biologist, the researcher should be familiar with not only the mean and variance, but also the form of the distribution and the range of values the data take on. The message: do not rush into throwing your data into some statistical "black box" without first understanding the nature of the data set. Of interest should be the distribution (e.g., normal, Poisson, bimodal) and identification of outliers.
2. *Sample size analyses*. Hopefully, sample size analyses will have been an ongoing process in the study. If not, then this is the time to determine if you did, indeed, gather adequate samples. Many researchers have found, when attempting to apply multivariate statistics, that they can only use several of the 10, 20, or even 100 variables they collected because of limited sample sizes.
3. *Testing assumptions*. Most researchers understand the need to examine their data for adherence to test assumptions associated with specific statistical algorithms, such as equality of variances between groups, and normality of data for each variable for a *t* test for example. In biological analyses, these assumptions are seldom met, thus rendering the formal use of standard parametric tests inappropriate. To counter violation of assumptions, a host of data transformation techniques is available (e.g., log, square root). However, we offer two primary cautions. First, remember that we are dealing with biological phenomena that may not correspond to a normal statistical distribution; in fact, they usually do not. Thus, it is usually far more biologically relevant to find a statistical technique that fits the data, rather than trying to force your biological data to fit a statistical distribution.

For example, nonlinear regression techniques are available, thus there is little biological justification for trying to linearize a biological distribution so you can apply the more widely understood and simple linear regression procedures. Second, if transformations are applied, they must be successful in forcing the data into a form that meets assumptions of the test. Most often, researchers simply state that the data were transformed, but no mention is made of the resulting distributions (i.e., were the data normalized or linearized?). Nonparametric and nonlinear methods are often an appropriate alternative to forcing data (especially when sample sizes are small) to meet parametric assumptions.

Once a specific α-value is set and the analysis performed, the P-value associated with the test is either "significantly different from α" or "not significantly different from α." Technically, a P-value cannot be "highly significantly different" or "not significant but tended to show a difference." Many research articles provide specific α-values for testing a hypothesis, yet discuss test results as if α were a floating value. Insightful papers by Cherry (1998) and Johnson (1999) address the issue of null hypothesis significance testing, and question the historical concentration on P-values (see Sects. 1.4.1 and 2.6.1). Although we agree with their primary theses, our point here is that, once a specific analysis has been designed and implemented, changing your evaluation criteria is not generally acceptable. The salient point is that the P-value generated from the formal test must be compared directly with the α-level set during study development; the null hypothesis is either rejected or not rejected. Borderline cases can certainly be discussed (e.g., a $P = 0.071$ with an α of 5%). But, because there should have been a very good reason for setting the α-level in the first place, the fact remains that, in this example, the hypothesis was not rejected. You have no basis for developing management recommendations as if it had been rejected. This is why development of a rigorous design, which includes clear development of the expected magnitude of biological effect (the effect size), followed by thorough monitoring of sampling methods and sample sizes must be accomplished throughout the study. Statistical techniques such as power analysis and model testing (where appropriate) help ensure that rigorous results were indeed obtained. Many articles and books have been written on hypothesis testing, so we will not repeat those details here (Winer et al. 1991; Hilborn and Mangel 1997).

8.2.10.1 Theme Example

Field sampling indicated that removing cowbird adults from the vicinity of nesting hosts, in combination with removal of cowbird nestlings and addling cowbird eggs caused a 54% increase in the number of young fledged (2.0 ± 0.8 young/nest on treatment plots vs. 1.3 ± 1.1 young/nest on nontreated plots; note that these data are preliminary and should not be cited in the context of cowbird–host ecology and management). According to our a priori calculations (see Sect. 8.2.6), we could not detect an effect size of <1 with 80% power and $P = 0.1$ given our sample sizes. However, these results for treatment effects did indicate that our findings were not significantly different ($P = 0.184$) and had an associated power of about 61%. Thus, the results we obtained were what we should have expected given our initial planning based on power analysis.

8.2.11 Step 11 – Interpretation

Conclusions must be drawn with reference to the population of inference only. A properly designed study will clearly identify the population of inference; few studies provide this statement. Rather, most studies are localized geographically, but seek to extend their results to a much wider geographic area in hopes of extending the utility of their results. Statements such as "if our results are applicable to a wider area..." are inappropriate. Without testing such an assumption, the researcher is helping to mislead the resource manager who needs to use this type of data. If the resource need identified before the study covers a "wider" geographic area, then the study should have been designed to cover such an area.

Estimates of population characteristics. Biologists rely on statistics to the point of often abandoning biological common sense when reporting and evaluating research results (see Sect. 1.5.3). We have become fixated on the mean, often ignoring the distribution of the data about that value. However, the mean value often does not occur, or at least seldom occurs, in nature. Populations are composed of individual animals that use various resource axes; they use one or more points along the available resource axis or continuum; an individual does not use the mean per se. For example, a bird foraging for insects on a tree trunk may move slowly up a tree, starting at 2 m in height and ending at 10 m in height. Simplistically this places the bird at a mean height of 6 m, a height at which it may have spent only 10–12% of its foraging times. Likewise, a bimodal distribution will have a mean value, yet from a biological standpoint it is the bimodality that should be of interest. Most of the parametric statistical analyses we use to help examine our data rely on comparisons of mean values, and consider the distribution about the mean only with respect to how they impact assumptions of the test.

Thus, when interpreting research results, biologists should examine both the distribution and the mean (or median or modal) values. Graphical representations of data, even if never published (because of journal space limitations), will be most instructive in data interpretation (e.g., the bimodality mentioned above).

Comparisons of estimated population characteristics with a published constant. A standard feature of most research reports is comparison of results with published data of a similar nature. This is, of course, appropriate because it places your results in context of published standards, and helps build a picture of how generalizable the situation may be. Such comparisons, however, should begin with a thorough analysis of the original papers under review. To be frank, just because a paper is published in a peer-reviewed outlet does not imply it is without weakness or even error. Thus, rather than simply wondering why "Smith and Jones" failed to reject the hypothesis that you have now rejected, you should examine Smith and Jones to see if (1) sample sizes were analyzed, (2) power analysis was conducted, (3) α was adhered to, and (4) their interpretation of results is consistent with accepted statistical procedures (e.g., assumptions tested?).

Comparisons of complementary groups with respect to the level of an estimated characteristic. "The proportion of adult males 3–4 years of age who had a parasite load was greater among residents of urban areas (47%) than among residents of non-urban areas (22%)." For this type of statement to be justified by data, the two estimates should meet minimum specifications of reliability (e.g., each estimate should have a CV of less than 25%), and they should differ from each other statistically.

8.2.12 Step 12 – Reporting

Publishing results in peer-reviewed outlets, or at least in outlets that are readily available in university libraries (e.g., certain government publications), is an essential part of the research process (see Sect. 1.3.1). Regardless of how well you think your study was designed, it must pass the test of independent peer review. The graduate thesis or dissertation, a final report to an agency, in-house agency papers, and the like, are not publications per se. These outlets are termed *gray literature* because they usually are not readily available, and typically do not usually receive independent peer review. Both the editor and the referees must be completely independent from your project for your work to be considered truly peer reviewed. We hasten to add that publication in a peer-reviewed outlet does not confer rigor to your study, nor should the papers published in such outlets be accepted without question. Rather, the reader can at least be assured that some degree of independent review was involved in the publication process.

Further, there are several tiers of journals available for publication. Although the division is not distinct, journals can be categorized roughly as follows:

- *First tier.* Publishes only the highest quality science; work that advances our understanding of the natural world through rigorously conducted studies. *Science, Nature.*
- *Second tier.* Similar to the first tier except not as broad in scope, especially work from ecology that has broad application to animals and/or geographic regions. *Ecology, American Naturalist.*
- *Third tier.* Journals that emphasize natural history study and application of research results, especially as related to management of wildlife populations: application usually to broad geographic areas. *Journal of Wildlife Management, Ecological Applications, Conservation Biology*, the "-ology" journals (e.g., *Condor, Auk, Journal of Mammalogy, Journal of Herpetology*).
- *Fourth tier.* As for third tier except of more regional application; work not necessarily of lower quality than second or third tier. Many regional journals, *Southwestern Naturalist, Western North American Naturalist, American Midland Naturalist.*

- *Fifth tier*. Similar to fourth tier except accepts papers of very local interest, including distribution lists and anecdotal notes. *Transactions of the Western Section of the Wildlife Society, Bulletin of the Texas Ornithology Society, Texas Journal of Science* (or similar).

We again hasten to add that there are numerous seminal papers even in fifth tier journals. Virtually any paper might help expand our knowledge of the natural world. Within reason, we should strive to get the results of all research efforts published. Some journals emphasize publication of articles that have broad spatial applicability or address fundamental ecological principles. In contrast, other journals publish articles that include those of relatively local (regional) and species-specific interest. Contrary to what some people – including journal editors – apparently think, the spatial or fundamental applicability of an article primarily concerns the service that a journal is providing to its readers (or members of the supporting society) rather than the quality of the article per se. Thus, publishing in a "lower tier" journal does not mean your work is not of as high a quality as that published in a high tier outlet.

The publication process is frustrating. Manuscripts are often rejected because you failed to explain adequately your procedures, expanded application of your results far beyond an appropriate level, provided unnecessary detail, used inappropriate statistics, or wrote in a confusing manner. In many cases, the comments from the editor and referees will appear easy to address, and you will be confused and frustrated over the rejection of your manuscript. All journals have budgets, and must limit the number of printed pages. The editor prioritizes your manuscript relative to the other submissions and considers how much effort must be expended in handling revisions of your manuscript. Your manuscript might be accepted by a lower tier journal, even though it may have received even more critical reviews than your original submission to a higher tier outlet. This often occurs because the editor has decided that your paper ranks high relative to other submissions, because the editor or an associate editor has the time to handle several revisions of your manuscript, and/or because the journal budget is adequate to allow publication.

The response you receive from an editor is usually one of the following:

1. Accept as is. Extraordinarily rare, but does occur.
2. Tentatively accepted with minor revision. This occurs more often, but is still relatively rare. The "tentative" is added because you still have a few details to attend to (probably of minor clarification and editorial in nature).
3. Potentially acceptable but decision will be based on revision. This is usually the way a nonrejected (see below) manuscript is handled. If the editor considers your manuscript of potential interest, but cannot make a decision because of the amount of revision necessary, he will probably send your revision back to the referees for further comment. The editor expects you to try to meet the criticisms raised by the referees. It is in your best interest to try to accommodate the referees, be objective, and try to see their point. A rational, detailed cover letter should explain to the editor how you addressed referees' concerns and suggestions, and why you did not follow certain specific suggestions. This is especially important if your revision is being sent back to the same referees!

4. Rejected but would consider a resubmission. In this case, the editor feels that your manuscript has potential, but the revisions necessary are too extensive to warrant consideration of a revision. The revisions needed usually involve different analyses and substantial shortening of the manuscript. In essence, your revision would be so extensive that you will be creating a much different manuscript. The revision will be treated as a new submission and will go through a new review process (probably using different referees).
5. Rejected and suggests submission to a regional journal. The editor is telling you that you aimed too high and should try a lower tier outlet. Use the reviews in a positive manner and follow the editor's suggestion. Your work likely does not have broad appeal but will interest a more local audience.
6. Returned without review. The editors of some journals might return your manuscript without submitting it to review. He or she has made the decision that your manuscript is inappropriate for this journal. Returning without review happens much more frequently in European ecology journals and in the first and second tier North American journals. Although it is understandable that an editor wants to keep the workload on the referees to a minimum, we think it is usually best to allow the referees to make an independent judgment of all submissions; otherwise, the editor functions as a one-person peer review.

Although the reviews you receive are independent from your study, they are not necessarily unbiased. We all have biases based on previous experience (see Sect. 1.2.4). Further, at times, personal resentment might sneak into the review (although we think most people clearly separate personal from professional views). Nevertheless, it is not uncommon for your initial submission to be rejected. It is not unusual for the rejection rate on the 1–3 tier journals to exceed 60%. Although disappointing, you should always take the reviews as constructive criticism, make a revision, and submit to another, possibly lower-tier, journal. Your "duty" as a scientist is to see that your results are published; you have no direct control over the review process or the outlet your work achieves.

8.3 Trouble Shooting: Making the Best of a Bad Situation

Once you have completed your field or lab study there is nothing you can change about your study design per se. That is, if you sampled birds in three treated and three untreated, 1 ha plots; you cannot suddenly have data on birds from 500 m long transects or from larger plots. But, let us say you have determined (perhaps through a hypercritical peer review) that you had insufficient samples to determine treatment effects; you needed more than three treated plots. Is your study ruined? If you are a student, what about completing your thesis? In the sections that follow, we discuss some general approaches to getting something meaningful out of the data you collected even when your design was inappropriate, or when a catastrophe struck your study.

8.3.1 Post-study Adjustments of Design and Data Handling

As noted in Sect. 8.3, you cannot change your study design after the study is completed. You can, however, legitimately change the way you group and analyze your data, which in essence changes your design. For example, say you have collected data in a series of eight plots with the intent of determining the impacts of thinning hardwoods on bird abundance. Because there were no pretreatment data, you were forced to use plots that had been thinned over a period of 4 years; plot size varied with size of the hardwood stands. These treatment plots were distributed over a minor elevation gradient (say, 200 m), and were located on slopes of varying slope and aspect. At once you will recognize several key problems with this design with regard to treatment effects: the treatments are confounded by differences in age, elevation, slope, and aspect. Perhaps analysis of covariance can assist with some of these confounding issues, but problems with sample size and plot size remain. An alternative to trying to move forward with a study of treatment effects would be to turn the study around and look at the response of birds to variations in a gradient of hardwood density and not the treatments per se. Yes, problems with confounding variables remain, but you will be held to a different standard during the peer review process by using a gradient approach rather than an experimental "treatment effects" approach. You will still be able to talk about how birds respond to hardwood density, and make a few statements about how "my results might be applicable to the design of hardwood treatments." Thus, you have a posteriori changed your method of data processing and analysis (from a two-group treatment vs. no treatment to a gradient approach).

Most field biologists recognize that the season of study and the age and sex of the study animals can have a profound influence on study results. As such, data are often recorded in a manner to separate effects of season (time), age, and sex from the desired response variable (e.g., foraging rate). However, dividing your sample into many categories effectively lowers the sample size. That is, if a priori power analysis indicates that you need a sample of 35 individuals (we discuss the issue of independence below) to meet project goals, this usually means a sample of 35 *each* of, say, adult males, subadult males, adult females, and subadult females. Thus, your sample has just been increased to about 140 and probably 140 per season. A standard rule of thumb is to always collect data in reasonable categories because you can always go back and lump, but often you cannot go back and split (i.e., if you did not record age you cannot divide your data into age categories). Thus, while lumping data certainly lessens your ability to tease out biologically meaningful relationships, it does remain an option when sample sizes are too small in your desired categories. You do run the risk of obscuring, say, age specific activities: e.g., the adults show a positive reaction to something you are measuring and the subadults a negative reaction, which expresses itself in your data as "no reaction." But, here we are talking about ways to make the best of a bad situation; we are not able to put insight into data that were collected in a manner that could obscure meaningful relationships.

Another rather common way to change a design is to add or remove nesting or blocking within an experimental framework (see Chap. 3 for discussion of experimental designs). We have often encountered situations where a referee has suggested that data could be more effectively analyzed by blocking across some environmental feature; for example, analyzing data on rodent abundance by elevational blocks or blocks based on vegetation type. Although seen more rarely than blocking, placing data into a nested organization might also enhance an analysis; for example, analyzing young within a burrow that are nested within a colony that is nested within a region. A posteriori *adding* blocking or nesting means that you managed to have an adequate sample size for such procedures. In most cases, you probably will lack sufficient data for blocking or nesting; instead, you move or *remove* blocks or nesting criteria. A related example would be the removal of paired plots into a nonpaired analysis (see Chap. 3).

The issue of independence of data (samples) is one of the central foundations of study design and statistics. Unfortunately, for many applications there are no absolute criteria upon which "independence" can be based. Repeatedly drawing blood samples from the same individual and calling the samples "independent" is likely a clear cut case of nonindependence (or pseudoreplication; see Sect. 2.3).

8.3.2 Adjusting Initial Goals to Actual Outcome

Related to and an integral part of our discussion and examples in Sect. 8.3.1 is the altering of your study goals in light of design inadequacies (or insufficient or inappropriate samples). Situations do arise that are outside of your reasonable control, such as natural or human-caused catastrophes. Fires and floods could virtually obliterate a study area. Perhaps serendipity will have left you with an ideal "treatment effects" study (i.e., pre- and postfire). More likely, you will be left with ashes. Given that a graduate student does not want to stay on for another 2–3 years to complete a different study, about all one might be left with in such circumstances is a brief "before the fire happened" look at the ecology of the animals that were under study. Alternatively, the study can be altered to select different study sites, abandon the initial study goals, and expand to look at the basic ecology of the target animal(s) over a wider geographic area.

As described above (Sect. 8.2.12), there are numerous opportunities to locate a journal that will welcome your manuscript. It might be that your study has been compromised in some way that prevents you from generating the type of ecological or management conclusions that had been intended when you began the work. Thus, you will likely save yourself, as well as referees and a journal editor, a lot of time by matching your manuscript with an appropriate journal (as described under Sect. 8.2.12 regarding tiers of journals). Again, your duty as a scientist is to get your work published. We are certainly not going to argue that a paper in *Science* or *Nature* would get you more accolades than a paper in the *Southwestern Naturalist*. Nevertheless, the fact remains that most of our work will be species and/or site specific. We recommend that you seek the advice from well-published individuals in selecting a journal for submission of your work.

Another option to consider when things have not gone as you anticipated with your study is conversion of your focus to that of a pilot study. If you are unable to draw reasonable conclusions based on your data, then a focus on hypothesis generating rather than hypothesis testing can be a reasonable approach. For example, say that your intended study goal was to make recommendations for management of a rare salamander based on marking and tracking individuals. But, because of various difficulties in first locating and then tracking the animals (e.g., marks could not be read), you failed to gather an adequate sample of individuals to address your initial goal when time and money expired. A reasonable approach, then, would be to focus your study as more methodological and report on solutions to these difficulties. Presentation of your ecological (e.g., habitat use) data would be appropriate, but only as preliminary findings; management recommendations would not likely be appropriate. While serving as co-Editors of *The Journal of Wildlife Management*, Block and Morrison frequently recommended that manuscripts be sent to a regional natural history journal because meaningful management recommendations could not be developed from the study results.

Related to focusing on a pilot study is focusing your work as a *case study*. There are situations in which your biological study was, perhaps, too localized or too brief in duration to warrant a full research article. For example, all of us have been contracted to conduct rather short-term (i.e., one season) assessments of the distribution of endangered species on a wildlife management area, military base, or a site proposed for development. Additionally, any study that results in a small data set collected over a short timeframe might be appropriate for addressing as a case study. While these are valuable data for the issue at hand, they usually have little interest to the general scientific readership of a journal. However, focusing on the *issue* underlying the reason for the study rather than the data collected is a viable way to pursue publication. For example, data collected on the distribution of endangered species on a military base that is slated for closure and potential economic development could serve as the basis for an article on the role of military bases in conservation; the story is the vehicle for carrying the data. Likewise, data collected on a few water catchments could be used to review and discuss the issue of adding water to the environment.

8.3.3 Leadership

The fundamental resource necessary for success in any scientific study is leadership. Regardless of the rigor of the design and the qualifications of your assistants, you must be able to train, encourage, critique – and accept criticism and suggestions, and overall guide your study throughout its duration. Leadership skills are required to develop and guide successful research teams. We have all read job advertisements in which a requirement reads something like "a proven ability to work well with others...". Employers are looking for people they can work with.

Yet, seldom has a study continued to completion in which no interpersonal problems have arisen. Thus, the skills necessary to properly select, train, and then guide a research team are an essential component of a successful study.

We cannot detail the steps needed to become a leader or how to successfully develop and manage research teams in this book. There are many resources, including books and workshops, which seek to develop leadership skills in project leaders. Probably of more importance, fundamentally, are workshops and other programs that seek to help you understand what drives you as an individual and what causes you to behave and react in the manner that you do under stressful situations.

8.4 Summary

In this chapter, we provide a step-by-step guide to conceptualizing, designing, implementing, analyzing, and publishing a wildlife study. Thoroughly developing the sequence of concept to publication means not only that you have a well thought out plan, but also provides a process to foresee potential problems that may strike a project. The essential steps to a research study are outlined in Fig. 8.1. These steps are the single-study application of the steps typically used in successful natural science research programs (Table 1.2).

We also discuss some general approaches to getting something meaningful out of the data you collected even when your design was inappropriate, or when a catastrophe struck your study. Although you cannot change your study design after the study is completed, you can legitimately change the way you group and analyze your data, which in essence changes your design. We also describe other changes in how data are handled after collection that might lessen your ability to tease out biologically meaningful relationships (assuming the study had been appropriately designed and nothing went wrong), but retain enough valuable information to warrant analysis and publication. There are journals that will welcome your manuscript even if your study was compromised in some way, preventing you from generating the type of ecological or management conclusions that you intended at the onset. We close the chapter with a brief reminder on the central role that leadership plays in developing and guiding a successful research team.

References

Alldredge, M. W., T. R. Simons, and K. H. Pollock. 2007. Factors affecting aural detections of songbirds. Ecol. Appl. 17: 948–955.

Block, W. M., K. A. With, and M. L. Morrison. 1987. On measuring bird habitat: Influence of observer variability and sample size. Condor 89: 241–251.

Cherry, S. 1998. Statistical tests in publications of The Wildlife Society. Wildl. Soc. Bull. 26: 947–953.

Cochran, W. G. 1983. Planning and Analysis of Observational Studies. Wiley, New York, NY.

Cook, C. W., and J. Stubbendieck. 1986. Range Research: Basic Problems and Techniques. Society for Range Management, Denver, CO.

Ehrlich, P. R., D. S. Dobkin, and D. Wheye. 1988. The Birder's Handbook: A Field Guide to the Natural History of North American Birds. Simon and Schuster, New York, NY.

Finch, D. M. 1983. Brood parasitism of the Abert's towhee: Timing, frequency, and effects. Condor 85: 355–359.

Garton, E. O., J. T. Ratti, and J. H. Giudice. 2005. Research and experimental design, in C. E. Braun, Ed. Techniques for Wildlife Investigations and Management, pp. 43–71, 6th Edition. The Wildlife Society, Bethesda, MD.

Harris, J. H., S. D. Sanders, and M. A. Flett. 1987. Willow flycatcher surveys in the Sierra Nevada. West. Birds 18: 27–36.

Hilborn, R., and M. Mangel. 1997. The Ecological Detective: Confronting Models with Data. Monographs in Population Biology 28. Princeton University Press, Princeton, NJ.

Johnson, D. H. 1999. The insignificance of statistical significance testing. J. Wild. Manage. 63: 763–772.

Kepler, C. B., and J. M. Scott. 1981. Reducing bird count variability by training observers. Stud. Avian Biol. 6: 366–371.

Kuhn, T. S. 1962. The Structure of Scientific Revolutions. University of Chicago Press, Chicago, IL.

Laymon, S. A. 1987. Brown-headed cowbirds in California: Historical perspectives and management opportunities in riparian habitats. West. Birds 18: 63–70.

Lehner, P. N. 1996. Handbook of Ethological Methods, 2nd Edition. Cambridge University Press, Cambridge.

Levy, P. S., and S. Lemeshow. 1999. Sampling of Populations: Methods and Applications, 3rd Edition. Wiley, New York, NY.

Martin, T. E. 1992. Breeding productivity considerations: What are the appropriate habitat features for management? in J. M. Hagan and D. W. Johnston, Eds. Ecology and Conservation of Neotropical Migrant Landbirds, pp. 455–473. Smithsonian Institution Press, Washington, DC.

Martin, P., and P. Bateson. 1993. Measuring Behavior: An Introductory Guide, 2nd Edition. Cambridge University Press, Cambridge.

Morrison, M. L., L. S. Hall, S. K. Robinson, S. I. Rothstein, D. C. Hahn, and J. D. Rich. 1999. Research and management of the brown-headed cowbird in western landscapes. Stud. Avian Biol. 18.

Morrison, M. L., B. G. Marcot, and R. W. Mannan. 2006. Wildlife–Habitat Relationships: Concepts and Applications, 3rd Edition. Island Press, Washington, DC.

Ralph, C. J., and J. M. Scott, Eds. 1981. Estimating numbers of terrestrial birds. Stud. Avian Biol. 6.

Ramsey, F. L., and J. M. Scott. 1981. Tests of hearing ability. Stud. Avian Biol. 6: 341–345.

Robinson, S. K., J. A. Gryzbowski, S. I. Rothstein, M. C. Brittingham, L. J. Petit, and F. R. Thompson. 1993. Management implications of cowbird parasitism on neotropical migrant songbirds, in D. M. Finch and P. W. Stangel, Eds. Status and Management of Neotropical Migratory Birds, pp. 93–102. USDA Forest Service Gen. Tech. Rpt. RM-229. Rocky Mountain Forest and Range Experiment Station, Fort Collins, CO.

Romesburg, H. C. 1981. Wildlife science: Gaining reliable knowledge. J. Wildl. Manage. 45: 293–313.

Steidl, R. J., J. P. Hayes, and E. Schauber. 1997. Statistical power analysis in wildlife research. J. Wildl. Manage. 61: 270–279.

Thompson, W. L., G. C. White, and C. Gowan. 1998. Monitoring Vertebrate Populations. Academic, San Diego, CA.

Welsh Jr., H. H., and A. J. Lind. 1995. Habitat correlates of Del Norte Salamander, *Plethodon elongatus* (Caudata: Plethodontidae), in northwestern California. J. Herpetol. 29: 198–210.

Winer, B. J., D. R. Brown, and K. M. Michels. 1991. Statistical Principles in Experimental Design, 3rd Edition. McGraw-Hill, New York, NY.

Chapter 9
Education in Study Design and Statistics for Students and Professionals

9.1 Introduction

There is a fear of statistics among the public, state and federal officials, and even among numerous scientists. The general feeling appears to be based on the convoluted manner in which "statistics" is presented in the media and by the cursory introduction to statistics that most people receive in college. Among the media, we often hear that "statistics can be used to support anything you want"; thus, statistics (and perhaps statisticians by implication) become untrustworthy. Of course, nothing could be further from the truth. It is not statistics per se that is the culprit. Rather, it is usually the way in which the data were selected for analysis that results in skepticism among the public.

Additionally, and as we have emphasized throughout this book, "statistics" and "study design" are interrelated yet separate topics. No statistical analysis can repair data gathered from a fundamentally flawed design, yet improperly conducted statistical analyses can easily be corrected if the design was appropriate. In this chapter we outline the knowledge base we think all natural resource professionals should possess, categorized by the primary role one plays in the professional field. Students, scientists, managers, and yes, even administrators, must possess a fundamental understanding of study design and statistics if they are to make informed decisions. We hope that the guidance provided below will help steer many of you toward an enhanced understanding and appreciation of study design and statistics.

9.2 Basic Design and Statistical Knowledge

As undergraduates, we usually receive a rapid overview of frequency distributions, dispersion and variability, and basic statistical tests (e.g., *t* tests). It is our opinion, built on years of teaching and discussing statistics with students, that few receive a solid foundation of basic statistical or mathematical concepts let alone study design. Even if the foundation was thorough, however, it is not reinforced through continual use after the single undergraduate course. In addition, this undergraduate exposure comes before the student is sufficiently versed in scientific methodology to see how study design and statistical analysis fit together. This makes it less likely that they will be motivated to retain what they learn. Typically, advanced statistical training and courses in study design do not come until graduate school. Thus, only those continuing on to advanced degrees and associated research projected are given the opportunity to put statistical learning into practice. Even here, such experiences typically are usually limited. At the MS level, there seldom is adequate time to take more than the basic statistical courses that covers analysis of variance (ANOVA) and linear regression and to learn the necessities for evaluating data collected for the MS thesis. At the PhD level, some students take additional courses in more advanced procedures such as multivariate statistics, nonparametric analyses, and perhaps experimental (ANOVA-based) design. However, even for PhD students, most statistical knowledge is focused what is needed to complete the dissertation analysis. Relatively few schools offer study design and statistics courses oriented specifically toward natural resources.

People who do not continue on to graduate school seldom receive any additional training in study design or statistics. Also, there is a serious misconception that those entering management, administrative, or regulatory positions have no need for statistics in their work. We counter, however, that nothing could be further from reality. In fact, a case can easily be made that managers, administrators, and regulators must fully understand the general principles of study design and statistics, including advanced design and analysis methods. Regulators and administrators are called upon to make decisions regarding permit applications that often directly or indirectly impact sensitive species and their habitats. These decisions are based, in part, on the best scientific data available regarding the proposed impact. Because proposed projects usually have proponents and opponents, the regulator–administrator is confronted with opposing opinions on the meaning of the data; remember, "statistics can be used to support anything you want." As such, the regulator–administrator that is naive regarding design and statistics has little hope of making a rational, informed decision. Likewise, managers must sift through myriad published papers, unpublished reports, and personal opinions to make decisions regarding a multitude of land-use practices. It boils down to this: Ecological systems we manage are only partially observable, through sampling. Appropriate study designs and statistical analyses are necessary to extract the signal (i.e., potential causal factors) from the noise (inherent variability) so that the information derived from the scientific method can be maximized and informed management decisions made.

All professionals have a responsibility for making informed decisions; the "buck" does stop somewhere. By analogy, if your accountant reports to you that the budget is in order, but later discovers an accounting error, you will be ultimately responsible for the budget debacle (which could result in disciplinary or even legal actions). A manager must have at least a fundamental understanding of the budget – income and expenditures, basic accounting practices – to make sure that the budget is not grossly out of balance; he will be held accountable. Likewise, if an wildlife administrator cannot adequately evaluate the rigor of an endangered species survey, for example, and cannot determine if appropriate statistics were used, then he or she would look rather foolish blaming a failure to protect the species on his or her staff. That is, how can you manage people if you do not know – at least fundamentally – what they are doing? As noted by Sokal and Rohlf (1995), there appears to be a very high correlation between success in biometry and success in the chosen field of biological specialization.

Wildlife professionals must, at a minimum, be able to ask the proper questions needed to interpret any report or paper. Such questions include issues of independence, randomization, and replication; adequacy of sample size and statistical power; pseudoreplication and study design; and proper extrapolation of results (as we developed in Chaps. 1 and 2). You do not, for example, need to know how to invert a matrix to understand multivariate analyses (see Morrison et al. (2006) for some examples). In legal proceedings, one must be clear on the reasons underlying a management or regulatory decision, but does not need to be able to create statistical software.

Thus, it is incumbent on all professionals to not only achieve an adequate understanding of study design and statistics (both basic and advanced), but also keep current on methodological advances. The field of natural resource management is becoming more analytically sophisticated (see Chap. 2). For example, it is now common to use rather complicated population models to assist with evaluation of the status of species of concern – simply plotting trends of visual counts on an *X–Y* graph no longer suffices for either peer-review or management planning. Below we outline what we consider adequate training in design and statistics for natural resource professionals, including university training, continuing education, and the resources available to assist with learning.

9.2.1 The Undergraduate

It is an axiom that all education must rest on a solid foundation. Otherwise, any hope of advancement of knowledge and understanding is problematic. Most universities require that undergraduates in the sciences (e.g., biology, chemistry, physics, and geology) have taken courses in mathematics through algebra, trigonometry, and often calculus. Beyond these basic courses, universities vary in their requirements for students specializing in natural resources and their management. Many popular *biostatistics* textbooks are written so as to not require mathematical education beyond elementary algebra (e.g., Hoshmand 2006; Sokal and Rohlf 1995; Zar 1998), or are written in a "nonmathematical" manner (e.g., Motulsky 1995). These are good books that impart a solid foundation of basic statistics – mentioning their requirements imply no criticism. Sokal and Rohlf (1995) noted that, in their experience, people with limited mathematical backgrounds are able to do excellently in *biometry*. They thought there was little correlation between innate mathematical ability and capacity to understand biometric methodologies. However, many more advanced mathematical and statistical methods in natural resources require an understanding of more advanced mathematics, including calculus (e.g., many modeling techniques, population estimators). All students planning on later receiving graduate degrees should take at least an introductory course in calculus as well as an additionally course in probability theory in preparation for advanced methods in natural resources. Otherwise, they will be limited in the types of courses they will be qualified to take in graduate school. In addition, statistical methods are better understood, or at least students are better motivated to understand them, when they have had or concurrently take a course in scientific methodology as applied to natural resources. Such courses should be part of the core curriculum for undergraduates in natural resources.

9.2.1.1 Biometry or Fundamental Statistics?

The heading for this subsection implies a dichotomy between biometric and other approaches to statistics. In reality, textbooks and the way statistics courses are taught vary widely, from very applied, "cookbook" approaches, to highly theoretical instruction into the underlying mathematics of statistical methods. As noted earlier in this chapter, we think that all resource professionals require, at a minimum, a good knowledge of the principles of study design and statistics. Thus, the applied approach, which minimizes formula and mathematics, is adequate in many cases for interpretation of research results. Knowing that ANOVA *somehow* looks for significant differences in two or more groups, that various rules-of-thumb are available to determine necessary sample size, and that pseudoreplication includes the inappropriate calculating of sample sizes will suffice for many biologists, managers, and administrators.

Thus, "biometrics" courses tend to sacrifice fundamentals of theory and inference for applications and interpretation. This is appropriate if the course is considered a self-contained survey of common procedures, and not considered a prerequisite to more advanced statistics courses. However, if your expectation is that you will be conducting independent research and writing and evaluating scientific publications, then a better understanding of the mathematical underpinnings of statistics is required. Using our previous examples, advancing from simple one-way ANOVA to tests of interactions and blocking, properly anticipating sample size requirements (e.g., through power analysis), and understanding the statistical basis of pseudoreplication all require that the mathematics of the procedures be understood at least in general terms. The ability to interpret an ANOVA computer printout is much different from being able to explain how the residuals (error) were calculated. The single-semester "biometrics" courses often offered in biology and natural resource programs do not provide these fundamentals.

When graduate school is the goal, it is probably better to sacrifice application (i.e., the single-semester biometrics course) for fundamentals. Many universities offer a two-semester "fundamentals" course within the statistics department; many also offer a version of these courses for nonstatistics majors. Such courses usually require, for example, that each step in an ANOVA can be interpreted – calculation of degrees of freedom, sources of variation, and interaction terms. Such understanding is necessary to properly analyze and interpret complicated data, and is fundamental to more advanced parametric techniques (e.g., multivariate analyses). It is unlikely that the undergraduate will have time to take additional statistics courses.

9.2.2 *The Graduate*

The first task of many new graduate students is to fulfill the courses they missed (or avoided) at the undergraduate level. Many universities offer graduate level statistics courses in the Statistics Department aimed at nonmajors to fill these gaps. Such courses often cover two semesters and offer a detailed coverage of the fundamentals of statistics. However, most frequently these courses focus only on application of statistical approaches, rather than delving into the theory behind those applications. This is where the advantage of solid, fundamental mathematical and statistical training during one's undergraduate training begins to show its advantages. Such courses usually allow graduate students to step directly into more advanced courses such as sampling, nonparametric statistics, categorical data analysis, multivariate statistics, and experimental design.

9.2.2.1 The Importance of Formal Experimental Designs

As outlined throughout this book, fundamental to study design is an understanding of experimental methodologies. Most ecological studies are complex, and are made all the more difficult by a host of confounding factors. Although we hold to the notion that natural history observations qualify as science, natural historians nevertheless need to understand experimental designs and their associated statistical analyses. Even when a hypothesis is not specified and an experiment not initiated, advanced statistical procedures are often needed to try and isolate the factors causing the response variable to behave in the manner observed. For example, it is often difficult to know at what scale a process may be operating, such as the ecological processes influencing abundance at several spatial scales. Thus, it can be confusing to know how to start a sampling procedure. Nested sampling designs are one method to help determine the spatial pattern (Underwood 1997, p. 275). However, implementing such a design – except by luck – requires knowledge of the more advanced area of nested (or hierarchical) ANOVA. Ignorance of such procedures naturally limits even how the natural historian could approach a problem. Thus, we recommend that, following the introductory two-semester statistics courses, students enroll in an ANOVA-based experimental design course. A popular textbook that concentrates on these procedures is Underwood (1997).

9.2.2.2 Parametric vs. Nonparametric Methodologies

There are often several tests available to analyze a data set; choosing the most appropriate test can often be tricky. A fundamental decision that must be made, however, involves choosing between the two families of tests: namely, parametric and nonparametric tests (e.g., see Motulsky (1995) for a good discussion). Many sampling problems in natural resources involve small populations and/or populations that do not exhibit a normal distribution, i.e., they are skewed in some fashion. Large data sets usually present no problem. At large sample size, nonparametric tests are adequately powerful, and parametric tests are often robust to violations of assumptions as expected based on the central limit theorem. It is the small data set that represents the problem. It is difficult to determine the form of the population distribution, and the choice of tests becomes problematic: nonparametric tests are not powerful and parametric tests are not robust (Motulsky 1995, p. 300).

The researcher is presented with two major choices when dealing with samples that do not meet parametric assumptions. The choice initially selected by most researchers is to perform transformations of the original data such that the resulting variates meet the assumptions for parametric tests. Transformations, in essence, "linearize" the data. To some, implementing transformations seems like "data grinding," or manipulation of data to try and force significance (Sokal and Rohlf 1981, p. 418). Further, most people have a difficult time thinking about the distribution of the logarithm of tree height, or the square root of canopy cover. Although it may take some getting use to, there is no scientific necessity to use common linear or arithmetic scales. For example, the square root of the surface area of an organism is often a more appropriate measure of the fundamental biological variable subjected to physiological and evolutionary forces than is the surface area itself (Sokal and Rohlf 1981, p. 418).

However, although attempting to transforms your data to meet assumptions of parametric tests might be statistically sound, such actions also likely obscure biological relationships. We go back once again to the fundamental importance of viewing your data graphically before applying any statistical tests. Visual examinations often reveal interesting biological properties of your data, such as nonlinear relationships and distinct thresholds in response variables. Further, applying transformations to data does not usually linearize biological data. Additionally, if data are transformed for analysis, they must be back transformed if biological interpretations to be valid.

The second choice involves the use of nonparametric tests. Most common parametric tests have what we could call nonparametric equivalents, including multivariate analyses (Table 9.1). Nonparametric tests are gaining in popularity as researchers become more familiar with statistics, and concomitantly, as nonparametric tests are increasingly being included on canned statistical packages. Because beginning and intermediate statistics courses spend little time with nonparametric statistics (concentrating primarily on chi-square tests), wildlife scientists are not as familiar with the assumptions or interpreting the results of nonparametric tests as they are with the parametric equivalents. This engenders a resistance among many to use of the nonparametric tests.

So, how do researchers handle the difficulties of small sample size and data that are in violation of assumptions of parametric tests? The procedures are many, although not necessarily always appropriate. In virtually any issue of a major ecology journal you can find studies that:

- Simply conduct parametric tests and say nothing about testing assumptions
- Conduct tests of assumptions but do not say if assumptions were met
- Conduct nonparametric tests without giving the rationale for their use or stating whether these tests met relevant assumptions

- Call parametric tests "robust" to violation of assumptions and conduct no transformations

Fortunately, by the 1990s, most journals insisted that statistical procedures be fully explained and justified; today, few papers lack details on the testing of assumptions. However, in our readings, it is quite common to read that transformations were performed, but no mention is given regarding the success of those transformations in normalizing data. Simply performing transformations does not necessarily justify using parametric tests. Thus, the graduate student would be advised to take a course in nonparametric statistics. There is no doubt that all researchers will have the need to use these tests, especially those listed in Table 9.1.

9.2.2.3 Categorical Data Analyses

Perhaps the most relevant advanced statistical course graduate students in wildlife sciences should consider is one that covers analysis of categorical data. Categorical data analysis, or analyses of data categorized based on a measurement scale consisting of a set of categories (Agresti 1996), has seen a considerable increase in applications to wildlife research. These measurement scales typically are either ordinal (data has a natural ordering such as age classes) or nominal (data has no natural ordering, such as names of different birds species located at a site). Thus, categorical data analysis makes use of both parametric and nonparametric statistical procedures.

For many wildlife studies, we deal with data that are either distributed binomially (0, 1; died or survived) or placed into a categorical framework (counts of individuals within a plot). Thus, fundamental understanding of binomial, multinomial, Poisson, and exponential distributions are necessary for a majority of statistical analyses used in wildlife ecology. For example, estimation of survival is often conducted using logistic regression, a form of a generalized linear model. Logistic regression relies on the logit link function, based on the binomial distribution, so that predictions of survival will be mapped to the range 0–1. Additionally, logit link functions can be used to evaluate proportional odds for ranked data (Agresti 1996) and underpin a host of the current capture–mark–recapture modeling approaches used wild wildlife science (Williams et al. 2002).

Frequently, many data of interest to wildlife ecologists are represented by discrete counts. The primary sampling model for count data is the Poisson regression, which is used to analyze count data as a function of various predictive variables, most frequently as a log-linear model, or a model where the log link function is used (Mood et al. 1974). Categorical data analysis is a field of statistics that has seen considerable research interest, ranging from simple contingency table analyses using chi-square tests to methods for longitudinal data analysis for binary responses. Although perhaps not obvious to many wildlife scientists, a majority of the statistical approaches used in wildlife ecology rely on categorical data analysis theory, thus highlighting its importance to wildlife students.

Table 9.1 Selecting a statistical test

	Type of data			
Goal	Measurement (from Gaussian population)	Rank, score, or measurement (from non-Gaussian population)	Binomial (two possible outcomes)	Survival time
Describe one group	Mean, SD	Median, interquartile range	Proportion	Kaplan–Meier survival curve
Compare one group to a hypothetical value	One sample t test	Wilcoxon test	Chi-square or binomial test	
Compare two unpaired groups	Unpaired test	Mann–Whitney test	Fisher's test (chi-square for large samples)	Log-rank test or Mantel–Haenszel
Compare two paired groups	Paired t test	Wilcoxon test	McNemar's test	Conditional proportional hazards regression
Compare three or unmatched groups	One-way ANOVA	Kruskal–Wallis test	Chi-square test	Cox proportional hazards regression
Compare three or matched groups	Repeated-measures ANOVA	Friedman test	Cochrane Q	Conditional proportional hazards regression
Quantify association between two variables	Pearson correlation	Spearman correlation	Contingency coefficients	
Predict value from another measured variable	Simple or linear regression or nonlinear regression	Nonparametric regression	Simple logistic regression	Cox proportional hazards regression
Predict value from several measured or binomial variables	Multiple linear regression or multiple nonlinear regression		Multiple logistic regression	Cox proportional hazards regression

Source: From *Intuitive Biostatistics* by Harvey Motulsky. Copyright © 1995 by Oxford University Press. Used by permission of Oxford University Press, Inc

9.2.2.4 Multivariate Analyses?

Beginning in the mid-1970s, multivariate analyses became a regular part of many studies of wildlife–habitat relationships (see Morrison et al. (2006) for review). Multivariate tests are used to analyze multiple measurements made on one or more samples of individuals. Multivariate analyses were applied to natural resource studies because many variables are typically interdependent, and because the many-dimensional concept of the niche and the many-dimensional sample space of multivariate analyses are analogous in many ways (Morrison et al. 2006). Thus, through the 1980s and 1990s, many graduate students chose a course in multivariate statistics as their advanced statistics course. The most commonly used parametric multivariate tests include multiple regression, principal component analysis, multivariate analysis of variance (MANOVA), and discriminant analysis.

Although multivariate analyses – including nonparametric forms – remain useful analytical tools, the emphasis on these methods was probably misplaced. The parametric methods carry assumptions that are similar to their univariate counterparts, but are even more difficult to test for and meet. For example, rather than having to achieve normality for a single variate, a typical multivariate analysis will use 5–10 or more variates, many of which will require different transformations. In addition, multivariate analyses require much large sample sizes than their univariate counterparts (Morrison et al. 2006). And as discussed above, nonparametric procedures are relatively more difficult to interpret given lack of attention they are given in most statistics courses.

9.2.2.5 Empirical vs. Model-Based Analyses

Most ecologists agree that the best decisions are those based on a solid database – the real stuff. However, there are numerous circumstances where the issue of the moment (e.g., management of endangered species) does not allow gathering of the data everyone would desire. For example, where the long-term persistence of a population in the face of development must be evaluated without the benefit of detailed demographic studies. Further, there are numerous situations where a good database exists, but the questions being asked concern the probability of population response to different management scenarios. For example, the influence of different livestock grazing intensities on the fecundity of deer. Model-based analyses are usually required to make such projections. Thus, we recommend that graduate students become familiar with basic modeling and estimation procedures, including analyses of population growth rates, and density estimators. These procedures require an understanding of matrix algebra and calculus.

9.2.2.6 Priorities

Obviously, any person would be well served by taking all of the courses described above. But, given the competing demands of other courses and fieldwork, what should the graduate student prioritize? We would like to see every MS student take a course in basic sampling design as well as the two-semester fundamental statistics courses. PhD students, on the other hand, should not only have Master's level coursework in sampling design, but also have additional courses such as probability theory, and a calculus-based math–stat course covering basic statistical theory and inference. Additional coursework at the PhD level would be simplified as theory and inference are the foundation of all other statistics courses.

9.2.3 The Manager

The natural resource manager must balance many competing issues when performing his or her duties. Many or most of the duties involve statistics, e.g., surveys of user preferences for services, designing and implementing restoration plans, managing and evaluating harvest records, monitoring the status of protected species, evaluating research reports, and budgetary matters. The statistical preparation outlined above for the undergraduate also applies here: A minimum of a general biometrics course. It is probably preferable to obtain a solid grasp of applications rather than the more fundamental statistics courses. Obviously, the more the better!

In addition to a basic understanding of statistics, managers need to understand the importance to statistics in making decisions. Personal opinion and experience certainly have a place in management decisions. [Note: we contrast personal opinion with expert opinion. *Personal opinion* implies a decision based on personal biases and experiences. In contrast, *expert opinion* can be formalized into a process that seeks the council of many individuals with expertise in the area of interest.] However, managers must become sufficiently versed in study design and statistics and avoid falling into the "statistics can be used to support anything you want" dogma. Falling back on personal opinion to render decisions because of statistical ignorance is not a wise management action. Using sound analyses avoids the appearance of personal bias in decision making, and provides documentation of the decision-making process; this is quite helpful in a legal proceeding.

Managers should also have an appreciation of statistical modeling and management science (e.g., adaptive resource management). We contend that every manager builds models in that every manager makes predictions (at least mentally) about how the system he or she is managing will respond to any given management action. Knowing the principles of management science will assist the manager in structuring the problem in his or her own thought processes, especially when the problem becomes very complex or other parties (e.g., stakeholders) must be brought into the decision process. These principles help to identify the sources of uncertainty (e.g., environmental variation, competing theories about system dynamics, partial controllability and observability of the system) that must be addressed, and how to manage in the face of them.

Managers require the same formal statistical training as outlined above for graduate students. Many students who were training as researchers – and thus received some statistical training – become managers by way of various job changes and promotions. However, many managers either never proceeded beyond the undergraduate level or completed nonthesis MS options. Unfortunately, most nonthesis options require little in the way of statistics and experimental design.

Thus, as professionals, they are ill-prepared to handle the aspects of their profession on which most management decisions are based (see also Garcia 1989; Schreuder et al. 1993; Morrison and Marcot 1995).

Thus, managers should be sufficiently motivated to obtain advanced training in statistics and design. This training can be gained through a variety of sources, including self-training, college courses, and professional workshops. Further, enlightened administrators could organize internal training workshops by contracting with statistical and design consultants.

9.2.4 The Administrator

The duties of manager and administrator – and sometimes even scientist – are often difficult to separate. Also, as discussed above for the manager, people often become administrators after stints as a manager or researcher. However, others become administrators of various natural resource programs through processes that involve little or no ecological – and especially statistical – training. Such individuals, nevertheless, need to be able to interpret the adequacy of environmental monitoring plans, impact assessments, research papers, personal opinion, and a host of other information. After all, it is the administrator who ultimately gives approval, and is often called upon to justify that approval. It is true that administrators (and managers) can hire or consult with statisticians. However, they must still be able to explain their decision-making process and answer questions that would challenge anyone with only a rudimentary understanding of technical matters.

We recommend that natural resource administrators be at least as knowledgeable as the managers under their supervision. Thus, administrators should possess the knowledge of statistics and design as outlined above for MS students.

9.3 Resources

9.3.1 Books

All natural resource administrators, managers, and researchers should have a personal library of books that are readily available for reference. This costs money, but the alternative is either ignorance or constant trips to a colleague's office or the library. Here, we provide some suggestions for assembling a small personal library that provides references for common study designs and statistical analyses. Fortunately, the basic designs and statistical procedures are relatively stable through time. As such, one does not need to purchase the latest edition of every text. In fact, the basic text used in most reader's undergraduate and graduate courses in statistics and study design certainly form the core of a personal library.

Kish (1987, p. vi) and Kish (2004) described the general area of statistical design as "ill-defined and broad," but described three relatively well-defined and specialized approaches (1) *experimental designs* that deal mostly with symmetrical designs for pure experiments, (2) *survey sampling* that deals mostly with descriptive statistics, and (3) *observational studies* including controlled investigations and quasiexperimental designs. There are numerous books that address each of these topics. As one's library grows and specific needs arise, we suspect that these specific topics will be added to the library.

Making recommendations for specific books is difficult because there are a multitude of excellent books available. Below we list some of our favorites, categorized by general analytical family. The fact we do not list a specific title by no means indicates our displeasure with its contents or approach; rather, these are books we have used and know to be useful. Each personal library should contain a book that covers each of the major categories listed below. Topics indented as subcategories provide more detailed coverage of the more common topics listed in the primary categories; these would be useful but not essential (i.e., could be reviewed as needed in a library, or added later as the need becomes evident).

9.3.1.1 Study (Statistical) Design

The books listed first are general overviews of two or more of the subtopics below:

- *Kish (1987)*. A well-written, although brief, review of the topics of experimental design, survey sampling, and observational studies. A good introduction to these topics. This book has been reprinted as Kish (2004). A related offering is Kish (1995), which is a reprinting of his original 1965 edition.
- *Manly (1992)*. An advanced coverage, emphasizing experimental designs, and including liner regression and time series methods.

Experimental Design (ANOVA)

- *Underwood (1997)*. This very readable book emphasizes application of ANOVA designs to ecological experimentation. We highly recommend this book.

Survey Sampling

- Survey sampling can be considered a subclass of the next subtopic, observational studies, but is separated because of its common use.
- *Levy and Lemeshow (1999)*. A popular book that presents methods in a step-by-step fashion. A nice feature of this book is the emphasis on determining proper sample sizes; also discusses statistical software.

Observational Studies (controlled investigations)

- *Cochran (1983)*. A short book that begins with some very useful material on planning observational studies and interpreting data.

- *Thompson* (*2002*). A sampling design book for an introductory level graduate student in natural resources or statistics. We highly recommend this book.
- *Rosenbaum* (*2002*). A detailed explanation of designing and analyzing observational studies.

9.3.1.2 Nonmathematical Approaches

These texts assume little or no mathematical knowledge. These texts are not the recommended stepping stone to more advanced statistical procedures:

- *Watt* (*1998*). A beginning text that explains basic statistical methods and includes descriptions of study design as applied to biology.
- *Fowler et al.* (*1998*). Another basic text that is easy to read and provides a good foundation with little use of mathematics for the field biologist.
- *Motulsky* (*1995*). A basic text that uses a minimal amount of mathematics to survey statistics from basics through more advanced ANOVA and regression. This is a good text for those not likely to advance immediately to more sophisticated procedures. It uses examples from the statistical software InStat (GraphPad Software, San Diego, CA), a relatively inexpensive program. The book and software would make a useful teaching tool for basic analyses of biological data.

9.3.1.3 Fundamentals

These texts assume knowledge of college algebra and incorporate fairly detailed descriptions of the formulas and structures of statistical procedures. This type of knowledge is necessary before advancing to more complicated statistical procedures:

- *Sokal and Rohlf* (*1995*). A widely used text that emphasizes biological applications. It covers primarily parametric tests from an elementary introduction up to the advanced methods of ANOVA and multiple regression.
- *Zar* (*1998*). A widely used text that provides introductory yet detailed descriptions of statistical techniques through ANOVA and multiple regression. Zar also provides very useful chapters on analysis of circular distributions.

Nonparametric and Categorical Data Analysis:

- *Agresti* (*2002*). Concentrates on two-way contingency tables, log-linear and logit models for two-way and multiway tables, and applications of analyses. *Le* (*1998*) presents a similar coverage and is readable.
- *Stokes et al.* (*2000*) presents a thorough development of categorical methods using SAS as the analytical system.
- *Hollander and Wolfe* (*1998*). A detailed and comprehensive coverage of nonparametric statistics.

- *Conover (1999)*. An authoritative, comprehensive, yet readable coverage of nonparametric statistics. This book is loaded with examples and is considered a classic.

Advanced

- *Draper and Smith (1998)*. Provides a detailed description of linear and nonlinear regression techniques. This book is considered a classic, and is well written and easy to interpret. Now includes a diskette containing data files for all the examples and exercises in the text. An understanding of fundamental (elementary) statistics is required.

9.3.1.4 Multivariate Methods

Dillon and Goldstein (1984), *Manly (2004)*, and *Afifi (2004)* are all very readable and thorough coverages of standard multivariate methods. We particularly recommend Afifi's text given the emphasis he places on interpretation of results. Included are examples using the more commonly used statistical packages.

Hosmer and Lemeshow (2000) details the use of logistic regression, which has become one of the most widely used multivariate procedures in wildlife science.

Kleinbaum (2005) is written in an understandable manner for the nonstatistician and is aimed at graduate students.

9.3.2 Web Resources

Here we present some of the many resources available over the internet that focus on design and statistical analyses. We usually provide the IRL for the home page of the organization sponsoring the Web page because the specific within-Web site links often change through time. Only sites offering free access to programs are provided; commercial sites (regardless of the quality of the products offered for purchase) are not listed:

- USGS Patuxent Wildlife Research Center (http://www.mbr-pwrc.usgs.gov/software.html): Contains an extensive list of programs focused on analyses of animal populations, including survival estimation and capture probabilities. Also contains or provides links to documentation of programs and literature sources.
- Illinois Natural History Survey (http://nhsbig.inhs.uiuc.edu): Manages the Clearinghouse for Ecological Software, which provides programs for density estimation, bioacoustics, home range analysis, estimating population parameters, habitat analysis, and more. For habitat analysis, programs such as Fragstats can be located.
- Colorado State University (http://www.warnercnr.colostate.edu): Offers the widely used program MARK (developed and maintained by Dr. Gary White), as well as other widely used programs such as CAPTURE and DISTANCE.

- The Eco-Tools (http://eco-tools.njit.edu/webMathematica/EcoTools/index.html): A Web-accessible means of performing many commonly used calculations; no special software is needed and all algorithms are open source. Programs available include life table calculations, count based PVA, estimating species diversity, and ordination; other programs are available.

9.4 Summary

We have emphasized throughout this book, "statistics" and "study design" are interrelated yet separate topics. No statistical analysis can repair data gathered from a fundamentally flawed design, yet improperly conducted statistical analyses can easily be corrected if the design was appropriate. In this chapter we provided specific guidance regarding the knowledge that we think all resource professionals should possess, including students, scientists, managers, and administrators. All resource professionals must possess a fundamental understanding of study design if they are to make informed decisions. Wildlife professionals must, at a minimum, be able to ask the proper questions needed to interpret any report or paper. Such questions include issues of independence, randomization, and replication; adequacy of sample size and statistical power; pseudoreplication and study design; and proper extrapolation of results.

Because many of the more advanced mathematical and statistical methods in natural resources require an understanding of more advanced mathematics, including calculus, we recommend that students planning on receiving graduate degrees should take at least a beginning course in calculus in preparation for advanced methods in natural resources. Otherwise, you will be limited in the types of courses you will be qualified to take in graduate school. Many "biometrics" courses tend to sacrifice fundamentals for specific applications and interpretation. When graduate school is the goal, it is probably better to sacrifice application (i.e., the single-semester biometrics course) for fundamentals. Graduate students must obtain a good understanding of experimental design, and take the opportunity to receive advanced statistical training in topics such as nonparametric, categorical, and multivariate analyses.

In addition to a basic understanding of statistics, managers and administrators need to understand the importance to study design and statistics in making decisions. Personal opinion and experience certainly have a place in management decisions, but all resource professionals must be able to grasp the strengths and weaknesses of various sampling approaches. Using sound analyses avoids the appearance of personal bias in decision making, and provides documentation of the decision-making process; this would be quite helpful in a legal proceeding.

We also provide guidance on classes to take, books to own and use as reference sources, and other ways in which you can obtain and maintain needed design and analytical skills. We also provide a list of Web sites where you may obtain extremely useful software to aid in ecological analyses.

References

Afifi, A. A. 2004. Computer-Aided Multivariate Analysis, 4th Edition. Chapman and Hall/CRC, Boca Raton, FL.
Agresti, A. 1996. An Introduction to Categorical Data Analysis. Wiley, New York, NY.
Agresti, A. 2002. Categorical Data Analysis, 2nd Edition. Wiley, New York, NY.
Cochran, W. G. 1983. Planning and Analysis of Observational Studies. Wiley, New York, NY.
Conover, W. J. 1999. Practical Nonparametric Statistics, 3rd Edition. Wiley, New York, NY.
Dillon, W. R., and M. Goldstein. 1984. Multivariate Analysis: Methods and Applications. Wiley, New York, NY.
Draper, N. R., and H. Smith. 1998. Applied Regresion Analysis, 3rd Edition. Wiley, New York, NY.
Fowler, J. L. Cohen, and P. Jarvis. 1998. Practical Statistics for Field Biology, 2nd Edition. Wiley, New York, NY.
Garcia, M. W. 1989. Forest Service experience with interdisciplinary teams developing integrated resource management plans. Environ. Manage. 13: 583–592.
Hollander, M., and D. A. Wolfe. 1998. Nonparametric Statistical Methods, 2nd Edition. Wiley, New York, NY.
Hoshmand, A. R. 2006. Design of Experiments for Agriculture and the Natural Sciences, 2nd Edition. Chapman and Hall/CRC, Boca Raton, FL.
Hosmer Jr., D.W., and S. Lemeshow. 2000. Applied Logistic Regression, 2nd Edition. Wiley, New York, NY.
Kish, L. 1987. Statistical Design for Research. Wiley, New York, NY.
Kish, L. 1995. Survey Sampling. Wiley, New York, NY (reprint of the 1965 edition).
Kish, L. 2004. Statistical Design for Research. Wiley, New York, NY (reprint of the 1987 edition).
Kleinbaum, D. G. 2005. Logistic Regression: A Self-Learning Text, 2nd Edition. Springer-Verlag, New York, NY.
Le, C. T. 1998. Applied Categorical Data Analysis. Wiley, New York, NY.
Levy, P. S., and S. Lemeshow. 1999. Sampling of Populations: Methods and Applications, 3rd Edition. Wiley, New York, NY.
Manly, B. F. J. 1992. The Design and Analysis of Research Studies. Cambridge University Press, Cambridge.
Manly, B. F. J. 2004. Multivariate Statistical Methods: A Primer, 3rd Edition. Chapman and Hall, Boca Raton, FL.
Mood, A. M., F. A. Graybill, and D. C. Boes. 1974. Introduction to the Theory of Statistics, 3rd Edition. McGraw-Hill, Boston, MA.
Morrison, M. L., and B. G. Marcot. 1995. An evaluation of resource inventory and monitoring program used in national forest planning. Environ. Manage. 19: 147–156.
Morrison, M. L., B. G. Marcot, and R. W. Mannan. 2006. Wildlife Habitat Relationships: Concepts and Applications, 3rd Edition. Island Press, Washington, DC.
Motulsky, H. 1995. Intuitive Biostatistics. Oxford University Press, New York, NY.
Rosenbaum, P. R. 2002. Observational Studies, 2nd Edition. Springer-Verlag, New York, NY.
Schreuder, H. T., T. G. Gregoire, and G. B. Wood. 1993. Sampling Methods for Multiresource Forest Inventory. Wiley, New York, NY.
Sokal, R. R., and F. J. Rohlf. 1981. Biometry, 2nd Edition. Freeman, New York, NY.
Sokal, R. R., and F. J. Rohlf. 1995. Biometry, 3rd Edition. Freeman, New York, NY.
Stokes, M. E., C. S. Davis, and G. G. Koch. 2000. Categorical Data Analysis in the SAS System, 2nd Edition. SAS Publishing, Cary, NC.
Thompson, S. K. 2002. Sampling, 2nd Edition, Wiley, New York, NY.
Underwood, A. J. 1997. Experiments in Ecology. Cambridge University Press, Cambridge.
Watt, T. A. 1998. Introductory Statistics for Biology Students, 2nd Edition. Chapman and Hall, Boca Raton, FL.
Williams, B. K., J. D. Nichols, and M. J. Conroy. 2002. Analysis and Management of Animal Populations. Academic, San Diego, CA.
Zar, J. H. 1998. Biostatistical Analysis, 4th Edition. Prentice-Hall, Englewood Cliffs, NJ.

Chapter 10
Synthesis: Advances in Wildlife Study Design

10.1 Introduction

In this chapter, we first briefly summarize our ideas on how to improve the way we pursue wildlife field studies through study design. We hope that our ideas, developed through the pursuit of many types of studies conducted under many different logistic and funding constraints, will serve to continue the discussion on improving scientific knowledge, conservation, and management of natural resources. We then provide the reader with a study guide for each chapter that serves as a reminder of the major points raised therein.

10.2 Suggestions for Improving Knowledge and Management

The underlying basis for wildlife research is the pursuit of knowledge about ecological systems. For this reason, researchers must understand the nature of the reality they study (ontology), the characteristics and scope of knowledge (epistemology), and what characterizes valuable and high quality research as well as value judgments made during the research process (axiology). Although there is no single prescriptive method of research in natural science, wildlife researchers employ certain intellectual and methodological approaches in common (see Chap. 1).

The goal of wildlife ecology research is to develop knowledge about wildlife populations and the habitats these populations use in order to benefit conservation. To attain this goal, wildlife ecologists draw from the fields of molecular biology, animal physiology, plant and animal ecology, statistics, computer science, sociology, public policy, economics, law, and many others disciplines when developing wildlife research studies. Using our knowledge of the species or system of interest, we ask important questions and generate hypotheses or statements about how we think the system works. We then draw on tools from many scientific disciplines to study, evaluate, and then refine our hypotheses about how ecological systems work, generate new hypotheses, ask new questions, and continue the learning process (see Table 1.2). It is critical that those implementing conservation, such as natural resource managers, also clearly understand the basics of sound methods of wildlife research; this knowledge is required to evaluate the quality of information available to them for making decisions.

Our review of wildlife study design and statistical analyses leads us to the following conclusions and suggestions for change. First, the field of ecology will fail to advance our knowledge of nature unless we ask important research questions and follow rigorous scientific methods in the design, implementation, and analysis of research and surveys. Natural resource management and conservation in general is ill served by poorly designed studies that ignore the necessity of basic concepts such as randomization and replication. More often than not, studies that ignore sound design principles produce flawed results.

Scientists must clearly elucidate study goals, and the spatial and temporal applicability of results, before initiating sampling. It is critical that managers determine how and where they will use study results so that results match needs. Researchers should carefully evaluate required sample size for the study before initiation of field sampling. Simple steps, such as sample size determination or power analysis, allow the researcher to evaluate the likely precision of results before the study begins. In this manner, researchers and natural resource managers alike can anticipate confidence in their decisions based on study results. Wildlife scientists require probabilistic samples and replication for all studies so that there is less chance that the results are biased and a greater likelihood that variation in the results can be attributable to treatment effects when they exist. Establishing replicates is often difficult in field situations, but scientists can usually achieve replication with planning. We must avoid pseudoreplication, however, so that natural resource managers do not make unsound decisions based on erroneous interpretations of data. If pseudoreplication is unavoidable (e.g., such as is often the case with isolated, rare groups of animals), we must acknowledge the implications of the sampling and account for it when interpreting results. Finally, we must interpret studies that do not employ probabilistic sampling and replication (all descriptive studies) critically. Although descriptive research can provide reliable data on such characteristics as typical clutch sizes for a given bird species, it generally cannot provide reliable data on more complex phenomena such as key factors limiting abundance of an endangered species.

In Sect. 10.3, we briefly summarize the primary points made in each of the previous chapters. We hope that these summaries will help flesh out the points made in Sect. 10.2 and refer readers back to the appropriate chapters more details where needed.

10.3 Summaries

10.3.1 Chapter 1 – Concepts for Wildlife Science: Theory

1. Wildlife scientists conduct research in the pursuit of knowledge, so they must understand what knowledge is and how it is acquired.
2. Ontologically, most wildlife scientists hold that there is a material reality independent of human thought and culture, while many social scientists and

humanists maintain that reality ultimately is a social construction because it is to some degree contingent upon human percepts and social interactions.

3. Several major perspectives toward the nature and the scope of knowledge, or epistemology, have developed in Western philosophy and each influence wildlife science to greater or lesser degrees. These include:

    - Empiricism
    - Rationalism
    - Pragmatism
    - Logical positivism
    - Postpositivism
    - Social constructionism

4. Regardless of the epistemological perspective one employs, logical thought, including inductive, deductive, and retroductive reasoning (Table 1.1), remains an integral component of knowledge acquisition.
5. At least three aspects of value or quality (axiology) influence wildlife science: scientific ethics, values researchers bring to their projects, and how both scientists and society determine the value and quality of scientific research.
6. Differences in ontological, epistemological, and axiological perspectives among natural scientists and many social scientists and humanists (e.g., postmodernists) have resulted in radically different interpretations of natural science since the 1960s as recently exemplified by "the science wars."
7. Because there is no single philosophy of science, there can be no single method of science either. Regardless, the natural sciences employ certain intellectual and methodological approaches in common (Table 1.2).
8. Because of the complex nature of scientific research, multiple researchers using a variety of methods often address different aspects of the same general research program.
9. Critiques of how natural science is conducted, written by scientists have helped wildlife researchers hone their study approaches; particularly regarding analytic methods.
10. Wildlife science commonly employs study designs (Fig. 1.1.) that are not consistent with Popper's (1959, 1962) falsification model of science (see postpositivism and the hypothetico-deductive model of science).
11. Epistemologically, wildlife science is better described by Haack's (2003) pragmatic model of natural science, where research programs are conducted in much the same way one completes a crossword puzzle, with warranted scientific claims anchored by experiential evidence (analogous to clues) and enmeshed in reasons (analogous to the matrix of completed entries). Under this pragmatic epistemology, truth, knowledge, and theory are inexorably connected with practical consequences, or real effects. The pragmatic model:

    - Permits any study design that can provide reliable solutions to the scientific puzzle (e.g., descriptive research, impact assessment, information-theoretic approaches using model selection, replicated manipulated experiments attempting to falsify retroductively derived research hypotheses, qualitative designs).
    - Does not imply that each of these study designs is equally likely to provide reliable answers to specific question in a given situation.

- Suggests that researchers must determine the best approach for each individual study given specific constraints; it does not provide or condone a rote checklist for excellent wildlife research programs.

12. Wildlife scientists use biological and statistical terms to represent various aspects of what they study. This sometimes can be confusing as the same word often is used in multiple contexts. Key biological and statistical concepts discussed in Chap. 1 and used in subsequent chapters include:

    - The term "significant" is particularly problematic as it can mean that something is biologically, statistically, or socially significant. Based on a particular study, not all statistically significant differences matter biologically, and just because we cannot find statistically significant differences does not imply that important biological differences do not indeed exist. Further, if wildlife scientists find something biologically significant does not imply that society will reach the same conclusion (and vice versa). Researchers must clearly stipulate what they mean by "significant."

13. Finally, we can divide wildlife studies into (1) those where the objectives focus on measuring something about individual animals or groupings of animals, (2) and those where the objectives focus on the habitat of the animal or group. This differentiation is critically important as appropriate study design hinges upon it.

10.3.2 Chapter 2 – Concepts for Wildlife Science: Design Application

1. Sound wildlife study design relies on the ability of the scientist to think critically when developing a study plan. Critical thought about the question of interest, the system under study, and potential methods for separating and evaluating sources of variation is necessary to ensure that we successfully define the causal and mechanistic relationships between variables of interest.
2. Disturbing variables limit our ability to examine the impacts of explanatory variables on the response variables of interest. Disturbing variables should be removed from the study through controlling for them by design using appropriate probabilistic sampling methods, or in the analysis by treating then as controlled variables or covariates.
3. Random selection of experimental study units permits us to use probability theory to make statistical inferences that extend to target population. Random assignment of treatments to study units helps to limit or balance the impacts of disturbing factors. Replication of experimental treatments is necessary to capture the full variability of treatment effects.
4. When developing a study, determine what type of design is most appropriate for the ecological question of interest. Determine whether a true experiment or quasiexperiment is feasible, whether a study is best suited to a mensurative approach, whether adaptive resource management is more appropriate, or whether the study is limited to description alone.

5. If conducting an observational study or sampling within experimental units, sampling design should account for variation over space or time and for the probability of missing subjects (e.g., animals, species, or plants) within a sampling unit.
6. Statistical inference methods are tools and should be treated as such. Generation of scientific hypotheses based on critical thought about the system and species of interest are paramount to developing defensible research studies. Estimation of population parameters, confidence intervals, tests for significance, and application of model selection should each be used, when appropriate, for evaluating scientific hypotheses. One should focus on statistical significance only when the observed biological effect is also deemed significant.
7. When sampling wildlife populations, the process of inference relies on using the appropriate sampling approach for the question of interest. The objective of inference is to extend the characteristics of the sample to the population from which it came by identifying the distribution of the estimator as it relates to the parameter of interest. We also should take additional properties of estimators into account when attempting to make inferences such as bias, precision, and accuracy.
8. Most wildlife research revolves around development of methods to assist with monitoring populations and evaluating those factors that influence population trajectories. Wildlife research requires not only well thought out questions, but also appropriate sampling designs that support the inference desired.
9. After project goals and data collection is accomplished, there is a wide variety of methods available for the analysis of ecological data. Programs for data storage, manipulation, and analysis are readily available and many are suitable for ecological data. However, the presentation of the results of ecological studies should be carefully considered, given the wide array of graphical methods available.
10. Finally, wildlife ecologists should identify and acknowledge to the extent possible any limitations on the strength or applicability of their inferences due to lack of randomization, replication, control, or violations of other statistical assumptions.

10.3.3 Chapter 3 – Experimental Designs

1. Wildlife studies may include manipulative experiments, quasiexperiments, or mensurative or observational studies. With manipulative experiments there is much more control of the experimental conditions; there are always two or more different experimental units receiving different treatments; and there is a random application of treatments. Observational studies involve making measurements of uncontrolled events at one or more points in space or time with space and time being the only experimental variable or treatment. Quasiexperiments are observational studies where some control and randomization may be

possible. The important point here is that all these studies are constrained by a specific protocol designed to answer specific questions or address hypotheses posed prior to data collection and analysis.

2. Once a decision is made to conduct research there are a number of practical considers including the area of interest, time of interest, species of interest, potentially confounding variables, time available to conduct studies, budget, and the magnitude of the anticipated effect.
3. Single-factor designs are the simplest and include both paired and unpaired experiments of two treatments or a treatment and control. Adding blocking, including randomized block, incomplete block and Latin squares designs further complicates the completely randomized design. Multiple designs include factorial experiments, two-factor experiments and multifactor experiments. Higher order designs result from the desire to include a large number of factors in an experiment. The object of these more complex designs is to allow the study of as many factors as possible while conserving observations. Hierarchical designs as the name implies increases complexity by having nested experimental units, for example split-plot and repeated measures designs.
4. ANCOVA uses the concepts of ANOVA and regression to improve studies by separating treatment effects on the response variable from the effects of covariates. ANCOVA can also be used to adjust response variables and summary statistics (e.g., treatment means), to assist in the interpretation of data, and to estimate missing data.
5. Multivariate analysis considers several related random variables simultaneously, each one being considered equally important at the start of the analysis. This is particularly important in studying the impact of a perturbation on the species composition and community structure of plants and animals. Multivariate techniques include multidimensional scaling and ordination analysis by methods such as principal component analysis and detrended canonical correspondence analysis.
6. Other designs are frequently used to increase efficiency, particularly in the face of scarce financial resources or when manipulative experiments are impractical. Examples of these designs include sequential designs, crossover designs, and quasiexperiments. Quasiexperiments are designed studies conducted when control and randomization opportunities are limited. The lack of randomization limits statistical inference to the study protocol and inference is usually expert opinion. The BACI study design is usually the optimum approach to quasiexperiments. Meta-analysis of a relatively large number of independent studies improves the confidence in making extrapolations from quasiexperiments.
7. An experiment is considered very powerful if the probability of concluding no effect when in fact effect does exist is very small. Four interrelated factors determine statistical power: power increases as sample size, α-level, and effect size increase; power decreases as variance increases. Understanding statistical power requires an understanding of Type I and Type II error, and the relationship of these errors to null and alternative hypotheses. It is important to understand the concept of power when designing a research project, primarily because such

understanding grounds decisions about how to design the project, including methods for data collection, the sampling plan, and sample size. To calculate power the researcher must have established a hypothesis to test, understand the expected variability in the data to be collected, decide on an acceptable α-level, and most importantly, a biologically relevant response level. Retrospective power analysis occurs after the study is completed, the data have been collected and analyzed, and the outcome is known. Statisticians typically dismiss retrospective power analysis as being uninformative and perhaps inappropriate and its application is controversial, although it can be useful in some situations.

8. Bioequivalence testing, an alternative to the classic null hypothesis significance testing reverses the burden of proof and considers the treatment biologically significant until evidence suggests otherwise; thus switching the role of the null and alternative hypotheses. The use of estimation and confidence intervals to examine treatment differences is also an effective alternative to null hypothesis testing and often provides more information about the biological significance of a treatment.
9. Regardless of the care taken, the best-designed experiments can and many will go awry. The most important characteristics of successful studies were (1) they trusted in random sampling, systematic sampling with a random start, or some other probabilistic sampling procedure to spread the initial sampling effort over the entire study area and (2) they used an appropriate field procedures to increase detection and estimate the probability of detection of individuals on sampled units. It seems clear that including good study design principles in the initial study as described in this chapter increases the chances of salvaging a study when things go wrong.
10. Study designs must be study-specific. The feasibility of different study designs will be strongly influenced by characteristics of the different designs and by the available opportunities for applying the treatment (i.e., available treatment structures). Other, more practical considerations include characteristics of study subjects, study sites, the time available for the study, the time period of interest, the existence of confounding variables, budget, and the level of interest in the outcome of the study by others. Regardless of the environment within which studies are conducted, all protocols should follow good scientific methods. Even with the best of intentions, though, study results will seldom lead to clear-cut statistical inferences.
11. There is no single combination of design and treatment structures appropriate for all situations. Our advice is to seek assistance from a statistician and let common sense be your guide.

10.3.4 Chapter 4 – Sampling Strategies: Fundamentals

1. Clearly define issues that influence sampling organisms in an ecological system, including study objectives, study area, range of the target population, and period of interest.

2. Probability sampling in wildlife studies is necessary to use inferential statistics and the resulting data are used to estimate those parameters for the population of interest such that those values can be generalized across the population under study and hopefully to the target population. Estimators represent the mathematical formula used to determine the parameters of interest in a population.
3. Clearly define the area of inference, the experimental unit, the sampling unit, and the sampling frame. Consider the species of interest when constructing plots for sampling, as species life history should influence plot shape and size.
4. Nonprobabilistic sampling, while common, results in potentially unbiased estimates for population parameters and the biases can seldom be estimated. Probabilistic sampling provides a process by which sampling units are selected at random, thus providing a basis for statistical inference.
5. Use a probability sampling plan for short-term studies and only stratify on relatively permanent features such as topography; use a systematic sampling plan for long-term studies and studies of spatial characteristics of the study area, spread sampling effort throughout area and time intervals of interest, and maximize sample size. Systematic sampling with a random start provides a close approximation of the statistical properties of a simple random sample.
6. Model-based approaches may provide less costly and logistically easier alternatives to large design-based field studies. Data analysis can improve the quality of the information produced by these studies; however, one should not ignore fundamentally flawed design issues and limited statistical inference.
7. In model-based analysis, have the model in mind as the sampling plan is developed. In a designed-based sampling plan, clearly define the parameters to measure, and in studies of impact or tests of effect, select response variables that are relatively uncorrelated to each other, measure as many relevant covariates as possible, and identify obvious biases.
8. Maximize sample size within budgetary and logistical constraints.
9. Use model-based sampling when enumeration of variables of interest is difficult and the risk of bias is outweighed by the desire for precision. Model-based sampling also can be used to identify and evaluate nuisance parameters (e.g., variability in detection rate).
10. Incorporate designed-based estimates of parameters as much as possible in model-based studies.

10.3.5 Chapter 5 – Sampling Strategies: Applications

1. Wildlife populations and ecologies typically vary in time and space. A study design should account for these variations to ensure accurate and precise estimates of the parameters under study.

2. Various factors may lend bias to the data collected and study results. These include observer bias, sampling and measurement bias, and selection bias. Investigators should acknowledge that bias can and does occur, and take measures to minimize or mitigate the effects of that bias.
3. A critical aspect of any study is development of and adherence to a rigorous quality assurance/quality control program.
4. Study plans should be regarded as living documents that detail all facets of a study, including any changes and modifications made during application of the study design. As a rule of thumb, study plans should have sufficient detail to allow independent replication of the study.
5. Sampling intensity should be sufficient to provide the information needed and the precision desired to address the study objectives. Anything less may constitute a waste of resources.
6. Plot size and shape are unique to each study.
7. Pilot studies are critical: "Those who skip this step because they do not have enough time usually end up losing time" (Green 1979, p. 31).

10.3.6 Chapter 6 – Impact Assessment

1. "Impact" is a general term used to describe any change that perturbs the current system, whether it is planned or unplanned, human induced or an act of nature and positive or negative.
2. There are several prerequisites for an optimal study design:
   - The impact must not have occurred, so that before-impact baseline data can provide a temporal control for comparing with after-impact data.
   - The type of impact and the time and place of occurrence must be known.
   - Nonimpacted controls must be available.
3. Impact assessment requires making assumptions about the nature of temporal and spatial variability of the system under study; assumptions about the temporal and spatial variability of a natural (nonimpacted) system can be categorized as in steady-state, spatial, or dynamic equilibrium.
4. Three primary types of disturbances occur: pulse, press, and those affecting temporal variance. Background variance caused by natural and/or undetected disturbances makes identifying the magnitude and duration of a disturbance difficult.
5. The "before–after/control–impact," or BACI, design is the standard upon which many current designs are based. In the BACI design, a sample is taken before and another sample is taken after a disturbance, in each of the putatively disturbed (impacted) sites and in an undisturbed (control) sites.
6. The basic BACI design has been expanded and improved to include both temporal and spatial replication (multiple controls; use of matched pairs).
7. Designs classified under "suboptimal" are designs without pretreatment data and most often apply to the impact situation where you had no ability to gather

preimpact (pretreatment) data or plan where the impact was going to occur. After-only impact designs also apply to planned events that resulted from management actions, but were done without any pretreatment data.

8. The gradient approach is especially applicable to localized impacts within homogeneous landscapes because it allows you to quantify the response of elements at varying distances from the impact and each gradient provides a self-contained control at the point beyond which impacts are detected.
9. A serious constraint in the design of wildlife impact studies is the limited opportunity to collect data before the disturbance. The before period is often short and beyond the control of the researcher, that is the biologist has not control over where or when the disturbance will occur. In some cases, it may be possible to improve our understanding of potential temporal variation without studying for multiple years by increasing the number of reference sites and spatial distribution of study sites such that the full range of impact response is sampled.
10. Because of the unplanned nature of most disturbances, pretreatment data are seldom directly available. Thus, the task of making a determination on the effects the disturbance had on wildlife and other resources is complicated by (1) natural stochasticity in the environment and (2) the unreplicated nature of the disturbance. To some extent, multiple reference areas can improve confidence in the attribution of impact by allowing a comparison of the condition in the impacted area to a distribution of conditions in the unimpacted (control) population.
11. Epidemiological approaches, by focusing on determining incidence rates, lend themselves to applications in impact assessment. The choice of the use factor, or denominator, is more important than the numerator. The choice arises from the preliminary understanding of the process of injury or death. The ideal denominator in epidemiology is the unit that represents a constant risk to the animal.
12. Obtaining information on the sensory abilities of animals is a key step in designing potential risk-reduction strategies.

10.3.7 Chapter 7 – Inventory and Monitoring Studies

1. Inventory and monitoring are key steps in wildlife biology and management; they can be done in pursuit of basic knowledge or as part of the management process.
2. Inventory assesses the state or status of one or more resources, whereas monitoring assesses population changes or trends.
3. Monitoring can be classified into four overlapping categories (1) implementation monitoring is used to assess whether or not a directed management action was carried out as designed, (2) effectiveness monitoring is used to evaluate whether a management action met its desired objective, (3) validation monitoring is used to evaluate whether an established management plan is working, and (4) compliance monitoring is used to see if management is occurring according to established law or regulation.

4. Selecting the appropriate variable to inventory or monitor is a key aspect of the study design, and direct measures (e.g., population numbers) are preferred over indirect measures (e.g., indices of population parameters).
5. The length of monitoring studies depends largely on the process or variable being studied, the magnitude and rate of change in the variable, and the natural variability in the variable and important covariates. The appropriate length for some variables may exceed available resources, necessitating alternative approaches, namely (1) retrospective studies, (2) substituting space-for-time, (3) modeling, and (4) substitutions of fast for slow dynamics.
6. Monitoring effects of management actions requires a clear and direct linkage between study results and management activities, often expressed as a feedback loop.

10.3.8 Chapter 8 – Design Applications

1. Studies should follow this process:
   - Establish questions
   - Develop hypotheses and predictions
   - Design research
   - Choose variables
   - Choose recording methods
   - Establish acceptable level of precision
   - Prepare a detailed protocol that clearly lays out the design, the data to be collected, the sampling plan, sample sizes, the methods for data collection, and the anticipated analysis
   - Collect preliminary data
   - Make necessary adjustments in the study protocol
   - Complete final data collection
   - Conduct quality/quantity assurance
   - Conduct analyses/hypothesis testing
   - Interpret results
   - Report and/or publish results
2. Questions are developed using literature, expert opinion, your own experiences, intuition, and guesswork. A thorough literature review is an essential cornerstone of all studies.
3. Simply stating a hypothesis in no way guarantees that knowledge will be advanced. What is required is careful and thoughtful evaluation of the predictive value of any proposed hypothesis.
4. Delineation of the study population is critical to deciding how to sample the population so that statistical inference can be made to the target population. Making statistical extrapolation beyond the study population requires replication in multiple populations.

5. Proper distribution of sampling locations (e.g., plots, transects) is a critical aspect of all studies.
6. Variables must be selected that are expected to respond to the treatment being tested or be closely linked to the relationship being investigated. Even purely descriptive, hypothesis-generating studies should focus sampling efforts on a restricted set of measurements.
7. All proposed recording methods should be thoroughly reviewed for potential biases and degree of precision attainable.
8. Calculating sample sizes necessary to achieve specified statistical precision or power are essential aspects of study design.
9. All studies should begin with a preliminary phase during which observers are trained to become competent in all sampling procedures, sample size calculations are conducted, and recording methods are refined.
10. It is important to the successful completion of the study that a formal protocol that includes a program of quality assurance/quality control is instituted on both the data collection and data processing components to ensure that the execution of the plan is in accordance with the study design.
11. A weakness of most studies is the lack of a detailed protocol that results in ad hoc implantation of the study details and a failure to enter, proof, and analyze data on a continuing basis.
12. Conclusions must be drawn with reference to the protocol by which the study is conducted and the population of inference only.
13. Although you cannot change your initial study design after the study is completed, you can legitimately change the way you group and analyze your data within the limitations of your design. Thus, you may be able to a posteriori change your method of data processing and analysis (e.g., from a two-group treatment vs. no treatment to a gradient approach).
14. Because of design inadequacies (or insufficient or inappropriate samples), you might have to revise your initial your study goals, such as narrowing the scope and applicability of the study.
15. The fundamental resource necessary for success in any scientific study is leadership. Regardless of the rigor of the design and the qualifications of your assistants, you must provide a detailed study protocol and you must be able to train, encourage, critique the implementation of the protocol – and accept criticism and suggestions, and overall guide your study throughout its duration.
16. Report your results, preferably in a peer-reviewed scientific journal when the results warrant.

10.3.9 Chapter 9 – Education in Design and Statistics for Students and Professionals

1. All wildlife professionals should, at a minimum, be able to ask the proper questions needed to interpret any report or journal article. Such questions include

issues of independence, randomization, and replication; adequacy of sample size and statistical power; pseudoreplication and study design; and proper extrapolation of results.
2. We think that all natural resource professionals require, at a minimum, a good knowledge of the principles and underlying theory of study design and statistics. The applied approach, which minimizes formulas and mathematics, is adequate in many cases for interpretation of research results.
3. All students planning to receive graduate degrees should take at least a beginning course in calculus in preparation for advanced methods in study design and evaluation.
4. We recommend that, following the introductory two-semester statistics course, graduate students enroll in an experimental design or sampling design course.
5. The graduate student, especially the PhD student, would also be advised to take more advanced courses in such topics as spatial analysis, time series, nonparametrics or resampling statistics, or multivariate analysis. The appropriate choice of courses will depend on the student's emphasis; however, we recommend that PhD students take a graduate course in statistical theory, as it will lay the foundation for all other courses.
6. We also recommend that graduate students at least become familiar with basic modeling, including analysis of population, community, or landscape dynamics, and the estimation of the parameters associated with these models. Estimation of these parameters often involves techniques for adjusting for imperfect detectability of subjects of interest (e.g., capture–recapture methodology).
7. We recommend that all natural resource professionals, including managers, administrators, and regulators, possess the statistical training outlined for the MS graduate student, or at least avail themselves of the advice of a good statistician.
8. All professionals should have a personal library of journal reprints and books that are readily available for reference. We provide some suggestions for assembling a small personal library that provides reference for common study designs and statistical analyses.

In closing, we wish all of you great success with your current and future research endeavors. All work is important; but appropriately designed, analyzed, and interpreted work has the greatest impact. We hope that our book assists in some positive way with the advancement of knowledge and conservation.

Index

N

T

Lightning Source UK Ltd.
Milton Keynes UK
UKOW06f2346040515
250874UK00005B/30/P